ELEMENTARY ALGEBRA
Part 1

About the Cover

The design on the cover, which is repeated in black and white on the title page, is a collage of symbols, formulas, instruments, and other devices indicating modern everyday uses of mathematics. The pairs of pictures which open chapters are designed to continue the same theme. They show achievements which have come about with the aid of mathematics in various fields such as science, communication, and transportation.

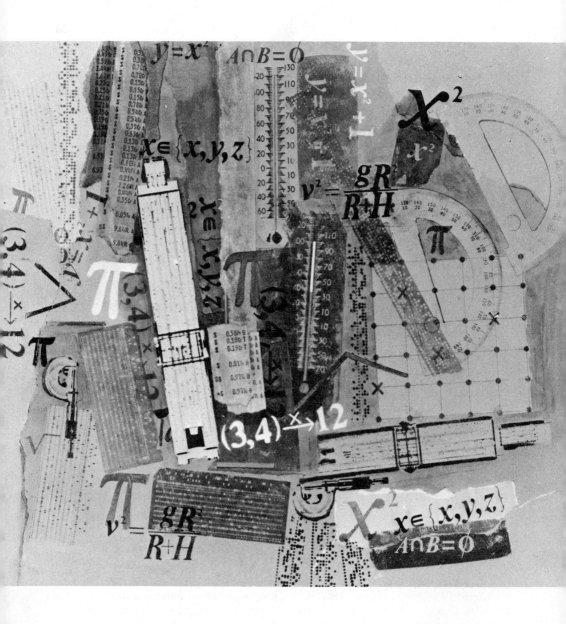

Elementary Algebra
Part 1

RICHARD A. DENHOLM

MARY P. DOLCIANI

GEORGE E. CUNNINGHAM

EDITORIAL ADVISERS   ANDREW M. GLEASON

ALBERT E. MEDER, JR.

HOUGHTON MIFFLIN COMPANY · BOSTON
NEW YORK   ATLANTA   GENEVA, ILLINOIS   DALLAS   PALO ALTO

ABOUT THE AUTHORS

Richard A. Denholm, Director of Curriculum and Instruction, Orange County, California, Department of Education. Previously, Dr. Denholm held the position of Coordinator of Mathematics and Physical Science Instruction for Orange County. He served as chairman of the research and writing team that developed a curriculum guide in modern mathematics for the California Association of County Superintendents of Schools.

Mary P. Dolciani, Professor and Chairman, Department of Mathematics, Hunter College of the City University of New York. Dr. Dolciani has been a member of the School Mathematics Study Group (SMSG) and a director and teacher in numerous National Science Foundation and New York State Education Department institutes for mathematics teachers.

George E. Cunningham, Mathematics Consultant, Anaheim Union High School District, Anaheim, California, and Instructor in Mathematics Education, California State College at Fullerton.

EDITORIAL ADVISERS

Andrew M. Gleason, Professor of Mathematics, Harvard University. Professor Gleason is prominently associated with curriculum changes in mathematics. He was chairman of the Advisory Board for SMSG as well as co-chairman of the Cambridge Conference which wrote the influential report, *Goals for School Mathematics.*

Albert E. Meder, Jr., Dean and Vice Provost and Professor of Mathematics, Emeritus, Rutgers University, the State University of New Jersey. Dr. Meder was Executive Director of the Commission on Mathematics of the College Entrance Examination Board, and has been an advisory member of SMSG.

COPYRIGHT © 1973, 1970 BY HOUGHTON MIFFLIN COMPANY

ALL RIGHTS RESERVED. NO PART OF THIS WORK MAY BE REPRODUCED OR TRANSMITTED IN ANY FORM OR BY ANY MEANS, ELECTRONIC OR MECHANICAL, INCLUDING PHOTOCOPYING AND RECORDING, OR BY ANY INFORMATION STORAGE OR RETRIEVAL SYSTEM, WITHOUT PERMISSION IN WRITING FROM THE PUBLISHER. PRINTED IN THE U.S.A.

STUDENT'S EDITION	ISBN: 0-395-14265-2
STUDENT'S EDITION W/ODD ANSWERS	ISBN: 0-395-14278-1
TEACHER EDITION	ISBN: 0-395-14264-4

Contents

1 Basic Concepts about Numbers and Numerals — 1

NUMBERS AND HOW TO REPRESENT THEM · **1-1** Numbers, Numerals, and the Number Line, 1 · **1-2** Number Relationships, 4 · **1-3** Numerical Expressions and Equality Statements, 7

FUNDAMENTAL ARITHMETIC OPERATIONS · **1-4** Addition and Its Inverse, 10 · **1-5** Multiplication and Its Inverse, 14 · **1-6** Graphs and the Number Line, 18 · **1-7** Operations on the Number Line, 20 · **1-8** Inequalities, 23

CHAPTER SUMMARY, 26 · CHAPTER TEST, 28
CHAPTER REVIEW, 29 · REVIEW OF SKILLS, 31
DIGITS AND NUMERALS, 32

2 The Language and Symbols of Mathematics — 35

FUNDAMENTAL IDEAS ABOUT SETS · **2-1** Sets and How to Represent Them, 35 · **2-2** Kinds of Sets, 39 · **2-3** Graphing a Set of Numbers, 42

SPECIAL SETS AND OPERATIONS ON SETS · **2-4** Sets and Subsets, 45 · **2-5** The Set of Whole Numbers — Some Special Subsets, 47 · **2-6** Sets and Operations, 50

CHAPTER SUMMARY, 54 · CHAPTER TEST, 55
CHAPTER REVIEW, 56 · REVIEW OF SKILLS, 58
TRIANGULAR NUMBERS, 59 · ARCHIMEDES, 61

3 Numbers, Functions, and Number Pairs — 63

REPRESENTING FUNCTIONS · **3-1** Functions: Rules and Machines, 63 · **3-2** Number Pairs and Graphs, 67 · **3-3** Functions and Graphs, 70

NOMOGRAPHS · **3-4** Special Graphs: Addition Nomographs, 75 · **3-5** Special Graphs: Multiplication Nomographs, 79

CHAPTER SUMMARY, 84 · CHAPTER TEST, 84
CHAPTER REVIEW, 85 · REVIEW OF SKILLS, 87
CARL FRIEDRICH GAUSS, 89

4 Variables, Expressions, and Sentences — 91

FUNDAMENTAL IDEAS OF GROUPING · **4-1** Expressions and Punctuation, 91 · **4-2** More about Expressions and Grouping Symbols, 94

THE USE OF VARIABLES IN EXPRESSIONS · **4–3** Expressions and Variables, 96 · **4–4** Factors and Coefficients in Algebraic Expressions, 102 · **4–5** Algebraic Expressions and Generalized Forms, 104

USING EXPONENTS · **4–6** Expressions Containing Exponents, 108 · **4–7** Exponents and Nomographs, 112

> CHAPTER SUMMARY, 115 · CHAPTER TEST, 117
> CHAPTER REVIEW, 118 · REVIEW OF SKILLS, 120
> EXPONENTS AND SCIENTIFIC NOTATION, 121 · PASCAL, 123

5 Open Sentences   125

EQUATIONS · **5–1** Open Equations and Inequalities, 125 · **5–2** Solving Equations, 128 · **5–3** Using Variables and Writing Expressions, 132

PROBLEM SOLVING · **5–4** Writing Equations and Solving Problems, 134 · **5–5** Using Formulas to Solve Problems, 138

INEQUALITIES · **5–6** Solving Inequalities, 140 · **5–7** Inequalities and the Relationship Symbols $\geq$ and $\leq$, 143 · **5–8** Writing Inequalities and Solving Problems, 146

> CHAPTER SUMMARY, 149 · CHAPTER TEST, 149
> CHAPTER REVIEW, 150 · REVIEW OF SKILLS, 153
> SPECIAL SETS OF NUMBERS, 154

6 Operations, Axioms, and Equations   157

BASIC PROPERTIES FOR EQUALITY AND EXISTENCE · **6–1** Basic Operations and Axioms, 157 · **6–2** Existence and Uniqueness; the Closure Property, 160

COMMUTATIVE, ASSOCIATIVE, AND DISTRIBUTIVE PROPERTIES · **6–3** Commutative and Associative Properties, 164 · **6–4** The Distributive Property, 168 · **6–5** Other Applications of the Distributive Property, 171

OTHER NUMBER PROPERTIES · **6–6** Properties of Zero and One, 174 · **6–7** Functions and Variables, 176

> CHAPTER SUMMARY, 181 · CHAPTER TEST, 183
> CHAPTER REVIEW, 184 · REVIEW OF SKILLS, 186
> PROPERTIES OF OPERATIONS, 186 · FIBONACCI, 189

7 Equations and Problem Solving   191

THE EQUALITY PROPERTIES · **7–1** Combining Similar Terms, 191 · **7–2** Addition and Subtraction Properties of Equality, 193 · **7–3** The Division Property of Equality, 197 · **7–4** The Multiplication Property of Equality, 200

Contents vii

WORKING WITH EQUATIONS · **7–5** More about Solving Equations, 202 · **7–6** Equations with the Variable in Both Members, 207

    CHAPTER SUMMARY, 209 · CHAPTER TEST, 210
    CHAPTER REVIEW, 211 · REVIEW OF SKILLS, 213
    POLYHEDRONS, 214

8 Negative Numbers 217

DIRECTED NUMBERS · **8–1** Directed Numbers and the Number Line, 217 · **8–2** Moves on the Number Line Using Directed Numbers, 221 · **8–3** Comparing Directed Numbers, 225 · **8–4** Representations of Directed Numbers, 227

INEQUALITIES AND DIRECTED NUMBERS · **8–5** Inequalities with Directed Number Solutions, 230 · **8–6** More about Inequalities and Directed Numbers, 232

    CHAPTER SUMMARY, 235 · CHAPTER TEST, 236
    CHAPTER REVIEW, 237 · REVIEW OF SKILLS, 239
    SYMMETRY, 240

9 Addition and Subtraction of Directed Numbers 243

ADDITION · **9–1** Number Line Addition and the Commutative Property, 243 · **9–2** Opposites of Directed Numbers, 247 · **9–3** Addition: Additive Inverses, 250 · **9–4** Using Additive Inverses to Simplify Expressions, 253 · **9–5** Number Line Addition and the Associative Property, 256

SUBTRACTION · **9–6** Subtraction of Positive Directed Numbers, 260 · **9–7** Subtraction of Negative Directed Numbers, 263

FUNCTIONS AND NOMOGRAPHS · **9–8** Functions and Directed Numbers, 266 · **9–9** Nomographs: Addition and Subtraction of Directed Numbers, 269

    CHAPTER SUMMARY, 273 · CHAPTER TEST, 274
    CHAPTER REVIEW, 275 · REVIEW OF SKILLS, 278
    PROBABILITY, 280

10 Multiplication and Division of Directed Numbers 283

MULTIPLICATION · **10–1** Products of Positive and Negative Numbers, 283 · **10–2** More about Products of Postive and Negative Numbers, 286 · **10–3** Products of Negative Numbers, 289 · **10–4** The Distributive Property and Algebraic Expressions, 293

Contents

DIVISION · **10–5** Division of Directed Numbers, 295 · **10–6** Directed Numbers and Reciprocals, 298 · **10–7** Functions: Multiplication and Division with Directed Numbers, 302

CHAPTER SUMMARY, 306 · CHAPTER TEST, 307
CHAPTER REVIEW, 308 · REVIEW OF SKILLS, 310
RATIONAL NUMBERS, 311

11 Solving Equations and Inequalities — 315

TYPES OF EQUATIONS · **11–1** Equations of Type $x + a = b$, 315 · **11–2** Equations of Type $ax = b$, 318 · **11–3** Equations of Type $ax + bx = c$, 321

PROPERTIES OF INEQUALITY · **11–4** The Addition Property of Inequality, 325 · **11–5** The Multiplication Property of Inequality, 329

FORMULAS · **11–6** Using Formulas to Solve Problems, 333

CHAPTER SUMMARY, 335 · CHAPTER TEST, 336
CHAPTER REVIEW, 337 · REVIEW OF SKILLS, 339
PAIRS OF DIRECTED NUMBERS, 340 · LEONHARD EULER, 343

12 Addition and Subtraction of Polynomials — 345

ADDING POLYNOMIALS · **12–1** Introduction to the Set of Polynomials, 345 · **12–2** Addition of Polynomials, 348 · **12–3** Polynomials and Addition Properties, 352

SUBTRACTING POLYNOMIALS · **12–4** Polynomials, Additive Identity, and Additive Inverse, 356 · **12–5** Subtraction of Polynomials, 359

USING POLYNOMIALS · **12–6** Problems Using Polynomials, 363 · **12–7** Equations and Polynomials, 365 · **12–8** Functions and Polynomials, 366

CHAPTER SUMMARY, 369 · CHAPTER TEST, 370
CHAPTER REVIEW, 371 · REVIEW OF SKILLS, 374
ABSOLUTE VALUE, 375 · THE EGYPTIANS, 377

13 Multiplication and Division of Polynomials — 379

MULTIPLICATION · **13–1** Polynomials and Exponents, 379 · **13–2** Multiplication of Monomials, 382 · **13–3** Power of a Product, 383 · **13–4** Multiplying a Polynomial by a Binomial, 386 · **13–5** Multiplying a Polynomial by a Polynomial, 389

SPECIAL PRODUCTS: PROPERTIES OF MULTIPLICATION · **13–6** Special Products of Polynomials, 394 · **13–7** Multiplication Properties for Polynomials, 397

DIVISION · **13–8** Division of Monomials, 400 · **13–9** Division of a Polynomial by a Monomial, 403 · **13–10** Division of a Polynomial by a Polynomial, 405

    CHAPTER SUMMARY, 407 · CHAPTER TEST, 408
    CHAPTER REVIEW, 409 · REVIEW OF SKILLS, 412
    ZERO AS EXPONENT, 413 · GALILEO GALILEI, 415

14 Geometric Figures in the Plane   417

LINES AND CURVES · **14–1** Basic Geometric Figures, 417 · **14–2** More about Lines and Line Segments, 421 · **14–3** Curves: Sets of Points, 424

ANGLES · **14–4** Rays and Angles in the Plane, 428 · **14–5** Angle Measurement and the Protractor, 431

TRIANGLES · **14–6** Kinds of Triangles, 436 · **14–7** Right Triangles; The Pythagorean Theorem, 440

    CHAPTER SUMMARY, 444 · CHAPTER TEST, 446
    CHAPTER REVIEW, 447 · SIMILAR TRIANGLES, 449
    CUMULATIVE TEST, 452 · TABLES, 460 · GLOSSARY, 461 · INDEX, 468

LIST OF SYMBOLS

		Page			Page
$=$	equals (is equal to)	8	$\ldots$	and so on	40
$\neq$	does not equal	23	$\sim$	is similar to	449
$>$	is greater than	24	$\geq$	is greater than or equal to	143
$<$	is less than	24	$\leq$	is less than or equal to	143
$\{\ \}$	set	35	$n(A)$	number property of set A	39
$\in$	is an element of	36	$P(e)$	probability of event e	280
$\notin$	is not an element of	36	$f(x)$	the value of function f for x	71
$\subset$	is a subset of	45	$(3,2) \overset{+}{\to} 5$	the sum of 3 and 2 is 5	11
$\not\subset$	is not a subset of	45	$(5,2) \overset{-}{\to} 3$	the difference of 5 and 2 is 3	11
$\emptyset$	the empty set	40	$(3,2) \overset{\times}{\to} 6$	the product of 3 and 2 is 6	14
$\cap$	intersection (of sets)	51	$(6,2) \overset{\div}{\to} 3$	the quotient of 6 and 2 is 3	14
$\cup$	union (of sets)	51	$\circ$	degree (angle measure)	432
π	pi	139	$\angle ABC$	angle ABC	428
a^n	a to the nth power	108	$m \angle A$	measure of $\angle A$	432
$\|a\|$	absolute value of a	376	$\triangle ABC$	triangle ABC	437
$-a$	opposite (additive inverse) of a	248	$\overleftrightarrow{AB}$	line AB	417
$^-2$	negative 2	218	$\overline{AB}$	line segment AB	417
$^+2$	positive 2	218	$\overrightarrow{AB}$	ray AB	428
(x, y)	ordered pair	64			

An early Roman sports arena . . .

A modern sports arena . . .

Basic Concepts about Numbers and Numerals

The famous Roman Colosseum is thought to be the finest surviving example of Roman architectural engineering. Completed in the year 80 A.D., it seated about 45,000 spectators, and was the scene of such spectacles as fights between gladiators or between men and wild animals. Awnings could be hung from the walls to protect the spectators from the sun. Originally, the arena could even be flooded for water spectacles. A modern and well-known sports arena is the Houston Astrodome, of which the interior is shown. This stadium seats 45,000 for a baseball game, 52,000 for football, and over 60,000 for a convention. With its plastic dome, which is 208 feet above the floor at its highest point, it can accommodate sports events and meetings in any kind of weather.

Numbers and How to Represent Them

1–1 Numbers, Numerals, and the Number Line

This first chapter contains many ideas already familiar to you. It is our purpose to present many of them from a fresh point of view. For example, consider the **whole numbers**, which include zero and the numbers that are used for counting. Perhaps you already know that each whole number can be matched with exactly one point on the **number line**.

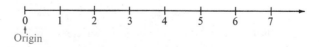

The point matched with 0 is called the **origin** of the number line. Notice that the other whole numbers are matched, in their natural order, with equally spaced points along the line. We say that they are in **consecutive order**.

Every whole number has a **successor** that is 1 greater than the number itself. Thus the successor of **5** is **5 + 1**, or **6**. The successor of **357** is **357 + 1**, or **358**. In general

The successor of any whole number **w** is the whole number **w + 1**.

Each of the symbols **3 × 6** and **10 + 8** names the whole number **18**. Can you think of other names for 18? The symbols **3 + 3 + 3** and $\frac{45}{5}$ name the whole number 9. These examples illustrate the very important idea that "a number has many different names."

In your earlier study of arithmetic you used whole numbers as well as numbers like $\frac{1}{2}$, 0.35, π, and $\frac{5}{8}$. You were using what are called the **numbers of arithmetic**. It is possible to match each number of arithmetic with exactly **one** point on the number line. The assignment of the numbers to points on the line is done in a systematic manner. The number line in Figure 1-1 shows that the distance between the points matched with $\frac{0}{3}$ and $\frac{1}{3}$ is the same as the distance between those matched with $\frac{1}{3}$ and $\frac{2}{3}$; the distance between points $\frac{4}{3}$ and $\frac{5}{3}$ is the same as between points $\frac{7}{3}$ and $\frac{8}{3}$, etc.

Figure 1-1

ORAL EXERCISES

Name the successor of each whole number.

SAMPLE: $\frac{10}{5}$

What you say: $\frac{10}{5}$ names the whole number 2; its successor is 2 + 1, or 3.

1. 39
2. 262
3. 100%
4. 200%
5. 4^2
6. 10^2
7. $\frac{12}{3}$
8. $\frac{18}{6}$
9. 7.0
10. 25.0
11. 2^3 (Recall that $2^3 = 2 \times 2 \times 2$)
12. 4^3

Basic Concepts about Numbers and Numerals 3

Name the numbers required to complete the part shown of each number line. The marks are equally spaced.

SAMPLE.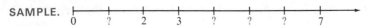
0 ? 2 3 ? ? ? 7

What you say: The numbers 1, 4, 5 and 6 are required.

13. ├──┼──┼──┼──┼──┼──→
   0 ? 2 ? ? 5

15. ←──┼──┼──┼──┼──┼──→
     ? 19 ? ? 22

14. ├──┼──┼──┼──┼──┼──→
   0 ? ? ? ? 5

16. ←──┼──┼──┼──┼──┼──→
     ? ? 1001 ? 1003

WRITTEN EXERCISES

Write the numbers described.

SAMPLE: Five consecutive whole numbers, the first of which is 6.

Solution: 6, 7, 8, 9, and 10.

A
1. Three consecutive whole numbers, the first of which is 4.
2. Five consecutive whole numbers, the first of which is 17.
3. Four consecutive whole numbers, the first of which is 9.
4. Six consecutive whole numbers, the first of which is 48.
5. Five consecutive whole numbers, the first of which is 4.

Answer each of the following questions.

SAMPLE: Which two consecutive whole numbers have a sum of 5?

Solution: Since $2 + 3 = 5$, the numbers are 2 and 3.

6. Which two consecutive whole numbers have a sum of 7?
7. Which two consecutive whole numbers have a sum of 13?
8. Which two consecutive whole numbers have a sum of 25?

B
9. Which two consecutive whole numbers have a sum of 1?
10. Which three consecutive whole numbers have a sum of 9?
11. Which three consecutive whole numbers have a sum of 6?
12. What is the sum of two consecutive whole numbers, the first of which is 8?
13. What is the sum of two consecutive whole numbers, the first of which is 19?

4 Chapter 1

C 14. What is the sum of three consecutive whole numbers, the first of which is 16?

15. What is the sum of three consecutive whole numbers, the last of which is 73?

Copy each number line picture and complete it by writing the missing numerals.

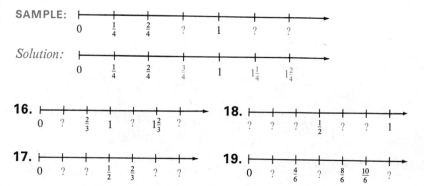

1-2 Number Relationships

You have seen how numbers can be matched with points on the number line. This is done according to the values of the numbers, with the point representing the **larger** of any two numbers lying to the **right** of the point representing the other number.

Capital letters are often used to label points on the number line. A number matched with a point is said to be the **coordinate** of the point. Thus in Figure 1-2 the **coordinate** of point F is $\frac{1}{3}$. What is the coordinate of T? of M? What is the name of the point whose coordinate is $\frac{5}{3}$?

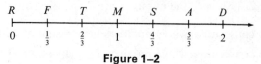

Figure 1-2

The word **between** has a particular meaning in mathematics. Suppose that on a given day the lowest thermometer reading indicated a temperature of 50 degrees and that the highest was 85 degrees. What can be said about a temperature of 65 degrees? of 77 degrees? of $84\frac{3}{4}$ degrees?

How would you answer the question "Which whole numbers are between 5 and 9?" Do you agree that the number line in Figure 1-3 shows that the numbers are 6, 7, and 8?

Basic Concepts about Numbers and Numerals

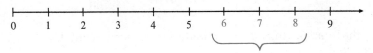

the whole numbers between 5 and 9

Figure 1–3

Notice that the numbers 5 and 9 are *not* included. However, if we were asked to name "the whole numbers between 5 and 9, inclusive" the word **inclusive** tells us to include 5 and 9 as is shown in Figure 1-4.

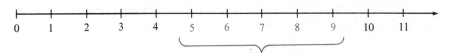

the whole numbers between 5 and 9, inclusive

Figure 1–4

ORAL EXERCISES

Name the coordinate of each point indicated.

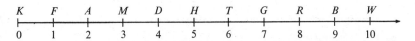

SAMPLE 1: *G*

What you say: The coordinate of *G* is 7.

SAMPLE 2: The point half the distance from 3 to 5.

What you say: The coordinate of the point half the distance from 3 to 5 is 4.

1. *K*
2. *M*
3. *D*
4. *H*
5. *W*
6. *A*
7. *B*
8. *F*
9. *T*

10. The point half the distance from 8 to 10.
11. The point half the distance from 2 to 4.
12. The point one-third of the way from 3 to 6.
13. The point one-fourth of the way from 2 to 6.
14. The point one-third of the way from 7 to 10.

6 Chapter 1

WRITTEN EXERCISES

Name the number that is the coordinate of each point that is labeled with a letter.

SAMPLE: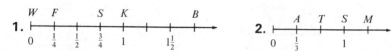

Solution: $H: \frac{1}{4}$ $T: \frac{1}{2}$ $M: \frac{3}{4}$ $E: 1\frac{1}{4}$

A 1. 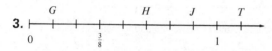 2.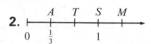

3.

4.

5.

Tell what letter on the number line below names the point described.

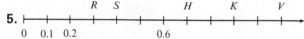

SAMPLE: The point that is one-third of the distance from M to S.

Solution: The distance from M to S is 3 units. The point described is $\frac{1}{3} \times 3$ (or 1) unit to the right of M. The point is named by the letter F.

6. The point that is one-fourth of the distance from E to F.
7. The point that is two-thirds of the distance from H to F.
8. The point that is half the distance from G to M.
9. The point that is one-third of the distance from H to R.
10. The point that is three-fourths of the distance from C to S.

Name the whole numbers described in each case. Refer to a number line, if necessary, to help you answer.

SAMPLE: The whole numbers between $1\frac{1}{3}$ and 6.

Solution: 2, 3, 4, and 5

Basic Concepts about Numbers and Numerals **7**

B 11. The whole numbers between 2 and 5.
12. The whole numbers between 4 and 9.
13. The whole numbers between π and 8. (Recall that π is about $3\frac{1}{7}$.)
14. The whole numbers between 1 and π.
15. The whole numbers between 3 and 7, inclusive.
16. The whole numbers between 6 and 7, inclusive.
17. The whole numbers between the origin and $6\frac{1}{2}$.
18. The whole numbers between the origin and $4\frac{9}{10}$.

Name the coordinate of the point described on the number line below.

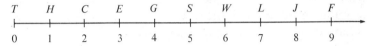

SAMPLE: The point two-thirds of the distance from *C* to *L*.

Solution: The distance from *C* to *L* is 5 units. So the point is $\frac{2}{3} \times 5$ (or $\frac{10}{3}$) units to the right of *C*. The coordinate of the point is $2 + \frac{10}{3}$, or $\frac{16}{3}$.

19. The point *S*.
20. The point *F*.
21. The point half the distance from *H* to *G*.
22. The point one-third of the distance from *G* to *J*.
23. The point one-fourth of the distance from *T* to *G*.
24. The point one-fifth of the distance from *E* to *J*.
25. The point 3 units to the right of the origin.
26. The point 7 units to the right of the origin.
27. The point $2\frac{1}{5}$ units to the right of *S*.
28. The point $3\frac{2}{3}$ units to the right of *W*.
29. The point 1.6 units to the left of *L*.
30. The point 2.8 units to the left of *S*.

C 31. The point one-fifth of the distance from *E* to *W*.
32. The point one-eighth of the distance from *C* to *W*.
33. The point two-thirds of the distance from *S* to *F*.
34. The point three-fifths of the distance from *Ḣ* to *G*.
35. The point five-eighths of the distance from *C* to *S*.

1–3 Numerical Expressions and Equality Statements

You know that in our decimal system we use the ten symbols 0, 1, 2, 3, 4, 5, 6, 7, 8, and 9 to represent numbers. These are called **digits**.

We also use symbols to represent the four basic operations: + for **addition**, − for **subtraction**, × for **multiplication**, and ÷ for **division**.

Each of the following symbols names the number twelve. They are called **numerals** or **numerical expressions**.

$$24 \div 2 \qquad 4 + 8 \qquad 15 - 3 \qquad 3 \times 4$$

Of course, "twelve" may be expressed by other numerical expressions. For example, $\dfrac{30 + 18}{4}$, $\dfrac{6 \times 6}{3}$, and $\dfrac{75 - 15}{5}$ are other ways of representing "twelve."

Two numerical expressions such as **4 + 8** and **3 × 4** may be used to write a number sentence in the form of an **equation**: **4 + 8 = 3 × 4**. An equation contains the symbol =, which is read "is equal to." An equation is called a **statement** if it is possible for you to decide whether it is **true** or **false**. Since both 4 + 8 and 3 × 4 name the same number, the equation **4 + 8 = 3 × 4** is a **true** statement. The equation **3 × 8 = 5 + 4** is a **false** statement.

If information is missing from a number sentence in equation form, the equation is *neither* true *nor* false until it is completed.

EXAMPLE.   Is the sentence **3 + 5 = 2 × ?** true or false?

$$3 + 5 = 2 \times 4 \text{ is true} \qquad 3 + 5 = 2 \times 6 \text{ is false}$$

Is it possible for a statement to be both true and false? Use the following statements to help you decide.

True Statements	False Statements
3 + 7 = 2 × 5	0 + 15 = 0
45 − 10 = 5 + 30	30 + 4 = 43
50 + 8 = 58	9 × 3 = 5 × 5

ORAL EXERCISES

Tell whether each statement is true or false.

SAMPLE 1:   12 + 4 = 4 × 4   SAMPLE 2:   30 − 7 = 40 − 6

What you say: True. 12 + 4 and 4 × 4 name the same number, 16.   *What you say:* False. 30 − 7 names 23, but 40 − 6 names 34.

1. 4 + 9 = 10 + 4
2. 5 + 7 = 3 + 10
3. 3 × 9 = 3 × 3 × 3

4. 2 × 3 × 5 = 40 − 10
5. 2 × 4 × 3 = 45 − 5
6. 6 × 1 = 2 × 1 × 3 × 1

Basic Concepts about Numbers and Numerals   9

7. $2 \times 4 = 2 \times 2 \times 2$
8. $8 + 0 = 7 + 1$
9. $3 \times 0 = 3 \times 1$
10. $6 \div 3 = 3 \div 6$
11. $15 \div 5 = 9 \div 3$

12. $1 \times 4 \times 5 = 3 \times 7$
13. $12 \div 3 = 2 \times 4$
14. $4 \times 3 \times 0 = 7 \times 9 \times 0$
15. $2 + 7 + 0 = 3 + 8 + 0$
16. $10 \div 5 = 5 \div 10$

WRITTEN EXERCISES

Tell whether each statement is true or false.

1. $0.03 + 2.1 = 1 + 1.13$
2. $3.50 + 0.17 = 4.77$
3. $12 \times \frac{1}{3} = \frac{1}{2} \times 8$
4. $\frac{1}{4} \times 16 = \frac{1}{3} \times 15$
5. $26 + 3 + 10 = 13 \times 2$
6. $15 + 19 + 11 = 3 \times 15$
7. $5\% + 12\% = 17\%$
8. $0.25 + 0.50 = 0.90 - 0.10$

9. $3\frac{1}{2} + 2\frac{1}{2} = 5 \times 1$
10. $32\% + 18\% = \frac{1}{2}$
11. $10^2 = 9^2 + 1$
12. $4\frac{3}{4} \times 2\frac{1}{4} = 14 \div 2$
13. $4 \times 4 \times 4 = 4 + 4 + 4$
14. $0.73 \times 0.4 \times 0 = 0.89 + 1$
15. $0.75 \div 3 = 7.5 \div 30$
16. $\frac{18}{3} + 6 = 2 + 2 + 8$

Determine the number named by each numerical expression.

SAMPLE: $\dfrac{16 + 8}{2 + 1}$ Solution: $\dfrac{16 + 8}{2 + 1} = \dfrac{24}{3} = 8$

17. $\dfrac{9 + 6}{5}$
18. $\dfrac{10 + 14}{8}$
19. $\dfrac{35 - 11}{12}$
20. $\dfrac{10 + 6 + 9}{5}$

21. $\dfrac{3 \times 4 \times 4}{6}$
22. $\dfrac{54 - 21}{7 + 4}$
23. $\dfrac{21 - 9}{6 - 2}$

24. $\dfrac{8 + 24}{10 - 6}$
25. $\dfrac{7 \times 5}{5 \times 7}$
26. $\dfrac{7 \times 6}{2 + 1}$

Copy and complete each equation to make a true statement.

SAMPLE: $4\frac{1}{2} + 5\frac{1}{2} = 7 + ?$ Solution: $4\frac{1}{2} + 5\frac{1}{2} = 7 + 3$

27. $7 + 10 + ? = 5 \times 5$
28. $23 - ? = 10 + 6$
29. $48 + 62 = ? + 95$
30. $96 \div 4 = 30 - ?$

31. $? \div 9 = 3 + 4$
32. $? \times 4 = 6 \times 6$
33. $5\frac{1}{2} + ? = 18 \div 2$
34. $3\frac{2}{3} + 8 = 15 - ?$

10   Chapter 1

[C] **35.** $3\frac{1}{4} - ? = 1\frac{1}{4} + \frac{1}{4}$
36. $0.27 \div 0.3 = 1 - ?$
37. $1.25 - ? = 0.64 + 0.39$

Show whether each statement is true or false.

SAMPLE: $\dfrac{2+8}{7-2} = \dfrac{6+8}{2}$ Solution: $\dfrac{2+8}{7-2} \stackrel{?}{=} \dfrac{6+8}{2}$

$$\begin{array}{c|c} \frac{\cancel{10}}{\cancel{5}} & \frac{\cancel{14}}{\cancel{2}} \\ 2 & 7 \end{array}$$

The statement is false.

38. $\dfrac{12+3}{5} = \dfrac{21-3}{6}$ **41.** $\dfrac{9 \times 8}{4 \times 3} = \dfrac{2+2+2}{4}$

39. $\dfrac{42 \div 7}{2+4} = \dfrac{3 \times 3}{2 \times 2}$ **42.** $\dfrac{4 \times 4 \times 4}{10-2} = \dfrac{3+6+7}{1+1}$

40. $\dfrac{0.7 \times 0.4}{2} = 1.0 + 0.4$ **43.** $\dfrac{7 \times 8}{8 \times 7} = \dfrac{5 \times 3}{3+5}$

Fundamental Arithmetic Operations

1–4 Addition and Its Inverse

Since you first began to study arithmetic, you have made use of the four basic operations of **addition, subtraction, multiplication,** and **division.** It is often helpful to think of these as pairs of **inverse** operations. Let us illustrate this for addition and subtraction.

Suppose that a barrel contains 25 gallons of water. If another 7 gallons are poured into the barrel today, and 7 gallons are drawn out tomorrow, the barrel will again contain 25 gallons.

After 7 gallons are poured in: $25 + 7 = 32$
After 7 gallons are removed: $32 - 7 = 25$

Since addition and subtraction seem to produce **opposite** effects they are called **inverse** operations.

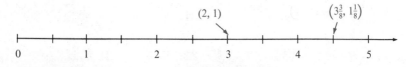

Figure 1–5

Basic Concepts about Numbers and Numerals   11

Addition can be thought of as the process of matching a pair of numbers, called **addends**, with a third number called the **sum**. Can you see, in Figure 1–5, why the number pair **(2, 1)** is matched with **3**? Is it correct to match the number pair $(3\frac{3}{8}, 1\frac{1}{8})$ with $4\frac{1}{2}$? The addition statement **2 + 1 = 3** could be written as the number pair statement **(2, 1)** $\xrightarrow{+}$ **3** similarly $(3\frac{3}{8}, 1\frac{1}{8})$ would be written as the number-pair statement $(3\frac{3}{8}, 1\frac{1}{8}) \xrightarrow{+} 4\frac{1}{2}$. In general the parts of an addition statement are named like this:

$$2 + 1 = 3$$
$$\downarrow \quad \downarrow \quad \downarrow$$
$$\text{addend} + \text{addend} = \text{sum}$$

Let us consider addition from another point of view. On the number line in Figure 1–6, (3, ?) is matched with 5. Using the method above we could write (3, ?) $\xrightarrow{+}$ 5. In this case one of the addends is missing, as shown by the question mark. Do you agree that the missing addend is **2**?

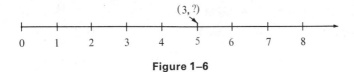

Figure 1–6

Since addition and subtraction are **inverse** operations, we have a choice of expressing this idea by any one of the following equations:

$$3 + ? = 5; \quad 5 - ? = 3; \quad 5 - 3 = ?$$

Each of the equations above becomes a true statement when the question mark is replaced by **2**; we say that they are **equivalent** statements. Their parts are named as shown below.

$$3 + ? = 5$$
$$\downarrow \quad \downarrow \quad \downarrow$$
$$\text{addend} + \text{addend} = \text{sum}$$

$$5 - ? = 3 \qquad\qquad 5 - 3 = ?$$
$$\downarrow \quad \downarrow \quad \downarrow \qquad\qquad \downarrow \quad \downarrow \quad \downarrow$$
$$\text{sum} - \text{addend} = \text{addend} \qquad \text{sum} - \text{addend} = \text{addend}$$

Thus, for the number pair statement (3, 2) $\xrightarrow{+}$ 5 we can write the two equivalent number pair statements (5, 2) $\xrightarrow{-}$ 3 and (5, 3) $\xrightarrow{-}$ 2.

Any addition statement of the form

$$a + b = c$$

has the equivalent subtraction forms

$$c - a = b \quad \text{and} \quad c - b = a.$$

ORAL EXERCISES

Tell how to complete each equation to make a true statement.

SAMPLE: Since $6 + 9 = 15$, we know $15 - 9 = ?$.

What you say: Since $6 + 9 = 15$, we know $15 - 9 = 6$.

1. Since $8 + 13 = 21$, we know $21 - 13 = ?$.
2. Since $12 + 7 = 19$, we know $19 - 12 = ?$.
3. Since $24 + 38 = 62$, we know $62 - ? = 38$.
4. Since $41 + 26 = 67$, we know $67 - ? = 41$.
5. Since $3.8 + 4.8 = 8.6$, we know $8.6 - 3.8 = ?$.
6. Since $18.3 + 5.9 = 24.2$, we know $24.2 - ? = 18.3$.

Describe each number in these statements as an addend or sum.

SAMPLE 1: $8 + 9 = 17$

What you say: 8 and 9 are addends. 17 is the sum.

SAMPLE 2: $45 - 17 = 28$

What you say: 17 and 28 are addends. 45 is the sum.

7. $26 + 35 = 61$
8. $88 = 58 + 30$
9. $67 - 25 = 42$
10. $31 - 16 = 15$
11. $108 = 40 + 68$
12. $325 + 42 = 367$
13. $39 = 75 - 36$
14. $10 = 53 - 43$

Tell what information is given in each equation and how to find the missing information.

SAMPLE 1: $(3, ?) \xrightarrow{+} 8$

What you say: 3 is an addend, 8 is the sum, so an addend is missing. Since $3 + 5 = 8$, the missing addend is 5.

SAMPLE 2: $(?, 7) \xrightarrow{+} 5$

What you say: 7 is an addend, 5 is an addend, so the sum is missing. Since $7 + 5 = 12$, the sum is 12.

15. $(8, 9) \xrightarrow{+} ?$
16. $(11, 3) \xrightarrow{+} ?$
17. $(6, ?) \xrightarrow{+} 19$
18. $(?, 7) \xrightarrow{+} 21$
19. $(16, ?) \xrightarrow{+} 23$
20. $(34, ?) \xrightarrow{+} 68$

21. $(?, 12) \xrightarrow{+} 16$
22. $(?, 28) \xrightarrow{+} 31$
23. $(?, \frac{2}{9}) \xrightarrow{+} \frac{3}{9}$
24. $(?, \frac{3}{4}) \xrightarrow{+} 6$
25. $(3.8, 1.2) \xrightarrow{+} ?$
26. $(15.75, 3.50) \xrightarrow{+} ?$

WRITTEN EXERCISES

Find the number needed to complete each number-pair statement.

SAMPLE 1: $(8\frac{1}{2}, 3) \xrightarrow{+} ?$

Solution: Since $8\frac{1}{2} + 3 = 11\frac{1}{2}$, the number needed is $11\frac{1}{2}$.

SAMPLE 2: $(8, 5\frac{1}{4}) \xrightarrow{+} ?$

Solution: Since $2\frac{3}{4} + 5\frac{1}{4} = 8$, we know $8 - 5\frac{1}{4} = 2\frac{3}{4}$, and the number needed is $2\frac{3}{4}$.

1. $(15, 8) \xrightarrow{+} ?$
2. $(8, 15) \xrightarrow{+} ?$
3. $(6\frac{3}{8}, 2\frac{1}{2}) \xrightarrow{+} ?$
4. $(?, 39) \xrightarrow{+} 25$
5. $(25, 39) \xrightarrow{+} ?$
6. $(91, 34) \xrightarrow{+} ?$
7. $(?, 7.4) \xrightarrow{+} 3.8$

8. $(?, 3.9) \xrightarrow{+} 14.2$
9. $(6\frac{1}{5}, ?) \xrightarrow{+} 9\frac{4}{5}$
10. $(?, 3\frac{1}{7}) \xrightarrow{+} 10\frac{6}{7}$
11. $(8.8, ?) \xrightarrow{+} 15.7$
12. $(68.2, ?) \xrightarrow{+} 23.5$
13. $(18\frac{7}{8}, 10\frac{1}{2}) \xrightarrow{+} ?$
14. $(?, 51.6) \xrightarrow{+} 89.2$

Complete the following to make equivalent statements.

SAMPLE: Since $? + 7 = 18$, we know $18 - 7 = ?$.

Solution: Since $11 + 7 = 18$, we know $18 - 7 = 11$.

15. Since $6 + ? = 23$, we know $23 - 6 = ?$.
16. Since $24 + ? = 36$, we know $36 - 24 = ?$.
17. Since $10 + ? = 18$, we know $18 - 10 = ?$.
18. Since $? + 17 = 42$, we know $42 - 17 = ?$.
19. We know $13 - 7 = ?$ since $7 + ? = 13$.
20. We know $85 - 14 = ?$ since $14 + ? = 85$.
21. We know $63 - 18 = ?$ since $? + 18 = 63$.
22. Since $4.9 + ? = 15$, we know $15 - ? = 4.9$.
23. Since $? + 18.4 = 26.2$, we know $26.2 - 18.4 = ?$.
24. Since $17 + ? = 55$, we know $? + 17 = 55$.

14 *Chapter 1*

Complete each of the following to make a true statement. Then write two equivalent subtraction statements.

SAMPLE: $6.2 + 8.9 = ?$

Solution: $6.2 + 8.9 = 15.1$; $15.1 - 6.2 = 8.9$; $15.1 - 8.9 = 6.2$

25. $3\frac{1}{4} + 2\frac{3}{8} = ?$
26. $10\frac{1}{5} + 3\frac{1}{2} = ?$
27. $72 + ? = 117$
28. $68 + ? = 155$
29. $36.5 + 27.7 = ?$
30. $62 = 45 + ?$

31. $52.5 + 19.7 = ?$
32. $? = 14\frac{3}{4} + 2\frac{1}{2}$
33. $? = 8\frac{5}{8} + 3\frac{1}{16}$
34. $429 + 375 = ?$
35. $642 + 109 = ?$
36. $83.9 = ? + 48.2$

Write each of the following in the form of an addition statement. Then complete the statement to make it true.

SAMPLE: $? - \frac{3}{8} = \frac{5}{8}$ *Solution:* $\frac{3}{8} + \frac{5}{8} = ?$
$\frac{3}{8} + \frac{5}{8} = 1$

|C|

37. $? - 15 = 35$
38. $? - 2\frac{1}{4} = 13\frac{1}{4}$
39. $? - 8\frac{1}{3} = 6\frac{1}{3}$
40. $? - 61 = 28$

41. $65 = ? - 20$
42. $41 = ? - 35$
43. $6\frac{3}{8} = ? - 2\frac{1}{2}$
44. $4.02 = ? - 3.27$

1–5 Multiplication and Its Inverse

Suppose that we begin with the number **8** and multiply it by **6**. The result is **48**. If **48** is divided by **6** the result is the number we began with, **8**.

$$8 \times 6 = 48 \quad \text{and} \quad 48 \div 6 = 8$$

Since multiplication and division produce opposite effects, just as was the case with addition and subtraction, these are also called **inverse** operations.

Multiplication can be thought of as the process by which a pair of numbers, called **factors**, is matched with a third number called the **product**. The number line picture in Figure 1–7 shows the number pair **(3, 4)** matched with **12** since $3 \times 4 = 12$. The number-pair statement to show this is $(3, 4) \xrightarrow{\times} 12$.

Do you see why **(2, 7)** is matched with **14**? This might be written as $(2, 7) \xrightarrow{\times} 14$.

Basic Concepts about Numbers and Numerals

Figure 1-7

In the equation 5 × ? = 35, the question mark stands for a missing factor. Since in each of the division equations 35 ÷ 5 = ? and 35 ÷ ? = 5, and in the multiplication equation 5 × ? = 35, the statement becomes true when the question mark is replaced by 7, we say that the three statements are **equivalent**. Their parts are named as follows:

$$5 \times 7 = 35$$
$$\downarrow \quad \downarrow \quad \downarrow$$
$$\text{factor} \times \text{factor} = \text{product}$$

$$35 \div 5 = 7 \qquad\qquad 35 \div 7 = 5$$
$$\downarrow \quad \downarrow \quad \downarrow \qquad\qquad \downarrow \quad \downarrow \quad \downarrow$$
$$\text{product} \div \text{factor} = \text{factor} \qquad \text{product} \div \text{factor} = \text{factor}$$

Written as equivalent number pair statements, we have:

$$(5, 7) \xrightarrow{\times} 35; \quad (35, 5) \xrightarrow{\div} 7; \quad (35, 7) \xrightarrow{\div} 5$$

Any multiplication sentence of the form

$$a \times b = c$$

has the equivalent division forms

$$c \div b = a \quad \text{and} \quad c \div a = b.$$

(*a* and *b* cannot equal 0)

ORAL EXERCISES

Tell how to complete each equation to make a true statement.

SAMPLE: Since 7 × 8 = 56, we know 56 ÷ 7 = ?.
What you say: Since 7 × 8 = 56, we know 56 ÷ 7 = 8.

1. Since 4 × 9 = 36, we know 36 ÷ 4 = ?.
2. Since 5 × 8 = 40, we know 40 ÷ 8 = ?.

3. Since 7 × 6 = 42, we know 42 ÷ ? = 7.
4. Since 13 × 5 = 65, we know 65 ÷ ? = 5.
5. Since 12 × 0.4 = 4.8, we know 4.8 ÷ 12 = ?.
6. Since 3.6 × 1.2 = 4.32, we know 4.32 ÷ ? = 3.6.

Describe each number as a factor or product.

SAMPLE 1: 15 × 7 = 105

What you say: 15 and 7 are factors. 105 is the product.

SAMPLE 2: 72 ÷ 9 = 8

What you say: 9 and 8 are factors. 72 is the product.

7. 16 × 3 = 48
8. 51 ÷ 17 = 3
9. 46 ÷ 2 = 23
10. 15 = 45 ÷ 3
11. 17 = 68 ÷ 4
12. 40 = 8 × 5

Tell what information is given and how to find the missing information.

SAMPLE 1: (6, ?) $\xrightarrow{\times}$ 54

What you say: 6 is a factor, 54 is the product, so a factor is missing. Since 6 × 9 = 54, the missing factor is 9.

SAMPLE 2: (?, 7) $\xrightarrow{\div}$ 5

What you say: 7 is a factor, 5 is a factor, so the product is missing. Since 7 × 5 = 35, the product is 35.

13. (12, 8) $\xrightarrow{\times}$?
14. (5, 12) $\xrightarrow{\times}$?
15. (10, ?) $\xrightarrow{\times}$ 30
16. (8, ?) $\xrightarrow{\times}$ 72
17. (3, ?) $\xrightarrow{\times}$ 27
18. (6, ?) $\xrightarrow{\times}$ 36
19. (28, 4) $\xrightarrow{\div}$?
20. (60, 20) $\xrightarrow{\div}$?
21. (?, 14) $\xrightarrow{\div}$ 3
22. (?, 2.9) $\xrightarrow{\div}$ 7.2
23. (?, $\frac{3}{4}$) $\xrightarrow{\div}$ $\frac{1}{2}$
24. (?, $\frac{2}{5}$) $\xrightarrow{\div}$ $\frac{1}{3}$

WRITTEN EXERCISES

Find the number needed to complete each number-pair statement.

SAMPLE 1: ($4\frac{1}{5}$, 3) $\xrightarrow{\times}$?

Solution: Since $4\frac{1}{5}$ × 3 = $12\frac{3}{5}$, the number needed is $12\frac{3}{5}$.

SAMPLE 2: (15, 3) $\xrightarrow{\div}$?

Solution: Since 3 × 5 = 15, we know 15 ÷ 3 = 5, and the number needed is 5.

Basic Concepts about Numbers and Numerals **17**

A
1. $(14, 7) \xrightarrow{\times} ?$
2. $(15, 12) \xrightarrow{\times} ?$
3. $(2\frac{1}{4}, 3) \xrightarrow{\times} ?$
4. $(?, 5) \xrightarrow{\div} 2\frac{1}{4}$
5. $(35, 7) \xrightarrow{\div} ?$
6. $(12, ?) \xrightarrow{\times} 60$
7. $(4.6, 7.1) \xrightarrow{\times} ?$
8. $(12.3, 8) \xrightarrow{\times} ?$
9. $(9, ?) \xrightarrow{\div} \frac{3}{4}$
10. $(15, \frac{5}{6}) \xrightarrow{\div} ?$

Complete the following to make equivalent statements.

SAMPLE: Since $12 \times ? = 36$, we know $36 \div 12 = ?$.

Solution: Since $12 \times 3 = 36$, we know $36 \div 12 = 3$.

11. Since $4 \times ? = 32$, we know $32 \div 4 = ?$.
12. Since $22 \times ? = 66$, we know $66 \div 22 = ?$.
13. Since $30 \times ? = 60$, we know $60 \div 30 = ?$.
14. Since $14 \times ? = 28$, we know $28 \div ? = 14$.
15. Since $25 \times ? = 75$, we know $75 \div ? = 25$.

B
16. Since $2 \times ? = 9.4$, we know $9.4 \div 2 = ?$.
17. Since $4 \times ? = 2.48$, we know $2.48 \div ? = 4$.
18. Since $\frac{4}{3} \times ? = \frac{8}{9}$, we know $\frac{8}{9} \div \frac{4}{3} = ?$.
19. Since $\frac{5}{2} \times ? = \frac{5}{6}$, we know $\frac{5}{6} \div ? = \frac{5}{2}$.

Complete each of the following to make a true statement. Then write two equivalent division statements.

SAMPLE: $16 \times 21 = ?$

Solution: $16 \times 21 = 336$
$336 \div 16 = 21$ and $336 \div 21 = 16$

20. $12 \times 25 = ?$
21. $14\frac{1}{2} \times 6 = ?$
22. $8 \times 4\frac{2}{3} = ?$
23. $13 \times ? = 195$
24. $? = 16 \times 23$
25. $? = 5.5 \times 1.7$
26. $4 \times ? = 33$
27. $9 \times ? = 17$

C
28. $78 = 13 \times ?$
29. $108 = ? \times 9$

Write each of the following in the form of a multiplication equation. Then complete the equation to make a true statement.

SAMPLE: $? \div 17 = 9$ *Solution:* $17 \times 9 = ?$
$17 \times 9 = 153$

30. $? \div 12 = 23$
31. $? \div 7 = 45$
32. $? \div \frac{3}{4} = \frac{5}{8}$
33. $? \div \frac{7}{3} = \frac{1}{2}$
34. $? \div 1.8 = 3.9$
35. $? \div 1.23 = 4.6$

1-6 Graphs and the Number Line

Many important ideas in mathematics can be illustrated through the use of **graphs**. When a number is matched with a point on the number line the point is said to be the **graph** of the number. Often large solid dots are used for emphasis, as in Figure 1–8. The graph of **all**

Figure 1–8

of the numbers (arithmetic numbers) between $\frac{1}{2}$ and $\frac{3}{2}$ is a solid line with an open dot at each end, as shown in Figure 1–9. The open dots

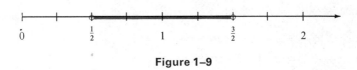

Figure 1–9

mean that the graph does **not** include the points $\frac{1}{2}$ and $\frac{3}{2}$. The graph of all the points between $\frac{1}{2}$ and $\frac{3}{2}$, **inclusive**, is shown in Figure 1–10.

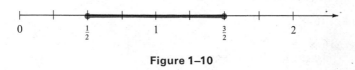

Figure 1–10

Do you agree that **solid dots** are needed at the ends of the graph in this case?

ORAL EXERCISES

Tell what numbers are graphed in each illustration.

SAMPLE:

What you say: 1, $1\frac{1}{2}$, and 2

1.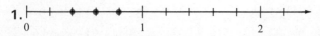

Basic Concepts about Numbers and Numerals **19**

2.

(Recall that the value of π is about 3.14.)

3.

4.

5.

6.

WRITTEN EXERCISES

Draw a number line for each exercise and graph the numbers given.

SAMPLE: 2, $3\frac{2}{3}$, and $4\frac{1}{3}$

Solution:

1. 1, 2 and 3
2. 0, 3 and $3\frac{1}{2}$
3. $\frac{1}{4}$, $\frac{1}{2}$ and $2\frac{3}{4}$
4. 0.5, 1.0 and 1.5

5. 1, $2\frac{1}{2}$ and π
6. $\frac{3}{5}$, $1\frac{4}{5}$ and 3
7. $\frac{6}{10}$, $1\frac{4}{10}$ and $3\frac{1}{10}$
8. $\frac{9}{8}$, $\frac{5}{8}$ and $2\frac{3}{8}$

9. 2, $\frac{9}{2}$ and $\frac{10}{2}$
10. 0.2, 0.7 and 1.6
11. 0.4, 1.5 and 2.9
12. 0.25, 0.75 and 2.0

Match each graph with its description.

13.

A. The numbers between 0 and 2, including 0.

14.

B. The numbers between $2\frac{1}{2}$ and $4\frac{1}{2}$, inclusive.

15.

C. The numbers between 3 and 9, inclusive.

16.

D. The numbers between 1 and $2\frac{1}{3}$.

20 Chapter 1

For each exercise, draw a number line and graph the number(s) described.

SAMPLE: The whole numbers between 3 and 7.

Solution: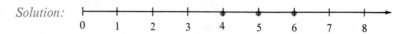

B 17. The whole numbers between 1 and 6.
18. The number halfway between 3 and 5.
19. The whole numbers between 2 and 7, inclusive.
20. Three consecutive whole numbers, the first of which is 4.
21. The numbers between 3 and 8.
22. The first three whole numbers.
23. The numbers between $2\frac{1}{2}$ and 7, inclusive.
24. The number halfway between 3 and 6.
25. The whole numbers between 0 and 1 and including 0.
26. The whole numbers between $\frac{1}{2}$ and $4\frac{1}{2}$.

C 27. The whole number two-thirds of the distance from the origin to 6.
28. The whole numbers less than π.
29. The numbers between $1\frac{1}{8}$ and 4, and including 4.
30. The numbers less than π.

1–7 Operations on the Number Line

Another use of number lines is to illustrate the basic operations. Do you see how the sum $\frac{1}{2} + \frac{3}{4} = ?$ is shown in Figure 1–11? We begin at **0** and draw an arrow $\frac{1}{2}$ unit long to the **right**; a second arrow begins at $\frac{1}{2}$ and is drawn $\frac{3}{4}$ unit long, also to the **right**. We finish at $\frac{5}{4}$. Therefore, $\frac{1}{2} + \frac{3}{4} = \frac{5}{4}$.

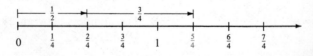

Figure 1–11

Do you see how the missing addend in $\frac{5}{4} - \frac{1}{2} = ?$ is found in Figure 1–12? We begin at **0** and draw an arrow $\frac{5}{4}$ units long to the **right**. Then an arrow $\frac{1}{2}$ unit long begins at $\frac{5}{4}$ and is drawn to the **left**.

Basic Concepts about Numbers and Numerals 21

We finish at $\frac{3}{4}$. Therefore, $\frac{5}{4} - \frac{1}{2} = \frac{3}{4}$.

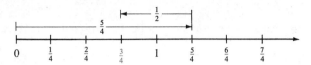

Figure 1–12

Writing the sentences $\frac{1}{2} + \frac{3}{4} = ?$ and $\frac{5}{4} - \frac{1}{2} = ?$ as number-pair statements might help you to decide in which direction to draw the second arrow. For $(\frac{1}{2}, \frac{3}{4}) \xrightarrow{+} ?$ the second arrow is drawn to the right. For $(\frac{5}{4}, \frac{1}{2}) \xrightarrow{-} ?$, in which direction is the second arrow drawn?

To show a problem like $4 \times \frac{1}{3} = ?$ on the number line, multiplication is treated as repeated addition. Do you see that $4 \times \frac{1}{3} = ?$ and $\frac{1}{3} + \frac{1}{3} + \frac{1}{3} + \frac{1}{3} = ?$ mean the same? Beginning at **0**, we draw four arrows, each $\frac{1}{3}$ of a unit long, to the right, as shown in Figure 1–13. We finish at $\frac{4}{3}$. Therefore, $4 \times \frac{1}{3} = \frac{4}{3}$.

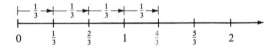

Figure 1–13

To show $\frac{4}{3} \div \frac{1}{3} = ?$, division may be thought of as repeated subtraction, or as the answer to the question: "How many $\frac{1}{3}$'s are in $\frac{4}{3}$?" Do you understand why we begin at $\frac{4}{3}$ and draw arrows, each $\frac{1}{3}$ unit long, to the **left**, as shown in Figure 1–14? **Four** arrows are required to finish at the origin. Therefore, $\frac{4}{3} \div \frac{1}{3} = 4$.

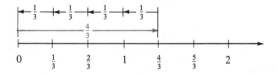

Figure 1–14

Can you suggest how the number pair statements could help you decide about the direction of the arrows? For $(4, \frac{1}{3}) \xrightarrow{\times} ?$, in which direction do you draw the arrows which have length equal to $\frac{1}{3}$? In which direction are the arrows of length $\frac{1}{3}$ drawn for $(\frac{4}{3}, \frac{1}{3}) \xrightarrow{\div} ?$

ORAL EXERCISES

Give a number statement suggested by each number-line picture.

SAMPLE:

What you say: $\frac{1}{2} + \frac{3}{2} = 2$

5.
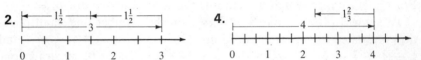

6.

WRITTEN EXERCISES

Make a number line representation of each number statement.

SAMPLE: $4 \times \frac{2}{3} = 2\frac{2}{3}$

Solution:

  1. $3 + 5 = 8$ 4. $4\frac{1}{4} + 2\frac{1}{2} = 6\frac{3}{4}$ 7. $7\frac{1}{2} - 4 = 3\frac{1}{2}$
2. $4 + 7 = 11$ 5. $6 - 2 = 4$ 8. $3 \times 5 = 15$
3. $2\frac{1}{2} + 3 = 5\frac{1}{2}$ 6. $10 - 3 = 7$

Write the number statement illustrated on each number line.

SAMPLE:

Solution: $\frac{3}{5} + \frac{6}{5} = \frac{9}{5}$

9.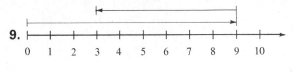

10.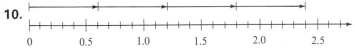

11.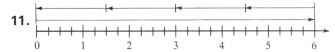

12.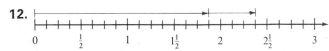

Illustrate each statement with a number line picture.

B
13. $4 \times 2\frac{1}{3} = 9\frac{1}{3}$
14. $10 \div 2\frac{1}{2} = 4$
15. $5 \times 1\frac{1}{2} = 7\frac{1}{2}$
16. $3 \times 0.4 = 1.2$

17. $4 \div \frac{4}{5} = 5$
18. $4.5 \div 1.5 = 3$
19. $6 = 4 \times 1.5$
20. $15 \div 2.5 = 6$

1–8 Inequalities

You have worked with number sentences, called equations, that describe the **equality** relation. Number sentences can be expressed in words or in mathematical symbols.

In words: Three, increased by six, is equal to nine.

In symbols: $3 + 6 = 9$

The equality relation is very important, but other relations are just as important. The subject of mathematics is often concerned with number sentences like those below. Do you agree that each statement is true?

EXAMPLE 1.

In words: The sum of three and five is not equal to ten.

In symbols: $3 + 5 \neq 10$

EXAMPLE 2.

    In words: Thirty-five is greater than twenty.

    In symbols: 35   >   20

EXAMPLE 3.

    In words: Fifteen is less than twenty-four.

    In symbols: 15   <   24

Do these sentences describe relations other than equality? They are not equations. A sentence that describes the "is not equal to," the "is greater than," or the "is less than" relation is called an **inequality**

Inequality statements of the "is less than" and "is greater than" type can be easily judged **true** or **false**. Simply think of the < and > symbols as arrowheads. In an equality statement that is **true**, the "arrowhead" always points toward the numeral for the **smaller** number.

Both of the statements $3 > 1\frac{1}{2}$ and $1\frac{1}{2} < 3$ are true. The number line in Figure 1–13 shows that 3 is to the **right** of $1\frac{1}{2}$, so 3 is **greater** than $1\frac{1}{2}$. Since $1\frac{1}{2}$ is to the **left** of 3, $1\frac{1}{2}$ is **less** than 3.

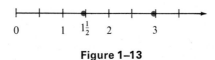

Figure 1–13

The statement "3 is between 2 and 5" means "3 is **greater** than 2 and 3 is **less** than 5."

    3 is greater than 2 and 3 is less than 5.
        $3 > 2$ and $3 < 5$

The statement $3 > 2$ **and** $3 < 5$ can be written $2 < 3 < 5$, and is read "2 is less than 3 and 3 is less than 5" or simply "3 is between 2 and 5."

ORAL EXERCISES

Tell the meaning of each relation symbol.

1. =   **2.** ≠   **3.** >   **4.** <

Express each of the following statements in words.

SAMPLE: $3 + 2 \neq 7$

What you say: Three plus two is not equal to seven.

5. $24 - 10 < 25$
6. $5 \times 3 < 18$
7. $1.2 \times 4 > 3$
8. $8 \times 9 > 16$
9. $10 < 3 + 9$
10. $18 > 10 + 5$
11. $3.8 < 3 + 1$
12. $1.3 + 2.4 \neq 7.3$
13. $575 \neq 5 \times 100$
14. $3 < 8 < 14$
15. $1\frac{1}{2} < 5 < 11$
16. $3.5 > 3.49$

WRITTEN EXERCISES

Show whether each statement is true or false.

SAMPLE 1: $3 + 9 \neq 3 \times 5$ Solution: $\dfrac{3 + 9 \neq 3 \times 5}{12 \mid 15}$

True, since $12 \neq 15$.

SAMPLE 2: $\dfrac{4 + 8}{2} > \dfrac{6 \times 4}{3}$ Solution: $\dfrac{\dfrac{4 + 8}{2} > \dfrac{6 \times 4}{3}}{\dfrac{12}{2} \mid \dfrac{24}{3}}$

$\dfrac{}{6 \mid 8}$

False, since $6 < 8$.

1. $12 + 0 > 5 \times 2$
2. $4 \times 5 < 7 + 5$
3. $\frac{1}{4} + \frac{3}{4} > 1$
4. $4 + 4 < 4 \times 4$
5. $3^2 < 3 \times 3 \times 3$
6. $\dfrac{1 + 5}{2} \neq 5$
7. $\dfrac{5 + 10}{3} \neq \dfrac{15}{3}$
8. $\dfrac{12 + 8}{5} > 2 \times 2$
9. $5 + 6 \neq \dfrac{12 + 18}{3}$
10. $3.6 + 2.3 < 5.91$
11. $3.29 \neq 3.299$
12. $\frac{125}{1} > 125$
13. $3\frac{1}{2} \times 0 = 0$
14. $16 \times 1 < 1$

Write each statement in symbols. Then tell whether it is true or false.

SAMPLE: Five is between one and eight.

Solution: $1 < 5 < 8$; True.

15. Six is less than fifteen.

16. The sum of five and four is greater than eight.
17. The product of one and nine is less than ten.
18. Three is not equal to the sum of seven and one.
19. Sixteen divided by one is not equal to sixteen.
20. Ten is between three and eight.

Replace each question mark with $<$, $>$ or $=$ to make a true statement.
SAMPLE: $3 + 6$? 5 *Solution:* $3 + 6 > 5$

21. $12 - 2$? 5×2
22. $3 \times 3 \times 3$? 3^3
23. 4.99 ? 4.98
24. $15 - 3$? 4×4
25. 25×0 ? 18×0
26. 16×1 ? $15 \div 1$

B
27. $\frac{18}{1}$? 19×1
28. $\frac{1}{5} + \frac{2}{5}$? $3 \times \frac{1}{5}$
29. $4 \times \frac{1}{3}$? $3 \times \frac{1}{3}$
30. 1.777 ? 1.778
31. $0.75 + 0.25$? $6 \div 6$
32. $2\frac{2}{3}$? 2.67
33. $1\frac{1}{3}$? 1.33
34. 8 ? $\frac{12 + 5}{2}$

Copy and complete each of the following to make true statements.
SAMPLE: $9 \times 4 \neq 12 \times$?
Solution: $9 \times 4 \neq 12 \times 2$ (or any number other than 3)

35. $12 - 7 = 2 +$?
36. $15 + 6 \neq 3 \times$?
37. $8 + 9 =$? $+ 8$
38. $\frac{2}{3} \neq \frac{1}{3} +$?

C
39. ? $\div 5 = 3 + 1$
40. $8 +$? $= 8$
41. $10 < 3 +$?
42. $4 \times 5 > 4 \times$?
43. $150 \neq$? $\times 50$
44. $\frac{15 + ?}{4} = 5$
45. $6 <$? < 8
46. ? $\div$? $= 1$
47. $3 \times$? > 30
48. ? $\times 13 > 0$

CHAPTER SUMMARY

Inventory of Structure and Concepts

1. **Whole numbers** can be matched with equally spaced points on the **number line**. Every whole number w has as its **successor** $w + 1$. Whole numbers have an **order property** which means they can be arranged according to size or value.

Basic Concepts about Numbers and Numerals   27

2. Each **arithmetic number** can be matched with a point on the number line.

3. **Addition, subtraction, multiplication,** and **division** can each be thought of as a process of matching a pair of numbers with a third number.

   For addition: addend + addend = sum
   For subtraction: sum − addend = addend
   For multiplication: factor × factor = product
   For division: product ÷ factor = factor

4. Addition and subtraction are a pair of **inverse** operations; that is, the effect of one operation is the opposite of the effect of the other. In the same way, the operations of multiplication and division are inverses of each other.

5. In graphing, the **coordinate of a point** on the number line is the number assigned to the point. The **graph of a number** is the point on the number line to which the number is assigned.

6. **Equations** are number sentences that contain the symbol = (is equal to). A number sentence that uses one of the following symbols is an **inequality**:

   $\neq$ (is not equal to)
   $<$ (is less than)
   $>$ (is greater than)

Vocabulary and Spelling

Pronounce, spell, and give the meaning of each of these words and expressions. The number refers to the page where the item is introduced.

origin (p. 1)
consecutive (order) (p. 1)
successor (p. 2)
numbers of arithmetic (p. 2)
coordinate (p. 4)
between (p. 4)
inclusive (p. 5)
digit (p. 7)
numeral (p. 8)
equation (p. 8)
numerical expression (p. 8)
statement (p. 8)

inverse (p. 10)
addend (p. 11)
number pair (p. 11)
sum (p. 11)
factor (p. 14)
product (p. 14)
equivalent statements (p. 15)
graph (p. 18)
repeated addition (p. 21)
repeated subtraction (p. 21)
inequality (p. 24)
$\neq, <, >$ (p. 24)

Chapter Test

1. Find two consecutive whole numbers whose sum is 41.
2. What is the sum of four consecutive whole numbers the first of which is 10?
3. Name the whole numbers between $4\frac{1}{2}$ and 9.
4. How many whole numbers are there between 7 and 8?

Refer to the following number line for Questions 5–7.

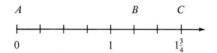

5. What is the coordinate of point A?
6. What is the coordinate of point B?
7. What is the coordinate of the point that is one-third of the distance from A to C?

In Questions 8–13, find the number needed to make each a true statement.

8. $(5.2, ?) \xrightarrow{+} 11$
9. $(?, 17) \xrightarrow{+} 49$
10. $(5\frac{1}{2}, 1\frac{1}{4}) \xrightarrow{-} ?$
11. $(3, 37) \xrightarrow{\times} ?$
12. $(20.5, 5) \xrightarrow{\div} ?$
13. $(?, 1) \xrightarrow{\times} 20$

14. On a number line, graph the whole numbers between 5 and 10.
15. On a number line, graph the whole numbers between 6.3 and 12, including 12.
16. Write the statement illustrated by the following number line.

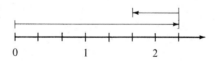

17. Illustrate $3\frac{1}{4} + 2\frac{1}{2} = 5\frac{3}{4}$ with a number line representation.
18. Illustrate $18 \div 3 = 6$ with a number line representation.

Replace each question mark with $=$, $<$, or $>$ to make a true statement.

19. $3\frac{3}{4} \times \frac{4}{15}$? 1
20. 2.555 ? 2.55

Basic Concepts about Numbers and Numerals   29

Chapter Review

1–1 Numbers, Numerals and the Number Line

1. List three consecutive whole numbers the first of which is 12.
2. Find two consecutive whole numbers whose sum is 37.
3. What is the sum of three consecutive whole numbers the first of which is 7?
4. Copy and complete the number line by writing the missing numerals.

1–2 Number Relationships

5. Name the number that is the coordinate of each lettered point.

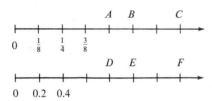

6. Name the whole numbers between π and $7\frac{1}{2}$.
7. Given the following number line:

   A B C D E F G H I
   ├──┼──┼──┼──┼──┼──┼──┼──┼──►
   0 1 2 3 4 5 6 7 8

   Name the point that is one-fourth the distance from F to G.
   Name the point that is one-third the distance from B to G.

1–3 Numerical Expressions and Equality Statements

8. Indicate whether each of the following statements is true or false.
   $6^2 - 5^2 = (6 - 5) \cdot (6 + 5)$
   $\dfrac{21 - 9}{6 - 2} = \dfrac{6 \times 4}{8}$

9. Complete each equation to make a true statement.
   $4\frac{1}{2} + ? = 7 + 3$
   $7 \times ? = 35 \times 3$

30 Chapter 1

1–4 Addition and Its Inverse

Find the number needed to make a true statement in each of Questions 10–14.

10. $(24, 15) \xrightarrow{+} ?$

11. $(19, ?) \xrightarrow{+} 24\frac{1}{2}$

12. $(?, 12.3) \xrightarrow{+} 15.6$

13. $(46, 15) \xrightarrow{-} ?$

14. $(37, 28.1) \xrightarrow{-} ?$

15. Write an equivalent addition statement for $49 - 15 = 34$.

1–5 Multiplication and Its Inverse

Find the number needed to make a true statement in each of Questions 16–22.

16. $(9, 12) \xrightarrow{\times} ?$

17. $(15, ?) \xrightarrow{\times} 75$

18. $(?, \frac{3}{4}) \xrightarrow{\times} 15$

19. $(105, 5) \xrightarrow{\div} ?$

20. $(23, 2) \xrightarrow{\div} ?$

21. $(18.6, 3) \xrightarrow{\div} ?$

22. Write $\frac{624}{13} = 48$ in the form of a multiplication statement.

1–6 Graphs and the Number Line

For each question, draw a number line and graph the numbers described.

23. The whole numbers between 3 and 7.

24. The whole numbers between $\frac{1}{2}$ and 6.2, inclusive.

25. The whole numbers between 0 and 5, including 5.

1–7 Operations on the Number Line

Write the number statement illustrated on each number line.

26.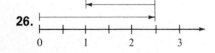

27.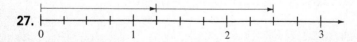

Make a number line representation of each number statement.

28. $4 \times 2.5 = 10$ **29.** $1\frac{1}{2} + 3 = 4\frac{1}{2}$ **30.** $15 \div 3 = 5$

1-8 Inequalities

Indicate whether each statement is true or false.

31. $15 + 1 < 15 \times 1$
32. $5^3 > 5 + 5 + 5$
33. $3\frac{1}{2} \times \frac{1}{4} < 1$

Replace each question mark with $>$, $<$, or $=$ to make a true statement.

34. $2 \times 2 \times 2 \times 2 \; ? \; 2^4$
35. $1.77 \; ? \; 1.777$
36. $5 \times \frac{1}{2} \; ? \; 5 \times \frac{1}{3}$

Review of Skills

Add.

1. $5 + 8$
2. $57 + 14$
3. $79 + 65$
4. $7.1 + 28.24$
5. $0.3 + 5.83$
6. $0.05 + 0.5 + 5$
7. $\frac{1}{4} + \frac{5}{8}$
8. $3\frac{3}{4} + 5\frac{7}{8}$

Subtract.

9. $25 - 15$
10. $53 - 28$
11. $90 - 24$
12. $2.7 - 1.9$
13. $5 - 2.6$
14. $7.1 - 0.24$
15. $1\frac{3}{4} - \frac{3}{8}$
16. $5\frac{1}{4} - 3\frac{3}{8}$

Multiply.

17. 3×10
18. 24×100
19. 345×1000
20. 24×0.4
21. 6.2×3.1
22. 2.25×3.5
23. $\frac{3}{4} \times \frac{1}{2}$
24. $4\frac{1}{2} \times \frac{7}{8}$

Divide.

25. $5 \overline{)35}$
26. $48 \div 16$
27. $\dfrac{105}{35}$
28. $0.2 \overline{)40}$
29. $5.5 \div 0.5$
30. $\dfrac{105}{0.35}$
31. $2 \div 1\frac{1}{2}$
32. $\frac{1}{4} \div 2$
33. $\frac{5}{2} \div \frac{1}{2}$

34. If 2 × 5 = 10, the numeral 2 is called a(n) __?__ of the product 10.
35. If 2 + 5 = 7, the numeral 2 is called a(n) __?__ of the sum 7.
36. The prime factored form of 12 is $2^2 \times 3$. Determine the prime factored form of: 18; 15; 28.
37. The divisors of 12 are 1, 2, 3, 4, 6 and 12. Name the divisors of: 10; 16; 24.

Make a prediction as to what numeral, or numerals, would continue each of the following patterns.

SAMPLE 1: 1, 5, 9, __?__, __?__   *Solution:* 13 and 17
SAMPLE 2: (12, 10), (7, 5), (10, 8), (3, __?__)   *Solution:* 1

38. 1, 3, 4, __?__, __?__
39. 2, 8, 32, __?__
40. 48, 24, 12, __?__, __?__
41. 15, 18, 12, __?__, __?__
42. 1, 8, 5, 12, 9, 16, __?__, __?__
43. 1, 1, 2, 3, 5, 8, 13, __?__, __?__
44. 27, 9, 3, __?__, __?__
45. 10, 5, $\frac{5}{2}$, __?__, __?__
46. (1, 2), (2, 3), (3, 4), (__?__, __?__)
47. (5, 10), (2, 4), (9, 18), (10, __?__), (__?__, 14)
48. (5, 0), (2, 0), (7, 0), (15, __?__), (23, __?__)
49. (2, 5), (3, 10), (5, 26), (7, 50), (6, __?__)
50. (12, 4), (18, 6), (33, 11), (39, 13), (54, __?__), (__?__, 5)

■ ■

CHECK POINT
FOR EXPERTS

Digits and Numerals

You know that we use ten symbols, called **digits**, to write numerals in our system of numeration. The digits are **0, 1, 2, 3, 4, 5, 6, 7, 8,** and **9**. For example, the numeral **1975** uses the four digits **1, 9, 7,** and **5**. In extended form, 1975 = 1000 + 900 + 70 + 5.

Basic Concepts about Numbers and Numerals 33

Of course, the same four digits can be arranged to form numerals that represent other numbers, such as **1759, 1957, 9715, 5791**, and so on. Do you see that the largest whole number that can be represented by using each of the digits 1, 9, 7, and 5 *exactly once* is **9751**, and that the smallest is **1579**?

Questions

1. What is the largest whole number that can be represented by a two-digit numeral?
2. What is the smallest whole number that can be represented by a three-digit numeral?
3. What is the largest whole number that can be represented by a three-digit numeral?
4. What is the largest even number that can be represented by a two-digit numeral?
5. What is the smallest odd number that can be represented by a two-digit numeral?
6. Using each of the digits 1, 2, 3, and 4 once and only once, write a four-digit numeral for:

    a. the smallest possible even number
    b. the largest possible even number
    c. the smallest possible odd number
    d. the largest possible odd number

7. Write in order of size, smallest first, all the three-digit numerals for whole numbers that can be represented by using each of the digits 2, 3, and 4, once and only once.
8. Write in order of size, largest first, all the three-digit numerals for odd numbers that can be represented by using each of the digits 1, 2, and 3, once and only once.
9. Write in order of size, largest first, all the three-digit numerals for even numbers that can be represented by using each of the digits 4, 5, and 6, once and only once.
10. What is the smallest odd number that can be represented by a five-digit numeral?

Martinette glider, 1909 . . .

Concorde supersonic transport . . .

The Language and Symbols of Mathematics

Man's first success in flying with wings was in gliders, which used air currents for soaring and gliding. Gliders were often launched from hilltops, or towed by automobiles, as in the picture of W. H. Martin's "Martinette" glider of 1909. In the early 1900's, the Wright brothers, experimenting with gliders, constructed reliable tables showing air pressure on wings. These provided valuable information for the construction of powered airplanes. The aircraft industry has advanced in a half-century to the spectacular achievement of supersonic aircraft like the "Concorde," built by the cooperation of French and British engineers. Designed to fly the Atlantic in three hours, it will travel at speeds up to 1400 miles an hour — twice as fast as sound.

Fundamental Ideas about Sets

2–1 Sets and How to Represent Them

The idea of sets is a very helpful one in the study of algebra. In this chapter we will investigate the notation and use of sets: the language and symbols of mathematics.

A **set** is commonly described as "a collection of things." You are probably familiar with different kinds of sets, such as a set of silverware, a set of dishes, a school of fish, or a collection of coins. The things that make up a set are called **members** or **elements** of the set.

In mathematics we use the notion of sets so often that we adopt a special symbolism for writing a set: we list its elements between braces. For example, for "the set whose elements are 3, 4, 5, and 7" we can write:

$$\{3,\ 4,\ 5,\ 7\}$$

For convenience, we often use a capital letter to represent some specified set. For example, if we write "$M = \{0, 2, 17, 85, 100\}$" we mean: "Set M is the set whose elements are **0, 2, 17, 85,** and **100.**" When a

35

36 *Chapter 2*

set is written in this way, the elements are separated by commas, and no element is named more than once.

A list of the members of a set is called a **roster** of the set. When a set contains a large number of elements, an abbreviated roster form can be used.

EXAMPLE 1. Abbreviated Roster: {0, 1, 2, ... , 99}
We read: "The set of whole numbers less than 100."

Another way to represent a large set is to use a **rule** that describes the members of the set.

EXAMPLE 2. Rule: {the books in the Congressional Library}
We read: "The set of books in the Congressional Library."

The order in which the elements of a set are listed does not change the set. That is, {1, 5, 7} is the same set as {5, 1, 7}.

A set may also be shown by enclosing its members in a ring marked with the letter name of the set. This sort of representation is called a **Venn diagram**.

This figure represents $M = \{4, 9, 15, 10\}$.

If we wish to state that an object k is a member of a set E, we may write $k \in E$. If we wish to state that an object m is not a member of set E, we may write $m \notin E$. The symbol $\in$ is read: "is an element of" or "is a member of"; similarly, $\notin$ is read: "is not an element of" or "is not a member of."

EXAMPLE 1. Symbols: $6 \in T$
We read: 6 is an element of set T.

EXAMPLE 2. Symbols: $10 \notin T$
We read: 10 is not an element of set T.

ORAL EXERCISES

Tell the meaning of each of the following:

SAMPLE 1: $K = \{1, 3, 8, 12\}$

What you say: Set K consists of the elements 1, 3, 8, and 12.

SAMPLE 2: $3 \in A$ *What you say:* 3 is an element of set A.

1. $Q = \{0, 1, 3\}$
2. $B = \{10, 20, 35, 40\}$
3. $J = \{4\}$
4. H
5. $9 \in W$
6. $M = \{\frac{1}{2}, \frac{1}{3}\}$
7. $19 \notin T$
8. K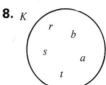

Give a rule to describe each set.

SAMPLE: $\{0, 1, 2\}$

What you say: {all the whole numbers less than 3}

9.
10. $\{1, 3, 5, 7\}$
11. $\{a, b, c\}$
12. $\{a, e, i, o, u\}$
13. {George Washington}
14. $\{0, 1, 2, \ldots, 15\}$
15. $\{0, 1, 2, \ldots, 9\}$
16.

WRITTEN EXERCISES

Match each item in Column 1 with the correct item in Column 2.

COLUMN 1

1. $\{0, 1\}$
2. $\{7, 9, 11\}$
3. $\{a, b, c, d\}$
4. $\{5\}$
5. $\{0, 2, 5, 8\}$
6. $\{9\}$
7. $\{1, 3, 5, \ldots, 999\}$
8. $\{27\}$
9.

COLUMN 2

A. $\{c, d, b, a\}$
B. {last three letters of the alphabet}
C. {odd numbers less than 1000}
D. {odd numbers between 6 and 12}
E. {sum of 9 and 18}
F. {whole numbers less than 2}
G. {digits in 5280}
H. {largest-valued digit}
I. {five}

Write the following sets in roster form.

SAMPLE: {whole numbers less than 9}

Solution: {0, 1, 2, 3, 4, 5, 6, 7, 8} or {0, 1, 2, ... , 8}

10. {whole numbers between 4 and 8}
11. {first five letters of the alphabet}
12. {odd numbers less than 10}
13. (all whole numbers between 2 and 10)
14. (all one-digit whole numbers)
15. {even numbers between 3 and 6}
16. {whole numbers between 2 and 6, inclusive}
17. {whole numbers less than 1000}
18. {months of the year with names that begin with the letter J}
19. {letters in the word *tomorrow*}

Tell whether each statement is true or false for the given set.

SAMPLE: $Q = \{8, 45, 61, 35\}$: $7 \in Q$; $45 \in Q$

Solution: $7 \in Q$ is **false**. $45 \in Q$ is **true**.

20. $K = \{3, 5, 11, 19, 35\}$: $5 \in K$; $6 \notin K$
21. $X = \{p, q, r, s\}$: $q \notin X$; $r \in X$
22. $T = \{0, 1\}$: $4 \in T$; $1 \notin T$; $3 \notin T$

B 23. $R = \{1, 2, 3, \ldots, 25\}$: $19 \in R$; $0 \in R$; $7 \in R$
24. $G = \{0, 2, 4, \ldots, 18\}$: $0 \in G$; $11 \in G$; $15 \notin G$
25. $B = \{a, b, c, d, \ldots, z\}$: $k \in B$; $m \in B$; $t \notin B$
26. $9 \in \{1, 3, 5, 7, 9, 11\}$
27. $Y = \{\frac{1}{2}, \frac{1}{3}, \frac{1}{4}, \frac{1}{5}, \ldots, \frac{1}{35}\}$: $\frac{1}{3} \notin Y$; $\frac{1}{18} \in Y$
28. $4 \in$ {whole numbers greater than 2}
29. $9 \notin$ {whole numbers greater than 10}
30. Thursday $\in$ {days of the week}

Write a rule or roster for each set.

SAMPLE 1: {months in the year}

Solution: {January, February, March, ... , December}

The Language and Symbols of Mathematics

SAMPLE 2: {5, 6, 7, 8}

Solution: {Whole numbers between 4 and 9}

31. {all days in the week}
32. {all school days in the week}
33. {3, 4, 5, 6, 7}
34. {13, 14, 15}
35. {8, 9, 10, 11, 12}
36. {0, 1, 2, . . . , 39}

 37. {all two-digit odd numbers}
38. {all one-digit odd numbers}
39. {odd numbers between 3 and 11}
40. {all numbers between 10 and 20 that are evenly divisible by 3}
41. {all two-digit numbers}
42. {all three-digit numbers}
43. {all even numbers between 99 and 301}

2–2 Kinds of Sets

The process called **counting** may be thought of as the matching of the counting numbers, 1, 2, 3, etc., with the objects in a given set. Any two sets are said to be in **one-to-one correspondence** when each member of one set is matched with one member of the other set, and no element in either set remains unmatched.

Sets in one-to-one correspondence:

$$\begin{array}{ccccc} a, & b, & c, & d, & e \\ \updownarrow & \updownarrow & \updownarrow & \updownarrow & \updownarrow \\ 1, & 2, & 3, & 4, & 5 \end{array}$$

Thus counting makes a one-to-one correspondence between the set to be counted and a part of the set of counting numbers.

The number that tells how many objects are in a set is called the **number property** of the set. For example, in $K = \{\triangle, \square, \bigcirc, \varheartsuit\}$ the set contains four members. So the number property of K is **4**.

EXAMPLE.   Symbols: $n(K) = 4$
   We read: The number property of set K is 4.

Suppose that you tried to write all of the whole numbers. Could you tell the number property of the set of whole numbers?

Set of whole numbers = {0, 1, 2, 3, 4, 5, 6, 7, 8, ...}

The three dots at the end tell us that the set continues on and on without ever ending. If there is no whole number that tells the number property of a set it is an **infinite set**. For example, do you agree that no whole number tells how many elements are in the set of **odd numbers**, **{1, 3, 5, 7, 9, 11, 13, ...}**?

A set whose number property can be named by a whole number is a **finite set**. For the sets shown below, the number property of set T is 4; the number property of set R is 75.

$$T = \{a, b, c, d\} \qquad n(T) = 4$$
$$R = \{1, 2, 3, 4, 5, \ldots 75\} \qquad n(R) = 75$$

How would you name the number property of the set of whole numbers between 5 and 6? Do you agree that there are **no** whole numbers between 5 and 6? Clearly, then, the number property of the set of whole numbers between 5 and 6 must be 0. If the number property of a set is 0 we call it **the empty set**. We also agree that there is only **one** empty set. For example, the set of whole numbers between 10 and 11 is the same set as the set of whole numbers between 5 and 6. Empty braces, { }, might be used to represent the empty set, but a more frequently used symbol is ∅.

The empty set is a **finite** set, because its number property can be named by a whole number, **0**. Thus, using symbols, we can write:

If $F = \{$**whole numbers between 5 and 6**$\}$, then
$$n(F) = 0 \quad \text{and} \quad F = \emptyset.$$

For two sets to be matched in one-to-one correspondence, they must have the same number property.

ORAL EXERCISES

Tell whether or not the members of the sets can be matched in one-to-one correspondence.

1. {0, 2, 4, 6} and {4, 6, 8}
2. {a, b, c} and {r, s, t}
3. {1, 2, 3, 4, 5} and {0, 1, 2, 3, 4}
4. {0} and { }
5. {5, 7, 9, 11} and {11, 9, 7, 5}

6. {0, 1} and {whole numbers less than 2}
7. {x, y, z} and {y, x, z}
8. {$, *, π, @} and {$, *, ¢}
9. {3, 5, 7, 9} and {3, 5, 7, 9, . . .}
10. {1, 2, 3, . . . , 10} and {a, b, c, d}

Tell whether each set is finite or infinite.

SAMPLE: {whole numbers}

What you say: The set of whole numbers is infinite.

11. {letters of the alphabet}
12. {1, 2, 3, 4, 5, 6, . . .}
13. {odd numbers less than 100}
14. {whole numbers between 4 and 6}
15. {whole numbers between 999 and 1000}
16. {odd numbers}
17. {whole numbers greater than 50}
18. {all two-digit numbers}

WRITTEN EXERCISES

Match each item in Column 1 with the correct item in Column 2.

COLUMN 1

1. {0}
2. {65, 66, 67, . . .}
3. {0, 1, 2, 3, 4, . . . , 366}
4. ∅
5. {97, 99, 101}

COLUMN 2

A. {whole numbers less than 367}
B. {odd numbers between 9 and 10}
C. {odd numbers between 96 and 102, inclusive}
D. {whole numbers less than 1}
E. {whole numbers greater than 64}

Describe the number property of each set.

SAMPLE: $B = \{21, 22, 23, 25, 29\}$

Solution: $n(B) = 5$

6. $R = \{3, 7, 11, 15\}$
7. $J = \{100\}$
8. $S = \{0, 1, 2, 3, 4, 5, 6\}$

42   Chapter 2

9. $C = \{1, 2, 3, \ldots, 15\}$
10. $T = \{7, 9\}$
11. $K = \{$odd numbers between 3 and 4$\}$
12. $F = \{$even numbers between 12 and 18$\}$
13. $W = \emptyset$
14. $D = \{w, x, y, z\}$
15. $P = \{$even numbers between 4 and 10, inclusive$\}$

Write the roster for each set and state whether the set is finite or infinite.

SAMPLE: $\{$letters in the word *set*$\}$

Solution: $\{s, e, t\}$; finite

B 16. $\{$days of the week whose names begin with $T\}$
17. $\{$months of the year whose names begin with $G\}$
18. $\{$even numbers greater than 80$\}$
19. $\{$odd numbers less than 91$\}$
20. $\{$whole numbers$\}$
21. $\{$women presidents of the United States$\}$

C 22. $\{$odd numbers evenly divisible by 3$\}$
23. $\{$whole numbers evenly divisible by 1$\}$
24. $\{$odd numbers evenly divisible by themselves$\}$
25. $\{$unit fractions for numbers less than 1$\}$ (Note: a unit fraction is a fraction with 1 as numerator)
26. $\{$unit fractions for numbers greater than 1$\}$
27. $\{$all the whole numbers between 2 and 3$\}$

2–3   Graphing a Set of Numbers

A set of numbers can be represented on the number line. The points on the number line that correspond to numbers in a set make up the **graph of the set.**

EXAMPLE: is the graph of
$\{$whole numbers between 1 and 7$\}$, or $\{2, 3, 4, 5, 6\}$.

The enlarged dots on the number line indicate the graph of $\{2, 3, 4, 5, 6\}$. Is this a finite or an infinite set?

Three graphs of infinite sets are shown in Figure 2–1. How do they differ from graphs of finite sets?

{the numbers between 4 and 5}

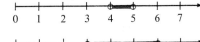

{the numbers between 3 and 6, inclusive}

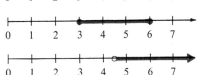

{the numbers greater than $4\frac{1}{2}$}

0 1 2 3 4 5 6 7
0 1 2 3 4 5 6 7
0 1 2 3 4 5 6 7

Figure 2–1

ORAL EXERCISES

Tell which graph in Column 2 should be matched with each set in Column 1.

COLUMN 1

1. {numbers greater than 2}

2. {numbers between 2 and 7}

3. {numbers equal to 3 + 1}

4. {2, 4, 5, 6}

5. {numbers between 2 and 6, inclusive}

COLUMN 2

A.
0 1 2 3 4 5 6 7
B.
0 1 2 3 4 5 6 7
C.
0 1 2 3 4
D.
0 1 2 3 4 5 6 7
E.
0 2 4 6

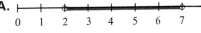

WRITTEN EXERCISES

Referring to the number line shown here, state the set of letters that describes the graph of each set of numbers.

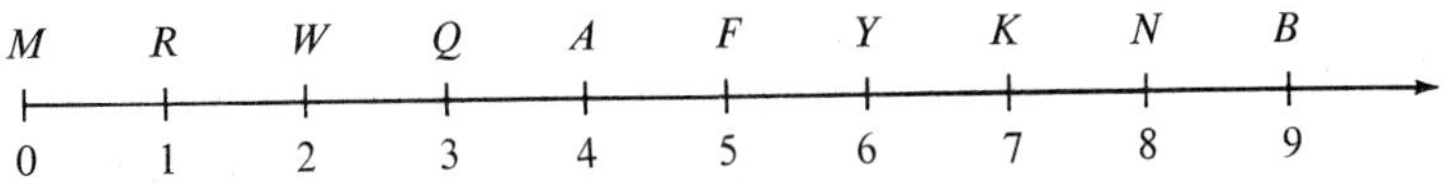

M R W Q A F Y K N B
0 1 2 3 4 5 6 7 8 9

SAMPLE: {even numbers between 2 and 8} *Solution:* $\{A, Y\}$

A

1. {3, 6, 9}

2. {0, 2, 4, 5}

3. {0, 9}

4. {1, 4, 6}

5. ∅

6. {8, 9, 3, 2, 5}

7. {0}

8. {1, 2, 3, . . . , 8}

44 Chapter 2

9. 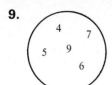 10. 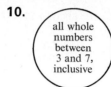 11.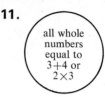

12. {whole numbers between 1 and 4, inclusive}
13. {whole numbers less than 5}
14. {whole numbers equal to the sum of 4 and 3}
15. {whole numbers equal to the product of 2 and 4}
16. {whole numbers greater than 5 but less than 9}

B 17. {whole numbers less than or equal to 3}
18. {whole numbers greater than 3 but less than 4}
19. {whole numbers between 1 and 4, including 1}

Draw the number line graph of each set.

SAMPLE: {numbers between 1 and 5, including 1}

Solution:

20. {numbers between 2 and 7, inclusive}
21. {whole numbers between 2 and 7, inclusive}
22. {numbers greater than 3.5}
23. {numbers greater than $1\frac{1}{3}$}
24. {numbers greater than or equal to 4}
25. {numbers greater than $2\frac{1}{2}$}
26. {whole numbers between 1 and π}
27. {even whole numbers between 3 and 7}

C 28. {numbers between $1\frac{1}{2}$ and $4\frac{1}{2}$, including $4\frac{1}{2}$}

29. {7, 2, 3, 9}
30. {0, 2, 4, ... , 10}
31.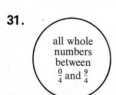
32. {0, 1, 2, 3, ... , 10}
33. {$\frac{4}{2}$, $\frac{6}{2}$, $\frac{8}{2}$, $\frac{10}{2}$, $\frac{12}{2}$}
34.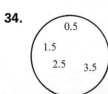

Special Sets and Operations on Sets

2-4 Sets and Subsets

Let us investigate the set of the first three odd numbers. If we call it set S, we have:

$$S = \{1, 3, 5\}$$

By using no numbers other than members of set S, we can form the following sets:

$$\{1\} \quad \{3\} \quad \{5\} \quad \{1, 3\}$$
$$\{3, 5\} \quad \{1, 5\} \quad \{1, 3, 5\} \quad \{\ \}$$

none of which contains any element that is not in set S.

Each of these sets is called a **subset** of S. To represent the relation "is a subset of," the symbol $\subset$ is used.

EXAMPLE. In words: $\{3, 5\}$ is a subset of $\{1, 3, 5\}$
In symbols: $\{3, 5\} \subset \{1, 3, 5\}$

Any subset of S that does not contain **every** element of S is called a **proper subset** of S. This includes the empty set.

Did you notice that every element of S is included in the subset $\{1, 3, 5\}$? A subset of S that contains every element of S is an **improper subset**.

For $R = \{1, 2, 3, 4\}$, the statement $\{2, 3\} \subset \{1, 2, 3, 4\}$ is true since every member of the subset is a member of R. However, the statement $\{4, 5\} \subset \{1, 2, 3, 4\}$ is false since at least one member of the subset is *not* a member of R. The statement could be made true by changing the "is a subset of" symbol $\subset$ to the symbol $\not\subset$, which means "is not a subset of."

EXAMPLE. In words: $\{4, 5\}$ is not a subset of $\{1, 2, 3, 4\}$
In symbols: $\{4, 5\} \not\subset \{1, 2, 3, 4\}$

ORAL EXERCISES

Tell which statements are true and which are false. In the case of a false statement, give a reason for your answer.

SAMPLE: $\{0, 1\}$ is a subset of $\{1, 2, 3, 4\}$.

Solution: False, since $0 \notin \{1, 2, 3, 4\}$.

1. {2, 4, 6} is a subset of {0, 1, 2, 3, 4, 5, 6, 7, 8}.
2. {0} is a subset of {0, 2, 4, 6, 8}.
3. {0} is a subset of {8, 12, 16, 25}.
4. {1, 2, 5, 7} is a subset of {7, 5, 1, 2}.
5. {p, q, r} is a subset of {a, b, c, d, ... , z}.
6. {39, 40} is not a subset of {1, 2, 3, 4, ...}.
7. {Abe Lincoln} is a subset of {U.S. presidents}.
8. {January} is not a subset of {days of the week}.
9. {the year 1971} is a subset of {leap years}.
10. ∅ is a subset of {9, 99}.

Give the meaning of each statement in words. Then tell whether the statement is true or false.

SAMPLE: {2, 6, 9} ⊄ {0, 2, 4, 6, ... , 10}

What you say: {2, 6, 9} is not a subset of {0, 2, 4, 6, ... , 10}; True.

11. {3, 9, 7} ⊄ {1, 3, 5, 7} 14. {0} ⊄ {0}
12. {0, 1} ⊂ {0, 10, 100} 15. ∅ ⊄ {0}
13. { } ⊂ {1971, 1972, 1973} 16. {8} ⊄ {2, 4, 6}

WRITTEN EXERCISES

Write a true statement using the given sets and the symbol ⊂.

SAMPLE: {1, 2, 3, 4, ...}; {3, 6, 9}

Solution: {3, 6, 9} ⊂ {1, 2, 3, 4, ...}

1. {1, 2, 3, ... , 10}; {7, 8}   4. {68}; ∅
2. {7, 8, 10}; {8}   5. ∅; {10, 20, 30, ...}
3. {3, 4}; {2, 3, 4, 5}   6. {3, 5, 7}; {whole numbers}

7. {0, 1, 2, 3, 4}; {whole numbers less than 3}
8. {odd numbers}; {69, 71, 73}
9. {odd numbers}; {whole numbers}
10. {0, 1, 2, 3}; { }

Write a true statement using the given sets and the symbol ⊄.

11. {2, 4, 6}; {2, 4, 6, 8, 10}
12. {3, 5, 7, 9}; {5, 9}

The Language and Symbols of Mathematics **47**

13. {3, 4, 5} ; {1, 3, 5, . . .}
14. {a, b, c, d} ; {a, b, c, d, e}
15. {1, 3, 5, 7, . . .} ; {whole numbers}
16. {3} ; ∅
17. { } ; {10, 12}
18. {0} ; { }
19. {$\frac{1}{2}$, $\frac{1}{3}$, $\frac{1}{4}$} ; {numbers between 0 and 1}
20. {states whose names begin with T} ; {Texas}

For each given set, list a proper subset containing four members.

SAMPLE: {presidents of the United States}

Solution: {Washington, Adams, Lincoln, Roosevelt} (Note: any four names of presidents can be used)

B 21. {oceans of the world}
22. {students in your algebra class}
23. {odd numbers less than 11}
24. {whole numbers between 4 and 10}
25. {two-digit numbers}
26. {arithmetic numbers between 0 and 1}

27.
28.

List the following subsets of set *U* if *U* = {3, 4}. Label each subset as proper or improper.

C 29. All those containing exactly one element.
30. All those containing no elements.
31. All those containing exactly two elements.

2–5 The Set of Whole Numbers — Some Special Subsets

Our work with whole numbers has also included some subsets of the set of whole numbers. Do you agree that every **even** number is a member of the set of **whole** numbers?

or, {even numbers} ⊂ {whole numbers}
{0, 2, 4, 6, 8, . . .} ⊂ {0, 1, 2, 3, 4, . . .}

You probably recall that an **even number** is a whole number that is evenly divisible by 2. That is, when the number is divided by two, the quotient is a whole number and there is no remainder. Is **0** an even number? What number will make the sentence "$0 \div 2 = ?$" a true statement? Since $2 \times 0 = 0$, we see that $0 \div 2 = 0$. Since there is no remainder, 0 must be an even number.

Let us consider whole numbers that are *not* evenly divisible by 2. These are called **odd numbers**. Do you agree that every odd number is also a member of the set of whole numbers?

or, $\quad$ {odd numbers} $\subset$ {whole numbers}
$\{1, 3, 5, 7, 9, \ldots\} \subset \{0, 1, 2, 3, 4, \ldots\}$

Suppose that you are asked to name "the set of factors of 18." What are the factors of 18?

We know that:
$18 = 1 \times 18 = 18 \times 1$
$18 = 2 \times 9 = 9 \times 2$
$18 = 3 \times 6 = 6 \times 3$

Thus each of the numbers 1, 2, 3, 6, 9, and 18 is a factor of **18**. Are there any other whole numbers that belong in the set? Try dividing 18 by each whole number that is less than 18 and you will see that there are no others than those listed that are factors of 18. Thus we can write:

The set of factors of 18 $= \{1, 2, 3, 6, 9, 18\}$.

As you may have discovered, a factor of a number is also a *divisor* of that number.

Number	Set of Factors (or Divisors)
15	{1, 3, 5, 15}
28	{1, 2, 4, 7, 14, 28}

A particularly important subset of the set of whole numbers is the set of **prime numbers**. A number is a prime number if its set of factors contains *exactly two* different members. Do you agree that the table on page 49 shows **2**, **3**, and **5** to be prime numbers? The set of prime numbers is an infinite set.

{Prime Numbers} $= \{2, 3, 5, 7, 11, 13, \ldots\}$

Why is 2 the only **even** prime number?

The Language and Symbols of Mathematics

Number	Factors	Set of Factors	Prime/Not Prime
1	1×1	$\{1\}$	Not prime
2	1×2	$\{1, 2\}$	Prime
3	1×3	$\{1, 3\}$	Prime
4	1×4 2×2	$\{1, 2, 4\}$	Not prime
5	1×5	$\{1, 5\}$	Prime
6	1×6 2×3	$\{1, 2, 3, 6\}$	Not prime

ORAL EXERCISES

Tell which statements are true and which are false. In the case of a false statement, give a reason for your answer.

SAMPLE: $\{2, 3, 9\} \subset \{\text{prime numbers}\}$

What you say: False; 9 is not a prime number, since its set of factors is $\{1, 3, 9\}$.

1. $\{2\} \subset \{\text{even numbers}\}$
2. $\{2\} \subset \{\text{prime numbers}\}$
3. $\{10\} \subset \{\text{prime numbers}\}$
4. $\{15, 17\} \subset \{\text{odd numbers}\}$
5. $\{\text{prime numbers}\} \subset \{\text{odd numbers}\}$
6. $\{\text{prime numbers}\} \subset \{\text{whole numbers greater than 1}\}$

WRITTEN EXERCISES

To each number in Column 1 match the correct set of factors in Column 2.

COLUMN 1

A
1. 22
2. 30
3. 65
4. 37
5. 20
6. 71

COLUMN 2

A. $\{1, 5, 13, 65\}$
B. $\{1, 71\}$
C. $\{1, 2, 4, 5, 10, 20\}$
D. $\{1, 2, 3, 5, 6, 15, 30\}$
E. $\{1, 37\}$
F. $\{1, 2, 11, 22\}$

Chapter 2

Write the set of factors of each number. Then tell whether or not the number is prime.

SAMPLE: 35   *Solution:* {1, 5, 7, 35}; 35 is not prime

7. 21	10. 17	13. 31	16. 43
8. 24	11. 25	14. 40	17. 55
9. 18	12. 15	15. 27	18. 48

Use the symbol ⊄ or ⊂ to replace each question mark and make a true statement.

SAMPLE: {2, 4, 3} ? {factors of 12}

Solution: {2, 4, 3} ⊂ {factors of 12}

B 19. {3, 8, 16} ? {factors of 18}
20. {prime numbers} ? {whole numbers}
21. {odd numbers} ? {prime numbers}
22. {31, 37, 43, 51} ? {prime numbers}
23. {19, 23, 29} ? {prime numbers}
24. {whole numbers} ? {prime numbers}

C 25. {factors of 6} ? {factors of 12}
26. {factors of 8} ? {factors of 12}

Write a roster of each set of numbers.

27. {whole numbers that have 2 as a factor}
28. {whole numbers that have exactly two different factors}
29. {prime numbers that have 2 as a factor}
30. {odd numbers that have 2 as a factor}

2–6 Sets and Operations

The Venn Diagram in Figure 2–2 consists of two sets of numbers. Notice that no member in set *K* is a member of set *S*. Sets that have no members in common are called **disjoint sets**.

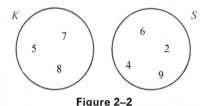

Figure 2–2   Figure 2–3

Suppose that we want to join K and S to form a third set. It will contain *every* element that is in K or in S. The new set, which we will call R, is shown in Figure 2-3.

The joining of two sets in this manner is the operation of **set union**. The symbol for this operation is ∪.

The union of the two sets K and S may be shown like this:

$$\{5, 7, 8\} \cup \{2, 4, 6, 9\} = \{2, 4, 5, 6, 7, 8, 9\}$$
$$K \cup S = \{2, 4, 5, 6, 7, 8, 9\}$$
$$K \cup S = R$$

The operation **union** can be applied to a pair of sets that are not disjoint (that is, have at least one element in common).

For $M = \{w, x, y, z\}$ and $N = \{u, v, w, x\}$
$$M \cup N = \{u, v, w, x, y, z\}$$

Note that, although w is an element of each set, it is listed only **once** in the union.

Another operation with two sets is called **intersection**.

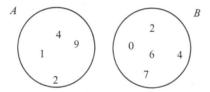

Figure 2-4

The **intersection** of sets A and B shown in Figure 2-4 is the set of elements which are common to *both* A and B. The shaded part of the diagram in Figure 2-5 shows the intersection of sets A and B.

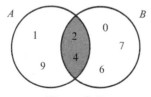

Figure 2-5

The symbol ∩ is used to indicate **intersection**. Hence the intersection pictured can be stated:

$$\{1, 2, 4, 9\} \cap \{0, 2, 4, 6, 7\} = \{2, 4\}$$
$$A \cap B = \{2, 4\}$$

The Venn diagram in Figure 2–6 illustrates the intersection of the disjoint sets $G = \{1, 3, 5, 9\}$ and $H = \{2, 4\}$.

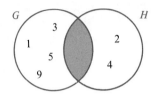

Figure 2–6

Do you agree that set G and set H have **no** common elements?

$$G \cap H = \emptyset$$

ORAL EXERCISES

Tell what set is indicated by each of the following.

SAMPLE: $\{3, 12, 15\} \cup \{7, 8\}$

What you say: The set containing the elements 3, 7, 8, 12, and 15.

1. $\{4, 5, 9\} \cup \{2, 3\}$
2. $\{r, s\} \cup \{x, y\}$
3. $\{2, 9, 6, 8\} \cap \{3, 7, 8\}$
4. $\{1, 3, 5, 7\} \cap \{5, 7, 8, 9\}$
5. $\{2, 4, 6\} \cup \{2, 3, 4, 5\}$
6. $\{p, q, r, s\} \cup \{s, t, w\}$
7. $\{0\} \cup \{0, 3, 5\}$
8. $\{\text{Tom, Bill}\} \cup \{\text{Bill, Ken, Al}\}$
9. $\{2, 7, 9, 5\} \cap \{3, 6, 8\}$
10. $\{a, b, c\} \cup \{b, c\}$
11. $\{\text{even numbers}\} \cap \{0, 1, 2, 3\}$

WRITTEN EXERCISES

Write the set indicated by each of the following. Use $M = \{3, 5, 7\}$, $N = \{1, 3, 5, 7, 9\}$, $R = \{2, 4\}$, and $S = \{0, 1, 8\}$.

SAMPLE: $M \cup R$ Solution: $M \cup R = \{2, 3, 4, 5, 7\}$

A 1. $M \cup N$ 2. $M \cap N$ 3. $R \cap S$

4. $N \cup S$
5. $S \cup R$
6. $N \cap S$
7. $M \cap R$
8. $S \cup M$
9. $M \cup S$
10. $R \cup M$
11. $N \cup R$
12. $R \cap N$

Write the set indicated by each union.

13. $\{1, 3, 5, 7\} \cup \{0, 1, 2, 3, 4, 5\}$
14. $\{1, 2, 3, 4, \ldots\} \cup \{0\}$
15. $\{0, 1, 2, 3, 4\} \cup \{0, 2, 4, 6\}$
16. $\{1, 3, 5, 7, \ldots\} \cup \{2, 4, 6, 8, \ldots\}$
17. $\{3, 5, 7, 11, 13, \ldots\} \cup \{2\}$
18. $\{3, 5, 7, 9\} \cup \{5, 7, 9\}$
19. $\{3, 6, 9\} \cup \{\ \}$
20. $\emptyset \cup \{1, 2, 3, 4\}$

Complete each of the following to make a true statement. Then draw a Venn diagram to illustrate it. Use $B = \{2, 4, 6, 8\}$, $C = \{0, 1, 2, 3\}$, $G = \{0, 2, 4\}$, and $R = \{1, 2, 6\}$.

SAMPLE: $C \cap B = ?$ Solution: $C \cap B = \{2\}$

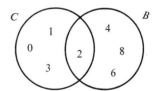

B
21. $B \cap G = ?$
22. $R \cap B = ?$
23. $C \cap R = ?$
24. $G \cap C = ?$
25. $R \cap G = ?$
26. $B \cap C = ?$

Write the union statement and the intersection statement suggested by each diagram.

SAMPLE: 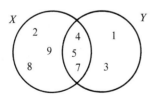 Solution: $X \cup Y = \{1, 2, 3, 4, 5, 7, 8, 9\}$
$X \cap Y = \{4, 5, 7\}$

C 27. 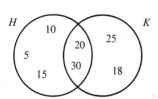   28.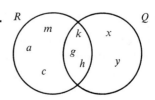

54 Chapter 2

29.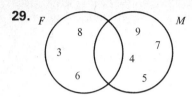

31.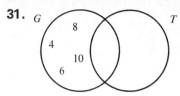

30.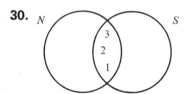

32.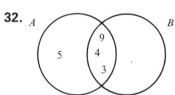

CHAPTER SUMMARY

Inventory of Structure and Concepts

1. A **set** is a collection of objects or things called members or **elements** of the set. A set may be represented by a list or **roster** of its members, by a **Venn diagram**, by a **graph**, or by a **rule** that describes the members of the set.

2. According to the **number property** of a set it may be either a **finite set** (including the **empty set**) or an **infinite set**.
   A finite set: The number property can be named by a whole number.
   The empty set (a special kind of finite set): The number property is named by the whole number 0.
   An infinite set: The number property cannot be named by a whole number.

3. Basic symbols used in set notation include:
   a. { } or ∅ (the empty set)
   b. ∈ (is an element of)
   c. ∉ (is not an element of)
   d. ⊂ (is a subset of)
   e. ⊄ (is not a subset of)

4. When each member of a set can be matched with a member of another set and no element in either set remains unmatched, the two sets are said to be in **one-to-one correspondence**.

5. A set whose members are also members of some other given set is a **subset** of the given set. If a subset does not contain all of the members of the given set it is a **proper subset**. If a subset does contain all of the members of the given set it is an **improper subset**. The empty set is a proper subset of every other set.

6. The set of factors of a given whole number has as its members all of the whole numbers that are **divisors** of the given number. A **prime number** is a number whose set of factors contains exactly **two** different numbers (1 and the number itself). The number 2 is the smallest prime number and is the only prime number that is also an even number.

7. The joining of two sets to form a third set which contains all of the elements in the original two sets is called set **union**. The set of all elements which are common to both of two given sets is called the **intersection** of the original sets.
   Disjoint sets are sets which have no members in common. The intersection of two disjoint sets is the empty set.

Vocabulary and Spelling

Pronounce, spell, and give the meaning of these words and expressions.

set (p. 35)
member (p. 35)
element (p. 35)
roster (p. 36)
Venn diagram (p. 36)
$\in$ and $\notin$ (p. 36)
one-to-one correspondence (p. 39)
number property (p. 39)
infinite set (p. 40)
finite set (p. 40)
empty set (p. 40)

{ } and $\emptyset$ (p. 40)
subset (p. 45)
proper subset (p. 45)
improper subset (p. 45)
$\subset$ and $\not\subset$ (p. 45)
factor (p. 48)
prime number (p. 48)
disjoint sets (p. 50)
set union (p. 51)
set intersection (p. 51)

Chapter Test

1. List in roster form the set of even numbers between two and six, including six.

Indicate whether each statement is true or false.

2. $\emptyset \in \{1, 2, 3\}$
3. The members of the set $\{0, 1, 2, \ldots, 25\}$ can be put in one-to-one correspondence with the members of the set {letters in the English alphabet}.
4. The set of even numbers is an infinite set.
5. $\{a, b, d\} \subset$ {letters in the English alphabet}

6. {the divisors of 48} = {the factors of 48}
7. The number property of {the numbers between 5 and 6} is zero.

For Questions 8–11, replace the question mark to make the resulting statement true.

8. If $A = \{2, 4, 6, 8\}$ and $B = \{1, 4, 8\}$ then in roster form $A \cup B = \{?\}$.
9. For the Venn diagram shown here, $\{3\} = A \; ? \; B$.
10. For the Venn diagram in Question 9, the number property of $A \cup B$ is ?.
11. If $M \subset N$ then $M \cap N = $?.

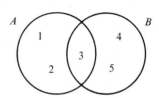

Draw a number line graph of each set described in Questions 12–14.

12. {the numbers between 5 and 7}
13. {the whole numbers less than 5 or equal to 5}
14. {the numbers greater than 4} ∩ {the numbers less than 5}
15. Copy and complete the Venn diagram, given: $A = \{1, 2, 3\}$, $A \cap B = \{2, 3\}$, and $A \cup B = \{1, 2, 3, 4, 5\}$.

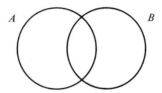

Chapter Review

2–1 Sets and How to Represent Them

1. Give in roster form all of the odd whole numbers less than 10.
2. Give in abbreviated roster form all of the whole numbers between 100 and 1000.
3. List the elements of set $A = $ {the even numbers between 9 and 15}.

For Questions 4–6, tell whether each statement is true or false.

4. If $B = \{2, 4, 6, 8\}$, then $4 \in B$.
5. If $C = $ {letters of the alphabet between a and p}, then $a \notin C$.

6. $1.5 \in$ {whole numbers}.
7. Draw a Venn diagram showing {the whole numbers between 3 and 8, including 8}.

2–2 Kinds of Sets

For Questions 8–11 tell whether the statement is true or false.

8. {1, 2, 3, 5} and {a, g, h, k} can be matched in one-to-one correspondence.
9. $\emptyset \in \{\ \}$
10. {1, 2, 3, . . . , 100} is an infinite set.
11. The number property of {the even whole numbers between 2 and 4} is 2.
12. What is the number property of {1000}?
13. Which of the following sets are finite?

    $A = \{2, 4, 6\}$ $C =$ {the even whole numbers}
    $B = \{1, 2, 3, \ldots\}$ $D =$ {the letters of the alphabet}

2–3 Graphing a Set of Numbers

Draw a number line graph for each of the following sets.

14. {8, 5, 7}
15. {whole numbers greater than 3 and less than 7}
16. {whole numbers greater than 4 and less than 5}
17. {numbers between 3 and 5}
18. {numbers greater than 2}
19. {numbers greater than or equal to 2}
20. {numbers between 5 and 8, including 5}

2–4 Sets and Subsets

Indicate whether each statement in Questions 21–24 is true or false.

21. {1, 2, 3} is an improper subset of {1, 3, 2}. ·
22. $\{a, b, c\} \subset \{a, b\}$
23. {even whole numbers} $\subset$ {whole numbers}
24. $\{2.4\} \subset$ {numbers between 3 and 4}
25. List all the subsets of {a, b}.
26. For {the whole numbers between two and six}, list all of the two-member subsets.

2–5 The Set of Whole Numbers — Some Special Subsets

Indicate whether each statement is true or false.

27. {the divisors of 15} = {3, 5}.
28. {the prime numbers} ⊂ {the even numbers}.
29. The set of factors of 39 is {1, 3, 13, 39}.

Write in roster form each of the following sets.

30. {whole numbers that do not have 2 as a factor}
31. {prime factors of 20}
32. {divisors of 28}

2–6 Sets and Operations

Give in roster form the set indicated by each of the following, if $A = \{1, 3, 5, 7, 9\}$, $B = \{0, 2, 4, 6, 8\}$, $C = \{1, 5, 9\}$, $D = \{2, 5, 7, 8\}$.

33. $A \cup B$ **35.** $A \cup D$ **37.** $B \cap A$
34. $C \cap B$ **36.** $C \cap D$ **38.** $C \cap A$

Draw a Venn diagram to illustrate each of the following, if $R = \{0, 3, 5\}$, $S = \{1, 2, 6\}$, $T = \{0, 2, 6\}$.

39. $R \cap S$ **40.** $S \cap T$ **41.** $R \cap T$

Review of Skills

For each equation, find the value of N which will make the statement true.

1. $\frac{7}{8} + \frac{3}{8} = N$
2. $2\frac{1}{2} - N = \frac{3}{4}$
3. $4 \cdot 9 = N$
4. $\frac{N}{6} = 9$
5. $12 \cdot N = 168$
6. $\frac{1001}{N} = 77$
7. $\frac{15}{4} - N = 3$
8. $3\frac{7}{8} - 2\frac{1}{2} = N$
9. $1\frac{1}{4} - \frac{1}{2} = N$
10. $\frac{9}{8} + N = \frac{9}{4}$
11. $5 \cdot N = 65$
12. $\frac{24}{N} = 3$
13. $\frac{N}{18} = 26$
14. $3\frac{7}{8} + N = 5$
15. $2\frac{1}{4} + 1\frac{7}{8} = N$
16. $3\frac{1}{2} \cdot 3 = N$

Replace each question mark with a number suggested by the pattern of number pairs.

17. (2, 5), (5, 8), (10, 13), (11, ?), (4, ?)
18. (2, 6), (5, 15), (0, 0), (4, ?), (8, ?)
19. (4, 2), (8, 4), (16, 8), (24, ?), (0, ?), (9, ?)
20. (2, 2.5), (7, 7.5), (1, ?), (0, ?)
21. (3, 2¼), (5, 4¼), (1, ¼), (8, ?), (2, ?)
22. (1, 3), (5, 11), (9, 19), (4, ?), (7, ?)
23. (2, 5), (5, 14), (7, 20), (1, ?), (6, ?)
24. (1, 1), (2, 4), (5, 25), (3, ?), (10, ?)
25. (3, 10), (2, 5), (6, 37), (4, ?), (1, ?), (0, ?)

■ ■

CHECK POINT FOR EXPERTS

Triangular Numbers

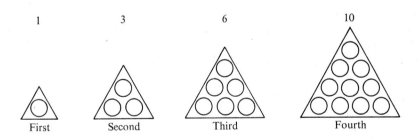

Study the triangular arrays of dots shown above. The number of dots in each array is a member of the set called **triangular** numbers. Can you find a pattern that will help you predict the next member of the set of triangular numbers?

Questions

1. What are the fifth, sixth, seventh, and eighth triangular numbers? Draw an array for each.

2. What are the first fifteen numbers in the set of triangular numbers? Use this pattern:

1st	2nd	3rd	4th	...	15th
1	3	6	10	...	?
1	(1 + 2)	(3 + 3)	(6 + 4)		

3. Copy this table and extend the pattern to include the tenth triangular number.

Triangular Number	Pattern
First	1 = 1
Second	3 = 1 + 2
Third	6 = 1 + 2 + 3
Fourth	10 = 1 + 2 + 3 + 4

4. The set of counting numbers is $\{1, 2, 3, 4, 5, \ldots\}$.
From your completed pattern in Question 3, notice that:
The **second** triangular number is the sum of the first **two** counting numbers.
The **third** triangular number is the sum of the first __?__ counting numbers.
The **fourth** triangular number is the sum of the first __?__ counting numbers.
The **fifth** triangular number is the sum of the first __?__ counting numbers.

5. In general, to find the *n*th triangular number, you find the sum of the first __?__ counting numbers.

A short cut for finding this sum is to find the value of $\dfrac{n \cdot (n + 1)}{2}$, where *n* stands for the number of counting numbers. Thus the sum of the first 5 counting numbers is $\dfrac{5 \cdot (5 + 1)}{2} = \dfrac{5 \cdot 6}{2} = 15$. Hence the **fifth** triangular number is **15**.

What is the fifteenth triangular number?
What is the twentieth triangular number?
What is the fifty-first triangular number?
What is the one-hundredth triangular number?

THE HUMAN ELEMENT

Archimedes

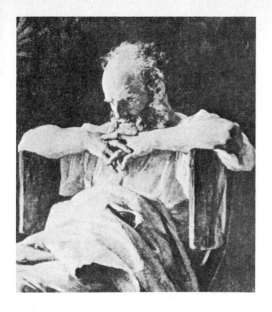

Archimedes of Syracuse (287–212 B.C.) was a great mathematician who devoted some of his talent to the solution of practical problems. He discovered a number of principles of physics, including the law of levers, and is said to have boasted: "If I have a lever long enough and somewhere to stand, I will move the whole earth."

One of the most famous stories about Archimedes concerns his solution of a difficult problem for King Hiero of Syracuse. The king had ordered a jeweller to make him a solid gold crown, and later suspected that the jeweller had cheated him by substituting another metal for part of the gold. He asked Archimedes to find out without damaging the crown. One day, as Archimedes was bathing, he noted that his body displaced a certain amount of water, causing the water level in the tub to rise. He realized that objects of the same weight but different densities would displace different amounts of water. If the crown was made of pure gold, it should displace the same amount of water as a solid gold bar of the same weight. Since it did not, Archimedes was able to calculate by exactly how much the jeweller had cheated the king.

Another service Archimedes performed for the king was the design and construction of machines to be used in war. Among the devices that he built were a machine which would hurl huge stones great distances, and a system of mirrors which used the sun's rays to set ships on fire. His machines were used to defend Syracuse against the Roman armies, and were so effective that the Roman soldiers were afraid to attack Syracuse.

In 212 B.C. the Roman general Marcellus finally conquered Syracuse by attacking on a holiday, when the citizens were not prepared to fight. Archimedes was killed during the attack. Legend says a Roman soldier killed the old mathematician as he contemplated a diagram which he had drawn in the sand to help him solve a problem in geometry.

Mesa Verde cliff dwelling . . .

Habitat apartment complex . . .

Numbers, Functions, and Number Pairs

In some parts of the southwestern United States, Indians once lived in homes built into cliffs, or along overhanging walls of canyons, for protection against other tribes. One of the best known of these is the Cliff Palace in Mesa Verde National Park, Colorado. This complex contained more than 200 living rooms on two, three, and four levels, as well as several *kivas,* or ceremonial chambers. Scientists believe that the Cliff Palace was built around 1066 A.D., and abandoned in the late 1200's. A modern counterpart is the Montreal "Habitat '67." Built for the Montreal World's Fair of 1967, it houses many people in modern style and comfort. The arrangement of the units is designed to insure privacy for each family.

Representing Functions

3–1 Functions: Rules and Machines

A **function**, as the term is used in algebra, may be described in terms of numbers, rules for carrying out operations, and number pairs. Suppose that the numbers **6, 7,** and **8** are read to you. As you hear each number (**input**) you reply (**output**) with a second number found according to a rule. Can you give the rule used by the man in this illustration? You probably found that his rule is "add ten to

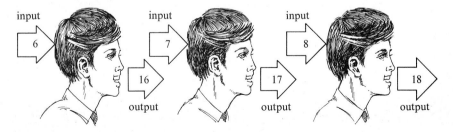

each input number." We can record the input and output information

Chapter 3

by using ordered number pairs, as shown in the last column of the following table. The rule "add ten" is represented by "+10." The three dots in each column indicate that we could extend the table indefinitely. From the table we can write the infinite set of ordered

	Input Number	Output Number	Number Pair
Rule: +10	6	16	(6, 16)
	7	17	(7, 17)
	8	18	(8, 18)
	⋮	⋮	⋮

number pairs: {(6, 16), (7, 17), (8, 18), ... }. These number pairs represent a function. Notice that in each number pair the **input** number is given **first** and the **output** number **second**. This is why they are called **ordered number pairs**.

Suppose that we create a simple "machine" to do our work and call it a **function machine**. Again the input and output numbers are used to write the ordered number pairs shown in the last column. Can

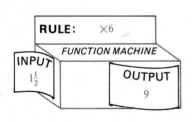

Input Number	Output Number	Number Pair
1½	9	(1½, 9)
2	12	(2, 12)
2½	? 15	(2½, 15)
3	? 18	(3, 18)

you tell how to complete the table? Notice that the rule given in this example is "×6". What does that tell you? The resulting set of number pairs represents a function:

$$\{(1\tfrac{1}{2}, 9), (2, 12), (2\tfrac{1}{2}, 15), (3, 18)\}$$

As you may have guessed, it is possible to make many kinds of rules. The next table is based on the rule "multiply by 2 and divide by 4." How is the rule represented? How would you complete the table? The number pairs in the third column represent a function: {(1, ½), (2, 1), (3, 1½), (4, 2), ...}.

Numbers, Functions, and Number Pairs

Rule: ×2 and ÷4

Input Number	Output Number	Number Pair
1	½	(1, ½)
2	1	(2, 1)
3	1½	(3, 1½)
4	2	?
5	?	?
⋮	⋮	⋮

A set of ordered number pairs, determined by a rule, represents a function. A very important idea about the number pairs in a function is that *no two* different number pairs may have the *same first number*. For example:

These are functions: {(0, 8), (1, 9), (2, 10), (3, 10), (4, 10)}
{(2, 4), (3, 9), (4, 16), . . .}
{(1, 2), (2, 3), (3, 4), (4, 5) . . .}

These are not functions: {(1, 3), (1, 5), (2, 4), (2, 6)}
{(1, 1), (2, 2), (1, 3), (1, 4)}
{(4, 3), (6, 3), (7, 4), (6, 5) . . .}

ORAL EXERCISES

Tell how to complete each table according to the given function rule.

Rule: Multiply by 10

	Input Number	Output Number	Number Pair
1.	1	10	
2.	3	30	
3.	5		(5, 50)
4.	7		(7, 70)
5.	?	90	(9, 90)
6.	?	?	(11, 110)

Rule: Add 2¼

	Input Number	Output Number	Number Pair
7.	3	5¼	
8.	3½	5¾	
9.	4		(4, 6¼)
10.	?	6¾	(4½, 6¾)
11.	5		(5, 7¼)
12.	?	?	(5½, 7¾)

WRITTEN EXERCISES

Use the given function rule to complete each set of number pairs.

SAMPLE: Add 3: {(4, 7), (7, ?), (2, ?), (10, ?), (3, ?)}

Solution: {(4, 7), (7, **10**), (2, **5**), (10, **13**), (3, **6**)}

[A] 1. Add 5: {(0, 5), (1, 6), (2, ?), (3, ?), (4, ?)}
2. Subtract 3: {(8, 5), (6, 3), (4, ?), (9, ?), (12, ?)}
3. Multiply by 8: {(0, 0), (2, 16), (4, ?), (7, ?), (11, ?)}
4. Divide by 6: {(36, 6), (54, 9), (0, ?), (48, ?), (72, ?), (132, ?)}
5. Multiply by $\frac{1}{10}$: {(8, $\frac{8}{10}$), (3, $\frac{3}{10}$), (5, ?), (9, ?), (13, ?), (10, ?)}
6. Add 9: {(18, 27), (24, ?), (28, ?), (35, ?), (40, ?), ($10\frac{1}{2}$, ?)}
7. Subtract 6: {(10, 4), (?, 12), (16, 10), (?, 21), (?, 32), (65, ?)}
8. Multiply by $\frac{1}{2}$: {(4, 2), (9, $4\frac{1}{2}$), (10, ?), (20, ?), (24, ?), (?, 6)}

Match each set of number pairs in Column 1 with the correct function rule in Column 2.

COLUMN 1	COLUMN 2
9. {(3, 1), (6, 2), (9, 3), (12, 4), (15, 5)}	A. ×7
10. {(10, 6), (15, 11), (19, 15), (8, 4), (4, 0)}	B. ×$\frac{1}{2}$
11. {(3, 21), (4, 28), (5, 35), (6, 42), (7, 49)}	C. ÷3
12. {(1, 1.3), (4, 4.3), (5, 5.3), (9.5, 9.8), (0, 0.3)}	D. +1 and ×2
13. {(4, 2), (5, $2\frac{1}{2}$), (6, 3), (7, $3\frac{1}{2}$), (8, 4)}	E. −4
14. {(4, 8), (5, 10), (6, 12), (7, 14), (8, 16)}	F. ÷$\frac{1}{2}$
15. {(0, 2), (1, 4), (2, 6), (3, 8)}	G. +0.3

Complete each table by using the given function machine.

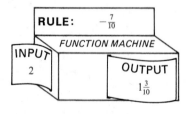

	Input Number	Output Number	Number Pair
[B] 16.	2	?	(2, $1\frac{3}{10}$)
17.	5	$4\frac{3}{10}$	?
18.	?	$6\frac{3}{10}$	?
19.	?	7	?
20.	9	?	?
21.	15	?	?
22.	?	?	(?, $12\frac{3}{10}$)

Numbers, Functions, and Number Pairs 67

	Input Number	Output Number	Number Pair
23.	?	?	(5, 4)
24.	3	3	?
25.	?	5	?
26.	10	?	?
27.	5.6	?	?
28.	11.4	?	?

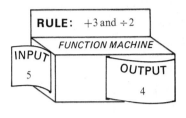

Name a function rule for each set of ordered pairs.

SAMPLE: {(2, 9), (3, 10), (4, 11), (5, 12)}   *Solution:* Add 7

[C] 29. {(3, 0), (5, 2), (7, 4), (9, 6), (12, 9), (10, 7)}
30. {(0, $2\frac{1}{2}$), ($1\frac{1}{2}$, 4), (2, $4\frac{1}{2}$), ($3\frac{1}{4}$, $5\frac{3}{4}$), (8, $10\frac{1}{2}$)}
31. {(3, 0), (4, 0), (5, 0), (6, 0), (7, 0), (8, 0), (9, 0)}
32. {($1\frac{1}{3}$, 2), ($2\frac{1}{3}$, 3), (5, $5\frac{2}{3}$), ($6\frac{2}{3}$, $7\frac{1}{3}$), (10, $10\frac{2}{3}$)}
33. {(2, 0.2), (3, 0.3), (6, 0.6), (4.8, 0.48), (9, 0.9), (10, 1)}

Tell whether or not each set of number pairs represents a function.

34. {(1, 7), (2, 8), (3, 9), (45, 51), (6, 19)}
35. {(3, 0), (0, 3), (5, 0), (0, 5), (6, 0), (0, 6)}
36. {(1, 2), (2, 3), (3, 4), (4, 5), (5, 6), (6, 7), . . .}
37. {(12, 21), (13, 31), (14, 41), (21, 12), (31, 13), (41, 14), . . .}
38. {(0, 1), (0, 2), (1, 1), (1, 2), (2, 1), (2, 3)}

Each set of ordered pairs represents a function. Name *two* different function rules for each set.

SAMPLE: {(10, 2), (15, 3), (20, 4), (25, 5), (30, 6)}

Solution: ÷5 or ×$\frac{1}{5}$

39. {(0, 0), (1, 1), (2, 2), (3, 3), (4, 4), (5, 5), (6, 6)}
40. {(0, 0), (1, $\frac{1}{10}$), (2, $\frac{2}{10}$), (3, $\frac{3}{10}$), (4, $\frac{4}{10}$), (5, $\frac{5}{10}$)}
41. {(4, 8), (9, 18), (14, 28), (19, 38), (24, 48)}

3-2 Number Pairs and Graphs

It is often difficult to visualize the complete meaning of a function when it is represented by a set of ordered number pairs. We can make a graph of pairs of whole numbers by using an array of points called a

68 Chapter 3

lattice. The array of points in Figure 3-1 forms a 6 × 6 (six by six) lattice.

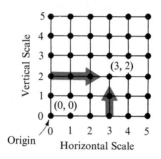

Figure 3-1

The large arrows drawn in the lattice indicate the point that is the **graph** of the number pair **(3, 2)**. Observe that this point is directly above **3** on the **horizontal** scale and opposite **2** on the **vertical** scale. Each point in the lattice is associated with exactly one ordered pair of whole numbers. The numbers of an ordered pair are called the **coordinates** of the point. The numbers in the ordered pair **(0, 0)** are the coordinates of the **origin** of the lattice. The **first** number in each ordered pair is the number on the **horizontal** scale. It is called the **first coordinate**. The **second** number in each pair is the number on the **vertical** scale and is called the **second coordinate**.

ORAL EXERCISES

Give the coordinates of each labeled lattice point.

SAMPLE: *Q*

What you say: The numbers in the ordered pair (3, 2) are the coordinates of point *Q*.

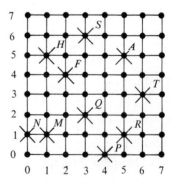

1. *M* (1, 1)
2. *T* (6, 3)
3. *R* (5, 1)
4. *H* (1, 5)
5. *P* (4, 0)
6. *A* (5, 5)

Numbers, Functions, and Number Pairs **69**

7. F (2,4) **8.** S (3,6) **9.** N (0,1)

Give the coordinates of each place named on this "map".

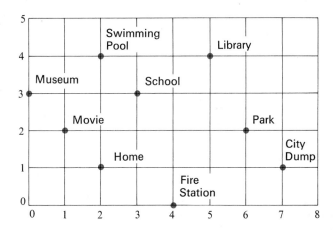

10. Home (2,1) **13.** City Dump (7,1) **16.** Swimming Pool (2,4)
11. Library (5,4) **14.** Museum (0,3) **17.** Park (6,2)
12. School (3,3) **15.** Movie (1,2) **18.** Fire Station (4,0)

WRITTEN EXERCISES

Use a piece of graph paper to make a lattice as indicated. Draw a cross (×) through each of the points named by the number pairs.

SAMPLE: 4 × 4 lattice;   *Solution:*
(2, 0), (2, 1), (1, 3), and (3, 3)

1. 4 × 4 lattice; (0, 3), (1, 3), (2, 0), and (0, 2).
2. 5 × 5 lattice; (0, 1), (1, 4), (4, 4), and (3, 2).
3. 5 × 5 lattice; (2, 2), (2, 4), (4, 0), and (1, 1).
4. 3 × 3 lattice; (0, 0), (0, 1), (1, 1), (1, 2), and (2, 2).
5. 6 × 6 lattice; (2, 3), (4, 4), (5, 0), (1, 5), (3, 5), and (2, 1).
6. 8 × 8 lattice; (4, 7), (2, 6), (5, 5), (6, 0), and (0, 7).
7. 4 × 4 lattice; (1, 1), (2, 1), (1, 2) and (2, 2).
8. 4 × 4 lattice; (0, 0), (3, 3), (3, 0), and (0, 3).

70 Chapter 3

Mark with a × each point in the set described. Use a 5 by 5 lattice for each part. Then make a roster of the members of the set.

SAMPLE: {All the points whose first coordinate is 2}

Solution: {(2, 0), (2, 1), (2, 2), (2, 3), (2, 4)}

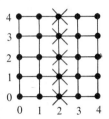

B 9. {All the points whose first coordinate is 4}
10. {All the points whose first coordinate is 0}
11. {All the points whose second coordinate is 2}
12. {All the points whose first and second coordinates are the same number}
13. {All the points whose second coordinate is 0}
14. {All the points whose first coordinate is 3}

C 15. {All the points whose first coordinate is greater than 2}
16. {All the points whose first coordinate is less than 3 and whose second coordinate is also less than 3}
17. {All the points whose first coordinate is either 2 or 3}

3–3 Functions and Graphs

You may have already guessed that we can use a function rule to name number pairs that can be graphed on a lattice. For example, the rule "**add 1**" gives the following results when the numbers **0, 1, 2,** and **3** are put into the function machine shown:

Input Number	Output Number	Number Pair
0	1	(0, 1)
1	2	(1, 2)
2	3	(2, 3)
3	4	(3, 4)

RULE: +1
FUNCTION MACHINE
INPUT 0
OUTPUT 1

The set of number pairs, {(0, 1), (1, 2), (2, 3), (3, 4)} represents a function. The corresponding points in the lattice shown in Figure 3–2 make up the graph of this function. Each point of the graph is marked with a ×.

Numbers, Functions, and Number Pairs   71

Graph of the function
{(0, 1), (1, 2), (2, 3), (3, 4)}

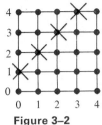

Figure 3–2

In algebra it is customary to use a letter such as *x* to represent the numbers to which the function rule is to be applied. Each value assigned to *x* becomes the first number in an ordered number pair. Then we simply replace *x* with one of the numbers from a given set, called the **replacement set,** and apply the function rule to find the second number in each number pair. This second number is called **the value of the function for *x*.** In symbols, "the value of the function for *x*" is written $f(x)$.

Suppose that we use {1, 3, 5, 7} as the replacement set for *x* in the function machine in Figure 3–3. Can you tell how to complete the table?

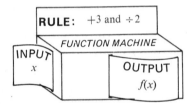

Figure 3–3

x	f(x)	Number Pair
1	2	(1, 2)
3	3	(3, 3)
5	4	(5, 4)
7	?5	(7 ?5)

+4 9 13 (9, 13)

Let us show the graph of this function in a lattice. Notice that the horizontal scale is the *x*-scale of Figure 3–4. The vertical scale is the $f(x)$-scale.

Graph of the function
{(1, 2), (3, 3), (5, 4), (7, 5)}

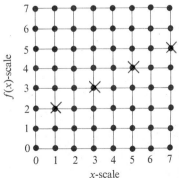

Figure 3–4

ORAL EXERCISES

According to the function rule on the machine, tell how to complete the table for the replacement set {0, 2, 4, 6, 8, 10}.

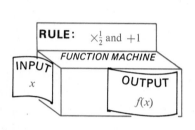

	x	f(x)	Number Pair
1.	0	1	(0,1)
2.	2	2	(2,2)
3.	4	3	(4,3)
4.	6?	4?	(6, 4)
5.	8?	5?	(8, 5)
6.	10	6	(10,6)

Tell how to complete each table according to the given function rule.

Rule: +1 and ×2

	x	f(x)	Number Pair
7.	0	2?	(0,2)
8.	1	4?	(1,4)
9.	2	6?	(2,6)
10.	3	8?	(3,8)
11.	4	10?	(4,10)
12.	5	12?	(5,12)

Rule: ÷2 and +1

	t	f(t)	Number Pair
13.	6	4?	(6,4)
14.	8	5?	(8,5)
15.	10	6?	(10,6)
16.	12	7?	(12,7)
17.	16?	8	(16,8)
18.	18?	9	(18,9)

WRITTEN EXERCISES

Complete each table according to the given function machine and function rule.

Numbers, Functions, and Number Pairs   73

SAMPLE:

m	f(m)	Number Pair
0	2	(0, 2)
3	3	(3, 3)
6	4	(6, 4)
9	5	(9, 5)

Solution:

A 1.

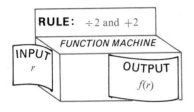

r	f(r)	Number Pair
12	8?	(12,8)
14	9?	(14,9)
16	10?	(16,10)
18	11?	(18,11)
20	12?	(20,12)

2.

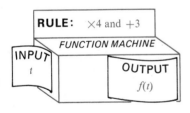

t	f(t)	Number Pair
0.5	5?	(.5,5)
1.0	7?	(1,7)
1.5	9?	(1.5,9)
2.0	11?	(2,11)
2.5	10?	(2.5,10)

3.

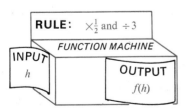

h	f(h)	Number Pair
2	?	?
4	?	?
6	1?	(6,1)
8	$1\frac{1}{3}$?	$(8, 1\frac{1}{3})$
10	$2\frac{1}{3}$?	$(10, 2\frac{1}{3})$

4.

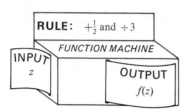

z	$f(z)$	Number Pair
$\frac{1}{2}$	?	?
$\frac{3}{2}$	?	?
$\frac{5}{2}$	1 ?	$(\frac{5}{2}, ?, 1)$
$\frac{7}{2}$	$1\frac{1}{3}$?	$(\frac{7}{2}, ?, \frac{1}{3})$
$\frac{9}{2}$	$2\frac{1}{3}$?	$(\frac{9}{2}, ?, 2\frac{1}{3})$

Use the given replacement set and function rule to write the ordered number pairs that represent the function.

SAMPLE: Replacement set for t: $\{10, 15, 20, 25, 30\}$; rule: $\times 2$ and $+5$

Solution:

t	$f(t)$	$(t, f(t))$
10	25	(10, 25)
15	35	(15, 35)
20	45	(20, 45)
25	55	(25, 55)
30	65	(30, 65)

5. Replacement set for x: $\{1, 4, 9, 16, 25\}$; rule: $+9$
6. Replacement set for t: $\{1, 3, 5, 7, 11, 13\}$; rule: $+3$ and $\div 2$
7. Replacement set for k: $\{10, 20, 30, 40, 50, 60\}$; rule: $\times \frac{1}{10}$
8. Replacement set for a: $\{3, 6, 9, 12, 15, 18\}$; rule: $\times 1$
9. Replacement set for y: $\{0, 1, 2, 3, 4, 5, 6, 7\}$; rule: $\times 2$ and $+3$
10. Replacement set for m: $\{1, 2, 3, 4, 5, 6\}$; rule: $+2$ and $\times 4$
11. Replacement set for h: $\{10, 8, 6, 4, 2, 0\}$; rule: $\div 2$ and $+5$
12. Replacement set for s: $\{5, 10, 15, 20, 25\}$; rule: $\times 6$ and $\div 10$

Complete each set of ordered pairs according to the given rule.

SAMPLE: Rule: $+4$; $\{(1, 5), (2, 6), (3, ?), (4, ?), (5, ?), (6, ?), (7, ?)\}$
Solution: $\{(1, 5), (2, 6), (3, \mathbf{7}), (4, \mathbf{8}), (5, \mathbf{9}), (6, \mathbf{10}), (7, \mathbf{11})\}$

B 13. Rule: $+7$; $\{(1, 8), (2, 9), (3, 10), (4, ?), (5, ?), (6, ?)\}$
14. Rule: -3; $\{(9, 6), (8, 5), (7, ?), (6, ?), (5, ?), (4, ?), (3, ?)\}$

15. Rule: ×2 and +1; {(0, 1), (1, 3), (2, ?), (3, ?), (4, ?), (5, ?)}
16. Rule: ×$\frac{1}{5}$; {(5, ?), (10, ?), (15, ?), (20, ?), (25, ?), (30, ?)}
17. Rule: ÷$\frac{1}{3}$; {(0, ?), (2, ?), (4, ?), (6, ?), (8, ?), (10, ?)}
18. Rule: ×1$\frac{1}{2}$; {(1, ?), (2, ?), (3, ?), (4, ?), (5, ?), (6, ?), (7, ?)}

Using the given function rule and replacement set, make a lattice graph of the function.

SAMPLE: Replacement set for x: {1, 3, 5, 7}; rule: +1 and ÷2

Solution:

x	$f(x)$	$(x, f(x))$
1	1	(1, 1)
3	2	(3, 2)
5	3	(5, 3)
7	4	(7, 4)

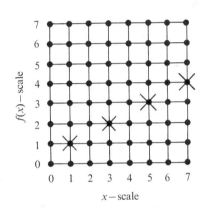

C

19. Replacement set for a: {0, 2, 4, 6}; rule: +4 and ÷2

20. Replacement set for n: {2, 3, 5, 7}; rule: ×$\frac{1}{n}$

21. Replacement set for s: {1, 3, 5, 7, 9}; rule: +5 and ×$\frac{1}{2}$

22. Replacement set for g: {0, 2, 4, 6, 8}; rule: ÷2 and ×0

23. Replacement set for x: {8, 4, 10, 2, 6}; rule: ÷$\frac{1}{2}$ and ÷4

Nomographs

3–4 Special Graphs: Addition Nomographs

The course of history shows that man has struggled for centuries to create devices to simplify his work in mathematics. A very simple but interesting type of mathematical device is the **nomograph**. The following

76   Chapter 3

illustration pictures a nomograph that can be used to solve simple problems in addition, and in its inverse, subtraction. The addition nomograph consists of three number lines or scales. Notice that two scales are labeled **addend** while the middle scale is labeled **sum**. Do you see that the length of a unit on the sum scale is **one-half** the length of a unit on the addend scales?

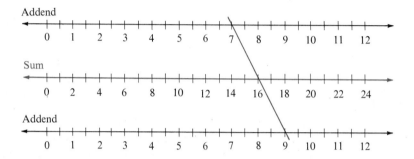

The familiar idea **addend + addend = sum** is important in using the nomograph for addition. A line joining the graphs of two addends intersects the sum scale at a point that corresponds with their sum. According to the nomograph above the unknown sum in the sentence 7 + 9 = ? is 16. So, 7 + 9 = 16.

Since subtraction is the **inverse** of addition, an addition nomograph can also be used to complete a sentence such as 15 − 5 = ?. Do you agree that **15** is the sum, and **5** is an addend, and that the other addend is unknown? The line drawn across the nomograph below through the addend **5** and the sum **15** locates the unknown addend **10**. So, 15 − 5 = 10.

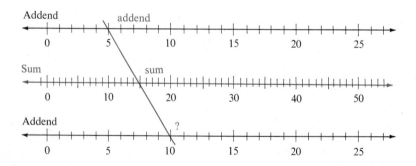

ORAL EXERCISES

Refer to the nomograph shown below. Give the addition statement and the two equivalent subtraction statements suggested by each line on the nomograph.

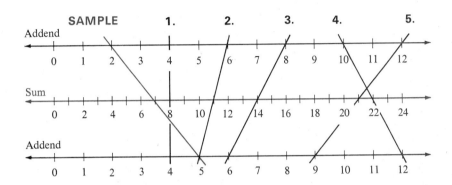

SAMPLE: (See nomograph above.)

What you say: $2 + 5 = 7$; $7 - 5 = 2$; $7 - 2 = 5$

1.–5. (See nomograph above.)

Answer each of the questions that refer to the nomograph below.

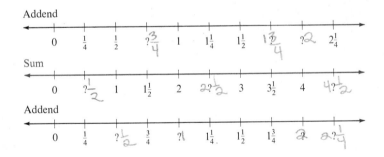

6. What numerals should replace the question marks on the first addend scale?
7. What numerals should replace the question marks on the sum scale?
8. What numerals should replace the question marks on the second addend scale?
9. Use the nomograph to complete the sentence $\frac{1}{4} + 1\frac{3}{4} = $?.
10. Use the nomograph to complete the sentence $3 - 2\frac{1}{4} = $?.

WRITTEN EXERCISES

Complete each sentence to make a true statement. Use the corresponding numbered line on the nomograph below.

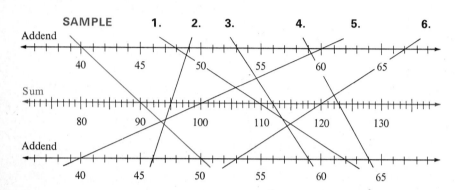

SAMPLE: 90 − 40 = ?   Solution: 90 − 40 = **50**

A
1. 48 + 62 = ?
2. 49 + 46 = ?
3. 53 + ? = 112
4. 123 − 59 = ?
5. ? − 40 = 60
6. 67 = 120 − ?

Use the following nomograph to complete each of Statements 7–12.

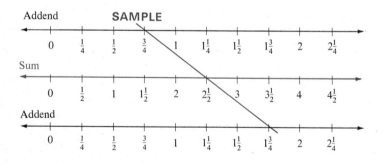

SAMPLE: $\frac{3}{4} + 1\frac{3}{4} = ?$ Solution: $\frac{3}{4} + 1\frac{3}{4} = 2\frac{1}{2}$

7. $\frac{1}{4} + \frac{1}{4} = ?$
8. $\frac{1}{2} + \frac{3}{4} = ?$
9. $1\frac{1}{2} + ? = 2$
10. $3 - 1\frac{1}{2} = ?$
11. $4\frac{1}{4} - ? = 2\frac{1}{4}$
12. $1\frac{1}{4} + ? = 3\frac{1}{2}$

Use the following nomograph to complete each of Statements 13–18.

Numbers, Functions, and Number Pairs

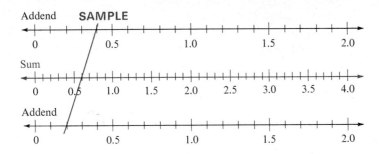

SAMPLE: $0.4 + 0.2 = ?$

13. $0.5 + 0.5 = ?$
14. $0.8 + 0.6 = ?$
15. $2.2 - 1.0 = ?$

Solution: $0.4 + 0.2 = \mathbf{0.6}$

16. $1.2 + ? = 2.5$
17. $1.4 = 3.3 - ?$
18. $? = 3.9 - 2.0$

Complete each of the following to make a true statement. Use the nomograph shown.

SAMPLE: $0.021 + 0.022 = ?$

Solution: $0.021 + 0.022 = 0.043$

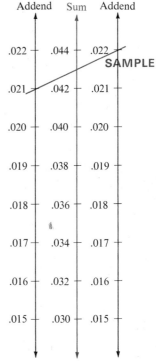

B 19. $0.021 + 0.019 = ?$
20. $0.017 + 0.015 = ?$
21. $0.021 + ? = 0.037$
22. $0.016 + ? = 0.038$
23. $0.030 - 0.015 = ?$
24. $0.038 - 0.017 = ?$
25. $0.036 - 0.018 = ?$
26. $0.018 + ? = 0.035$
27. $0.032 = 0.017 + ?$
28. $0.018 = 0.034 - ?$

3–5 Special Graphs: Multiplication Nomographs

You probably recall that the multiplication of two numbers may be thought of as **factor × factor = product**. Division, the inverse of multiplication, is indicated as **product ÷ factor = factor**. These ideas allow the use of multiplication nomographs to solve both division and multiplication problems. The **multiplication nomograph** shown below consists of two scales labeled **factor**, and a middle scale labeled **product**. It is especially important to note how numbers are matched with points on these scales, which mathematicians call **logarithmic scales**. You may study logarithms in another mathematics course, some day.

The line drawn across the nomograph represents the sentence $3 \times 5 = ?$. It intersects the product scale at the graph of **15**. So, $3 \times 5 = 15$.

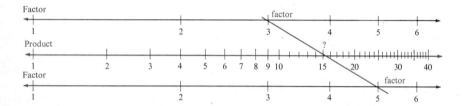

Notice that the same line across the nomograph might help us to complete the sentences $15 \div 5 = ?$ and $15 \div 3 = ?$.

When writing a multiplication sentence and the equivalent division sentences it helps to keep in mind that an expression like $10 \div 2$ may be written as the fraction $\frac{10}{2}$. For example:

$15 \div 5 = 3$ has the same meaning as $\frac{15}{5} = 3$,   and

$36 \div 2 = 18$ has the same meaning as $\frac{36}{2} = 18$.

ORAL EXERCISES

Give the multiplication statement and the two division statements suggested by each line on the nomograph below.

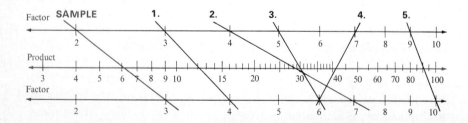

Numbers, Functions, and Number Pairs **81**

SAMPLE: (See nomograph.)

What you say: $2 \times 3 = 6$; $\frac{6}{3} = 2$; $\frac{6}{2} = 3$.

1.–5. (See nomograph.)

Name each of the numbers in the statements below as a product or as a factor.

SAMPLE: $\frac{20}{5} = 4$

What you say: 20 is the product; 5 and 4 are factors.

6. $36 \div 4 = 9$
7. $\frac{36}{4} = 9$
8. $24 \div 8 = 3$
9. $\frac{32}{4} = 8$
10. $4 \times 8 = 32$

11. $12 = 48 \div 4$
12. $12 = \frac{48}{4}$
13. $65 \div 13 = 5$
14. $\frac{27}{6} = 4\frac{1}{2}$
15. $\frac{65}{10} = 6\frac{1}{2}$

WRITTEN EXERCISES

Write an equivalent multiplication statement or two equivalent division statements in fractional form for each of the following.

SAMPLE 1: $6 \times 7 = 42$ Solution: $\frac{42}{6} = 7$; $\frac{42}{7} = 6$

SAMPLE 2: $\frac{39}{3} = 13$ Solution: $39 = 3 \times 13$

1. $9 \times 8 = 72$
2. $5 \times 11 = 55$
3. $\frac{45}{9} = 5$
4. $\frac{51}{17} = 3$
5. $57 = 3 \times 19$
6. $63 = 9 \times 7$

7. $15 = \frac{75}{5}$
8. $6 = \frac{42}{7}$
9. $\frac{13}{4} = 3\frac{1}{4}$
10. $\frac{21}{9} = 2\frac{1}{3}$
11. $\frac{0.36}{0.4} = 0.9$
12. $\frac{0.24}{0.6} = 0.4$

13. $28 \times 7 = 196$
14. $42 \times 5 = 210$
15. $2\frac{1}{2} = \frac{15}{6}$
16. $8\frac{1}{3} = \frac{50}{6}$
17. $228 = 12 \times 19$
18. $966 = 23 \times 42$

Complete each sentence to make a true statement. Use the corresponding numbered line on one of the nomographs.

82 Chapter 3

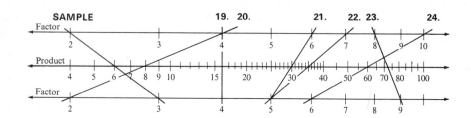

SAMPLE: 2 × 3 = ?   *Solution:* 2 × 3 = **6**
19. 4 × 4 = ?   **22.** ? × 5 = 35
20. 4 × 2 = ?   **23.** 8 × ? = 72
21. 6 × 5 = ?   **24.** $\frac{60}{6}$ = ?

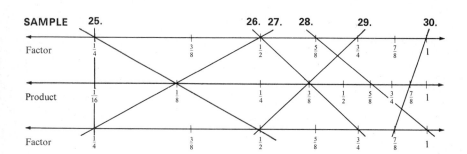

SAMPLE: $\frac{1}{4} \times \frac{1}{2}$ = ?   *Solution:* $\frac{1}{4} \times \frac{1}{2} = \frac{1}{8}$

B **25.** $\frac{1}{4} \times \frac{1}{4}$ = ?   **28.** ? × 1 = $\frac{5}{8}$
26. $\frac{1}{2} \times \frac{3}{4}$ = ?   **29.** $\frac{3}{8} \div \frac{3}{4}$ = ?
27. $\frac{1}{2} \times$? = $\frac{1}{8}$ **30.** $\frac{7}{8} \div 1$ = ?

Complete each sentence to make a true statement.

SAMPLE 1: $9 = \frac{?}{3}$ *Solution:* Since $9 \times 3 = 27$, $9 = \frac{27}{3}$.

SAMPLE 2: $\frac{4 \times 8}{2}$ = ?   *Solution:* $\frac{4 \times 8}{2} = \frac{32}{2}$
Since $2 \times 16 = 32$, $\frac{4 \times 8}{2} = 16$.

31. $6 = \frac{?}{7}$ **33.** ? ÷ 5 = 9
32. $\frac{?}{9} = 8$ **34.** ? ÷ 7 = 5

35. $? \div \frac{1}{3} = \frac{2}{3}$

36. $\frac{3}{5} = ? \div \frac{1}{2}$

37. $\frac{6 \times 9}{3} = ?$

38. $\frac{5 \times 8}{10} = ?$

39. $0.4 = \frac{?}{0.6}$

40. $0.3 = \frac{?}{0.9}$

41. $? \div 0.8 = 0.3$

42. $? \div 0.25 = 3$

SAMPLE: $.2 \times .4 = ?$

Solution: $.2 \times .4 = .08$

C **43.** $0.2 \times 0.1 = ?$

44. $0.3 \times 0.2 = ?$

45. $? = 0.4 \times 0.5$

46. $0.5 \times ? = 0.30$

47. $0.5 \times ? = 0.15$

48. $0.36 \div 0.6 = ?$

49. $\frac{0.49}{0.7} = ?$

50. $\frac{0.40}{?} = 0.8$

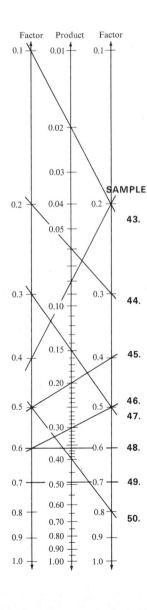

CHAPTER SUMMARY

Inventory of Structure and Concepts

1. A **function** is a set of ordered pairs. These may be shown in roster form, or given by a rule, or by the graph of the set. No two different ordered number pairs in a set that is a function can have the same **first** number.

2. A **lattice** is an array of points used to give a graphical representation of ordered number pairs. The numbers of an ordered pair are called the **coordinates** of the point. The **first** number of the pair, or **first coordinate**, is the number on the **horizontal** scale. The **second** number, or **second coordinate**, is the number on the **vertical** scale.

3. **Nomographs** are devices that can be used to carry out mathematical operations. The **addition** nomograph consists of three appropriately arranged number line scales, with units of uniform length.

    Since **subtraction** is the inverse operation of addition, both **addition** and **subtraction** may be done on the same nomograph.

    The **multiplication** nomograph consists of three **logarithmic** number scales. Since **multiplication** and **division** are inverse operations, both can be done on the same nomograph.

Vocabulary and Spelling

Pronounce, spell, and give the meaning of these words and expressions.

function (*p. 63*)
lattice (*p. 68*)
first coordinate (*p. 68*)
second coordinate (*p. 68*)

x-scale (*p. 71*)
$f(x)$-scale (*p. 71*)
nomograph (*p. 75*)

Chapter Test

Indicate whether each statement is true or false.

1. The set of ordered pairs {(1, 2), (5, 2), (7, 2), (9, 2)} defines a function.
2. The infinite set of ordered pairs {(1, 2), (2, 3), (3, 4), ...} defines a function.

3. The set of nine ordered pairs indicated by points marked × defines a function.
4. The sum of the first coordinates of the nine ordered pairs is 16.

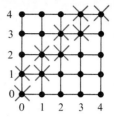

In Questions 5–8, replace each question mark to make the resulting statement true.

5. $1\frac{3}{4} - ? = 1\frac{3}{8}$

6. $\frac{504}{?} = 3$

7. $\frac{?}{9} = 71$

8. $? + 1\frac{1}{4} = 2\frac{7}{8}$

Draw lattice graphs for Questions 9 and 10.

9. All points whose first coordinate is twice its second coordinate. (Use a 5 × 5 lattice.)
10. All points whose second coordinate is a divisor of its first coordinate. (Use a 4 × 4 lattice.)
11. If the rule is $+4$ and $\div 2$, then $f(h) = ?$ when $h = 3$.
12. If the rule is $\times 5$, then $h = ?$ when $f(h) = 7$.
13. If the rule is -2 and $\times \frac{5}{2}$, then $f(h) = ?$ when $h = 8$.

Complete the ordered pairs so that some function rule relates the first number to the second number.

14. (2, 7), (5, 16), (10, 31), (4, ?), (8, ?)
15. (4, 3), (8, 5), (16, 9), (10, ?)

Chapter Review

3–1 Functions: Rules and Machines

Which of the following sets of ordered pairs represent functions?

1. {(3, 5), (4, 6), (5, 7)}
2. {(3, 5), (4, 5), (5, 5)}
3. {(3, 5), (3, 6), (5, 5)}
4. {(3, 9), (4, 1), (8, 3), (5, 2)}

86 Chapter 3

For each given function rule, write the set of ordered pairs with input numbers 1, 3, 5, and 6.

5. +5   **6.** ×3   **7.** −0   **8.** ÷2

Complete the ordered pairs so that for each set some function rule relates the first number to the second number.

9. {(2, 3), (4, 7), (10, 19), (6, ?), (12, ?), (?, 9)}
10. {(2, 4), (6, 36), (8, ?), (10, ?), (?, 49)}

3–2 Number Pairs and Graphs

For each question, construct a 4 × 4 lattice and mark the points described.

11. (0, 3), (3, 1), (2, 2)
12. All the points whose first coordinate is 1
13. All the points whose first coordinate is one less than its second coordinate

List the set of ordered pairs represented by the points marked by a X in each lattice.

14. 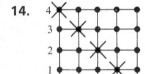   **15.**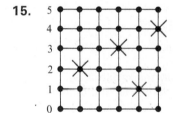

3–3 Functions and Graphs

16. If the rule is +3, determine $f(x)$ when x is 5.
17. If the rule is ×2 and −5, determine $f(t)$ when t is 7.
18. If the rule is +1 and ×2, make a lattice graph of the function $(x, f(x))$ when $x \in \{0, 1, 2\}$.

3–4 Addition Nomographs

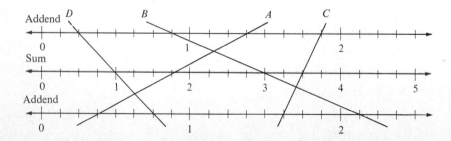

Using the nomograph on page 86, match each of the following equations with the corresponding line (A, B, C, or D) from the nomograph.

19. $\frac{7}{8} + 2\frac{1}{8} = 3$

20. $1\frac{5}{8} + 1\frac{7}{8} = 3\frac{1}{2}$

21. $1 - \frac{1}{4} = \frac{3}{4}$

22. $\frac{7}{8} = 3 - 2\frac{1}{8}$

Using the same nomograph, complete each of the following equations.

23. $1\frac{1}{8} + 2\frac{3}{8} = ?$

24. $\frac{7}{8} + ? = 1\frac{3}{4}$

25. $2\frac{3}{4} - ? = 1\frac{1}{8}$

26. $3\frac{3}{4} - 1\frac{1}{2} = ?$

3–5 Multiplication Nomographs

Write an equivalent multiplication statement or two equivalent division statements (fractional form) for each of the following.

27. $4 \times 6 = 24$

28. $\frac{35}{7} = 5$

29. $168 = 8 \times 21$

30. $\frac{2.10}{0.6} = 3.5$

Complete each sentence to make a true statement.

31. $105 \div ? = 15$

32. $\frac{?}{12} = 106$

33. $0.5 \times ? = 210$

34. $0.8 = \frac{?}{0.6}$

35. $\frac{3}{4} \div ? = 3$

36. $? \div \frac{1}{4} = 5$

Review of Skills

Simplify each expression.

1. $13 + 5 - 6$

2. $2 \cdot 5 \cdot 6$

3. $2 \cdot 5 + 3$

4. $2 \cdot (5 + 3)$

5. $2 + 5 \cdot 6 - 4$

6. $(2 + 5) \cdot (6 - 4)$

7. $\frac{17 + 3}{2}$

8. $24 - 13 + 5$

9. $3 \cdot 7 \cdot 2$

10. $10 + 3 \cdot 4$

11. $(10 + 3) \cdot 4$

12. $15 - 3 \cdot 4 + 1$

13. $(15 - 3) \cdot (4 + 1)$

14. $\frac{11 - 5 + 2}{2}$

15. $29 - 24 - 3$

16. $10 \cdot 10 \cdot 10$

17. $15 - 2 \cdot 5$

18. $(15 - 2) \cdot 5$

19. $4 + 2 \cdot 1 - 1$

20. $(4 + 2) \cdot (1 - 1)$

21. $\frac{18 + 18}{2}$

88 Chapter 3

22. $3 \cdot 4$
23. 5^2
24. $(2 + 3)^2$
25. $2 \cdot 3^2$
26. $\dfrac{5^2}{5}$
27. $\dfrac{8 + 8}{8}$
28. $2^2 \cdot 2^3$
29. $3^1 \cdot 3^2$
30. $3 + 2^2$
31. $5 \cdot 5 \cdot 5$
32. $2^3 \cdot 3^2$

33. 3^4
34. 2^5
35. $2^2 + 3^2$
36. $2^2 \cdot 3$
37. $\dfrac{5^2}{5^1}$
38. $\dfrac{3 + 6}{3}$
39. 2^6
40. 3^3
41. $(2 + 3)^3$
42. 5^3
43. $2^2 \cdot 3^2$

44. $5 \cdot 2$
45. $2 \cdot 5$
46. $(2 \cdot 3)^2$
47. $\dfrac{3^3}{3^2}$
48. $\dfrac{8 \cdot 8}{8}$
49. $\dfrac{3 \cdot 6}{3}$
50. $3 \cdot 3^2$
51. $2 + 3^2$
52. $(3 + 2)^2$
53. $2 \cdot 2 \cdot 2 \cdot 3 \cdot 3$
54. $(2 \cdot 5)^2$

Given the following diagrams representing a rectangle and a square:

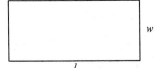

55. Compute the area of the rectangle if $l = 8$ feet and $w = 3$ feet.
56. Compute the perimeter of the rectangle if $l = 8$ feet and $w = 3$ feet.
57. Compute the area of the square if $s = 5$ inches.
58. Compute the perimeter of the square if $s = 5$ inches.

To each name in Column 1, match the correct figure from Column 2.

COLUMN 1   COLUMN 2

59. Cone   A.    D.

60. Cube

                 B.   E.

61. Sphere

62. Square   C.   F.

63. Cylinder

THE HUMAN ELEMENT

Carl Friedrich Gauss

Carl Friedrich Gauss (1777–1855) was born into the family of a laborer and gardener in Brunswick, Germany. From this poor background he rose to be one of the greatest mathematicians of all times, ranking with Archimedes and Newton.

Gauss's genius showed itself at an early age. Watching his father add a long column of figures, Gauss, who was then three years old, told him the answer was wrong, and gave another answer. When the father checked, he discovered that his son's answer was the correct one.

At the age of seven, Gauss began his formal education in the local grammar school. One day the teacher assigned the problem of adding all the numbers from one through a hundred, hoping to keep the class busy for some time. As soon as the teacher had finished stating the problem, Gauss handed in his slate, with nothing written on it but the correct answer. He had discovered for himself how to find quickly the sum of a sequence of numbers. The diagram below will help clarify his reasoning.

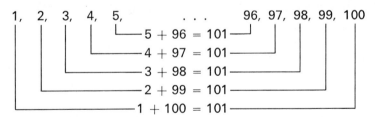

There are 50 pairs of numbers, and the sum of each pair is 101, so the sum of all the numbers in the sequence is $50 \times 101 = 5050$.

Eventually Gauss's genius came to the attention of the Duke of Brunswick, who financed his university education, and made it possible for Gauss to develop his great talents. By the time he was 25 years old, this son of a poor laborer had solved problems that had waited centuries for a solution. His most significant work was in number theory, the branch of mathematics which deals with the properties of integers. He referred to this as *arithmetic* when he said: "Mathematics is the queen of the sciences, and arithmetic is the queen of mathematics."

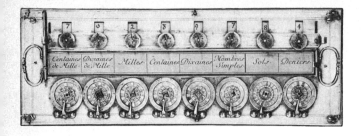

To compute taxes in the year 1642 . . .

For modern tax computation . . .

Variables, Expressions, and Sentences

The French mathematician Blaise Pascal devised one of the earliest calculating machines in 1642, in order to avoid the drudgery of bookkeeping for his tax-collector father. Figures entered in the calculator were transmitted through connecting gears, with the results of the addition and subtraction appearing at the top of the machine. Modern computers are used for checking income tax returns at the Internal Revenue Service Center, Andover, Massachusetts. In the section of the computer room shown, the Director of the Center stands at a high speed printer unit. Behind him is a central processing control which includes memory control and arithmetic units. When a tax return is processed, the computer checks the arithmetic. If an error is found, the return moves to a printout for a notice to be sent to the taxpayer concerned.

Fundamental Ideas of Grouping

4–1 Expressions and Punctuation

An important goal in mathematics is that of clarity. For example, in the case of the expression "$5 + 7$" we can agree without argument that it names the number 12. However, the meaning of the expression "$10 - 5 + 2$" might easily cause confusion. Depending upon the order in which the indicated operations are completed, the symbol $10 - 5 + 2$ could name either the number 7 or the number 3, as shown below.

$$\underbrace{10 - 5}_{5} + 2 = ? \qquad \text{but} \qquad 10 - \underbrace{5 + 2}_{7} = ?$$
$$5 + 2 = 7 \qquad\qquad\qquad 10 - 7 = 3$$

When we write words, the possibility of misunderstanding is avoided by the use of punctuation marks. Do you see that the following statements contain the same words, but that each has a different meaning?

Please call Tom.   Please call, Tom.

In mathematics, grouping symbols such as **parentheses**, **brackets**, and the **fraction bar** serve as punctuation for making the meaning clear. Such symbols tell us which operation is to be completed first.

Study these illustrations:

Expression	Meaning of the Expression
$2 \times [4 + 5]$	First 4 and 5 are added; then the result is multiplied by 2. Thus $2 \times [4 + 5] = 2 \times 9 = 18$.
$[2 \times 4] + 5$	First 2 and 4 are multiplied; then 5 is added to the result. Thus $[2 \times 4] + 5 = 8 + 5 = 13$.
$(18 \div 6) + 3$, or $\frac{18}{6} + 3$	First 18 is divided by 6; then 3 is added to the result. Thus $(18 \div 6) + 3 = 3 + 3 = 6$.
$18 \div (6 + 3)$ or, $\frac{18}{6+3}$	First 6 and 3 are added; then 18 is divided by the result. Thus $18 \div (6 + 3) = 18 \div 9 = 2$.
$7 + 6(8 - 3)$	First 3 is subtracted from 8; then 6(5) is replaced by the numeral 30, a simpler form of the number it represents. Finally, 30 is added to 7. Thus $7 + 6(8 - 3) = 7 + 6(5) = 7 + 30 = 37$.

Note that, in the last illustration, $7 + 6(5)$ was treated as if it were $7 + [6(5)]$. The symbol of grouping around a multiplication may be omitted. However, the multiplication is performed before an addition or subtraction unless there are parentheses or brackets to indicate otherwise. In general, to **simplify** an expression one performs the operations in the order indicated by grouping symbols, written or understood.

ORAL EXERCISES

Tell how to find the number named by each expression.

SAMPLE. $(8 + 10) - 6$

What you say: First add 8 and 10; then subtract 6. The number named is 12.

Variables, Expressions, and Sentences

1. $4 + (5 \times 2)$
2. $10 + (6 \div 3)$
3. $(3 \times 4) + 6$
4. $(16 \div 4) + 5$
5. $\dfrac{5 + 9}{7}$
6. $10 - (3 \times 3)$
7. $(5 \times 7) - 10$
8. $(4 \times 5) + 9$
9. $(8 \div 2) - 3$
10. $5 \times [3 + 4]$
11. $[3 \times 8] \div 6$
12. $\dfrac{24}{3 + 1}$
13. $\dfrac{10 + 8}{5 + 4}$
14. $(3 + 6) + \tfrac{3}{4}$
15. $(2 + 6) + (3 \times 5)$

WRITTEN EXERCISES

Match each expression in Column 1 with the correct item in Column 2.

COLUMN 1

1. The sum of 5 and 7, decreased by 9.
2. The product of 5 and the sum of 3 and 7.
3. The quotient of 25 and the sum of 8 and 2.
4. The sum of 8 and 13 divided by 7.
5. 9 decreased by the product of 4 and 7.
6. The quotient of 25 and the product of 6 and 2.
7. The product of 5 and the quotient of 3 and 7.
8. The product of 4 and 7, decreased by 9.

COLUMN 2

A. $5 \times \tfrac{3}{7}$
B. $(4 \times 7) - 9$
C. $(5 + 7) - 9$
D. $9 - (4 \times 7)$
E. $\dfrac{25}{6 \times 2}$
F. $5 \times (3 + 7)$
G. $25 \div [8 + 2]$
H. $\dfrac{8 + 13}{7}$

Copy each statement and supply the punctuation that will make it true.

SAMPLE. $16 + 4 \div 2 = 10$ Solution: $(16 + 4) \div 2 = 10$

9. $5 \times 2 + 7 = 17$
10. $5 \times 2 + 7 = 45$
11. $5 + 4 \div 2 = 7$
12. $17 + 3 \div 5 = 4$
13. $8 \div 2 + 7 = 11$
14. $4 \times 7 - 3 = 25$
15. $\tfrac{3}{2} + \tfrac{5}{2} \div 2 = 2$
16. $3\tfrac{1}{2} \times 2 \times 7 = 49$
17. $20 \div 2 \div 5 = 2$
18. $16 - 8 - 2 = 6$

Simplify each expression.

SAMPLE. $(9 \times 6) \div 3$ Solution: $(9 \times 6) \div 3 = ?$
$54 \div 3 = 18$

B 19. $(15 - 3) - 12$ 24. $\dfrac{15 + 1}{10 - 2}$ 29. $\dfrac{21 \div 7}{4 \times \frac{1}{2}}$

20. $8 - (3 - 2)$ 25. $(6 + 9) + (7 + 5)$ 30. $3 \times [4 + 2 + 1]$

21. $[48 \div 8] \div 2$ 26. $(10 - 6) \div (8 \times \frac{1}{4})$ 31. $(6 + 9) \div (15 \div 3)$

22. $20 \div (3 + 2)$ 27. $\dfrac{15 - 10}{10 - 2}$ 32. $\dfrac{16 \times 2}{2 \times (4 + 4)}$

23. $\dfrac{18 + 4}{11 + 7}$ 28. $\dfrac{9 - 5}{40 + 5}$ 33. $[3 \times 2] \div 4$

34. $(27 \div 9) + (4 \times 3) + (18 \div 9)$

35. $(3 \times 5 \times 7) + (9 + 6 + 10 + 4)$

36. $\dfrac{36 + 4}{5 \times 8} + \dfrac{36 \div 4}{3} + 35$

4–2 More about Expressions and Grouping Symbols

As we move ahead in the study of algebra we shall try to eliminate any symbolism that can be spared. For example, the use of the multiplication sign, ×, can be discontinued if we agree that the expression **3(4 + 10)** has the same meaning as **3 × (4 + 10)**. Furthermore, the expression **5 × 9** may be written with a dot to indicate multiplication, or may be written with grouping symbols. Thus each of the following is equal to **45**:

$$5 \cdot 9; \quad 5(9); \quad (5)9; \quad (5)(9).$$

In working with expressions you will often find both parentheses and brackets used in the same expression. Notice that the following expression has parentheses inside of the brackets.

$$3[24 \div (2 + 4)]$$

To find the meaning of such an expression, in which one grouping symbol is inside another, it is best to begin with the innermost grouping symbol and simplify that expression first. Then, the expression in the outer grouping symbol can be simplified. For example:

$$3[24 \div (2 + 4)] = 3[24 \div 6]$$
$$= 3[4]$$
$$= 12$$

ORAL EXERCISES

Tell what number is named by each expression.

SAMPLE: $4(2 + 8)$

What you say: $4(2 + 8) = 4(10) = 40$

1. $(3 + 7)5$
2. $9(6 - 2)$
3. $(8)(8)$
4. $2(3)(5)$
5. $(4)(3)10$
6. $(2 + 4)(5 + 6)$
7. $(3 + 5 + 10)2$
8. $10(6 + 8 + 12)$
9. $3 \cdot (5 - 4)$
10. $2 + (2 \cdot 3 \cdot 4)$
11. $3[2 + (3 + 4)]$
12. $5[10 - (3 + 4)]$

WRITTEN EXERCISES

Simplify each expression.

1. $(3 + 8 + 9) \div 4$
2. $(3 + 9) \div (4 - 1)$
3. $(3)(5)(4 + 1)$
4. $10 + 3(2 + 9)$
5. $12 + 2(4 + 6) + 3$
6. $\dfrac{8 + 4(6 - 2)}{4}$
7. $\dfrac{4(9 + 3) - 6}{7}$
8. $(5 + 4)(6)(3)(0)$
9. $(8)(\tfrac{6}{2})(1)3$
10. $7 + 3(5 - 1)$
11. $(2 + 1)(3 + 1)10$
12. $7 + \dfrac{4(7 - 2)}{5}$
13. $25 - 3(6 - 2)$
14. $\dfrac{(36 \div 9) + 3}{8}$
15. $\dfrac{4(7) + 16}{3 + 8}$
16. $\dfrac{12 + (3 \cdot 6)}{2(3)}$
17. $4[18 \div (6 + 3)]$
18. $5[2 + (3 \cdot 5)]$

Rewrite each expression, using a fraction. Do not simplify.

SAMPLE 1. $36 - (15 \div 4)$ *Solution:* $36 - \tfrac{15}{4}$

SAMPLE 2. $(11 - 3) \div (5 + 1)$ *Solution:* $\dfrac{11 - 3}{5 + 1}$

19. $45 - (17 \div 6)$
20. $(13 \div 2) + 8$
21. $(2 + 6) \div (10 - 2)$
22. $72 \div (3 \cdot 35)$
23. $(19 + 3) \div (4)2$
24. $[(4)(8)] \div 5$

Simplify each expression.

B 25. $\dfrac{48}{\left[\frac{8}{2}\right]} + 3$

26. $\dfrac{45}{\left(\frac{6}{2}\right)} \cdot 5$

27. $\left(\tfrac{3}{4}\right)\left(\tfrac{1}{2}\right)2$

28. $(16 + 10) \div [(3 \cdot 8) \div 12]$

29. $[28 \div (8 \cdot \tfrac{1}{2})] + 6$

30. $(36 \div 9) + [(15 + 10) \div 10]$

31. $(28 \div 4) - [(6 + 8) \div 7]$

32. $\dfrac{\left(\frac{45}{3}\right)}{5} - 2$

33. $6[(2 + 3) + 4] + 10$

34. $(10 + 4) \div [(3 \times 6) - 11]$

35. $(49 - 25) \div [(7 + 5)2]$

36. $\left[\dfrac{72 - 30}{4 + 3}\right]8$

37. $3\left[\dfrac{100 - 64}{3 + 15}\right] + 10$

38. $[(5 \times 20) + (6 \div 3)] \div 3$

C 39. $\left[\dfrac{12 + 8}{(11)(3)} - \dfrac{5}{33}\right] + \dfrac{2}{33}$

40. $\left[\dfrac{9 + 8}{13(5)} + \dfrac{4}{65}\right] + \dfrac{1}{65}$

41. $(3)(9) + [(9 \cdot 3) + 2]$

42. $4(3) + [(3 + 15)(2)]$

The Use of Variables in Expressions

4–3 Expressions and Variables

The idea of using letters or other symbols to represent numbers is probably not new to you. To calculate the area of any rectangle the product of the length and width is found. This can be written as (length × width), or as $l \cdot w$. Symbols such as l and w are called **variables** and each may be replaced with a number from some indicated set of numbers called the **replacement set**. Thus, the value of the expression $l \times w$ will depend upon which element of the indicated replacement set is used to replace each of the variables l and w. When an expression contains at least one variable it is called a **variable expression**. Any variable expression or numerical expression is known as an **algebraic expression**.

Suppose that r in the variable expression $r + 6$ is replaced by each member of the replacement set $\{1, 2, 3, 4\}$. Then, $r + 6$ would have the following values:

$1 + 6$, or 7 $\qquad$ $3 + 6$, or 9

$2 + 6$, or 8 $\qquad$ $4 + 6$, or 10

Variables, Expressions, and Sentences

If a variable expression represents a product, the use of the multiplication symbol, ×, may often be avoided in either of the following ways:

(1) The two factors may simply be written with no symbol between them; thus

$$lw \text{ means } l \times w$$
$$4m \text{ means } 4 \times m$$

(2) Either or both of the factors may be enclosed in parentheses. For example:

$$l(w) = (l)w = l \times w$$
$$4(m) = (4)(m) = 4 \times m$$

How might an algebraic expression be written for "6 times the sum of 3 and some number"? Since we do not know the exact meaning of "some number," our algebraic expression must contain a variable. If we choose the letter *n* to stand for "some number," then "**6 times the sum of 3 and some number**" becomes

$$6(3 + n).$$

Study these examples of algebraic expressions.

a. the difference of 8 and 5, increased by *m* $(8 - 5) + m$
b. 3 times the sum of 6 and *k* $3(6 + k)$
c. the sum of *x* and *y*, divided by 7 $\dfrac{x + y}{7}$
d. the product of 3 and *n*, decreased by 8 $3n - 8$
e. the product of 5, *r*, and *s* $5(r)(s)$

ORAL EXERCISES

Give the value of each expression if $r = 3$, $k = 5$, and $m = 6$.

SAMPLE. $2r + km$ *Solution:* $2(3) + 5(6) = 6 + 30 = 36$

1. $r + 5$
2. $3k$
3. $m - 2$
4. $m \div 3$
5. $2(r)$
6. rk
7. $2kr$
8. $m + k + r$
9. $(r)(r)$
10. $\frac{1}{2}m$
11. $\dfrac{m}{2}$
12. $\dfrac{r + k}{2}$

13. $r + 3 + 8$ **15.** $r + m + k$ **17.** $2r + k$

14. $10m$ **16.** $\dfrac{10}{k}$ **18.** $\dfrac{m + r}{k}$

Match each item in Column 1 with the correct algebraic expression in Column 2.

COLUMN 1

19. the quotient of ten and the sum of 3 and b
20. 8 less than the sum of v and w
21. m divided by 3, and then increased by 2
22. the product of 5 and g, decreased by 8
23. 5 more than the sum of x and y
24. the difference of 15 and z, increased by the sum of a and b

COLUMN 2

A. $\dfrac{m}{3} + 2$
B. $(15 - z) + (a + b)$
C. $10 \div (3 + b)$
D. $5g - 8$
E. $\dfrac{m + 2}{3}$
F. $(x + y) + 5$
G. $(v + w) - 8$
H. $8 - 5g$

WRITTEN EXERCISES

Find the value of each algebraic expression, if $a = 7$, $b = \frac{1}{3}$, $r = 12$, $m = 9$, and $n = 5$.

SAMPLE. $\dfrac{6b + r}{a}$ Solution: $\dfrac{(6 \cdot \frac{1}{3}) + 12}{7} = \dfrac{2 + 12}{7}$
$= \dfrac{14}{7} = 2$

1. $r + b$
2. $\dfrac{m + r}{3}$
3. $5a + 3m$
4. $4(m + n)$
5. $ar + mr$
6. $\dfrac{r + m}{a}$
7. $5n \div a$
8. $\frac{1}{4}(a + m)$

9. $r \div \frac{1}{3}m$
10. $\dfrac{ar + 6}{m}$
11. $4(2a + 3n)$
12. $\dfrac{r}{a} + \dfrac{n}{a}$
13. $(an)(an)$
14. $(aa)(nn)$
15. $2a + 3b + 5r + n$
16. $4n\left(2a + \dfrac{10}{n}\right)$

B 17. $(ar - m) + n$
18. $(a + r)(a + m)$
19. $(m + n)(m - n)$
20. $\dfrac{5a + m}{11}$
21. $(abr) + \dfrac{a}{b}$
22. $\dfrac{5nr}{bnr}$
23. $\dfrac{2a(2n + m)}{3n(r - a)}$

24. $(a + b)(m - n)$
25. $(br)(mb)(n)(n)$
26. $4(ar) + \dfrac{m + n}{a}$
27. $\dfrac{r + r + r}{(br)(br)}$
28. $\dfrac{5a - r}{5a + r}$
29. $\dfrac{n(6b + a)}{n(6a - n)}$
30. $3(n)(n) + mn - 4m$

Find all the values of each algebraic expression if the replacement set for *n* is {2, 4, 6, 8} and the replacement set for *x* is {3, 6, 9}.

SAMPLE. $2n + 10$ Solution: $2(2) + 10 = 14$
$2(4) + 10 = 18$
$2(6) + 10 = 22$
$2(8) + 10 = 26$

31. $\dfrac{n + 12}{2}$
32. $(n)(n) + n$
33. $2x + x$
34. $\dfrac{2n + 3n}{n}$

35. $\tfrac{1}{3}(x + 6)$
36. $\dfrac{x}{3} + (x)(x)$
37. $2n + 3n + 4n$
38. $\dfrac{2n}{n} + \dfrac{3n}{n} + 6$

Write an algebraic expression for each of the following.
39. the sum of 3 and *n*, decreased by 5
40. 5 times the sum of *g* and *h*
41. the quotient of *k* and the sum of 5 and *r*
42. the product of 5, *n*, *m*, and *r*
43. *m* less than the sum of *x* and 15
44. the sum of *a* and *b* decreased by the sum of *s* and *t*

Find the value of each expression, if $x = 24$, $y = 5$, and $a = 4$.
SAMPLE. $3[x \div (y + 3)]$ Solution: $3[24 \div (5 + 3)] = 3[24 \div (8)]$
$= 3[3]$
$= 9$

C 45. $x + [a(y + 3)]$
46. $[2(a + y)] \div x$

47. $a + [ay + (a + y)] + x$
48. $(y + a)[y(a + 3)]$

100 Chapter 4

49. $\left[\left(\dfrac{x}{y}\right)+y\right]+10$

50. $\left[\left(\dfrac{y+a}{y-a}\right)+3\right]\dfrac{x}{8}$

51. $\left[\left(\dfrac{x-y}{x+y}\right)+3\right]+y$

52. $\left[\left(\dfrac{x-a}{10}\right)+\left(\dfrac{ax}{3}\right)\right]ay$

PROBLEMS

Calculate the value of each of the following algebraic expressions for the suggested values of the variables.

SAMPLE. Volume of a cylinder: Bh; let $B = 120$ sq. ft. and $h = 15$ feet.
Solution: $120(15) = 1800$; the volume is 1800 cu. ft.

1. Perimeter of a rectangle: $2l + 2w$; let $l = 10.5$ feet and $w = 6.3$ feet.
2. Distance traveled by a car: $(r)(t)$; let $r = 52$ m.p.h. and $t = 3\frac{1}{2}$ hours.
3. Interest charged for a loan: prt; let $p = \$800$, $r = 15\%$, and $t = 1\frac{1}{2}$ years.
4. Volume of a cylinder: Bh; let $B = 138.2$ sq. in. and $h = 15.7$ in.
5. Volume of a cone: $\frac{1}{3}(Bh)$.
   Let $B = 234$ sq. in. and $h = 45$ in.

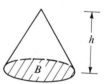

6. Area of a triangle: $\frac{1}{2}bh$; let $b = 36.48$ in. and $h = 10.02$ in.
7. Circumference of a circle: $2\pi r$; let $\pi = \frac{22}{7}$ and $r = 84$ in.
8. Distance around a semi-circle: $\dfrac{\pi d}{2} + 2r$.
   Let $\pi = 3.14$. $d = 29$ in. and $r = 14.5$ in.

9. Area of each of these quadrilaterals:

   a. $\dfrac{l+l}{2} \cdot w$; let $l = 9\frac{1}{2}$ in. and $w = 7$ in.

   rectangle

   b. $\dfrac{s+s}{2} \cdot s$; let $s = 11$ in.

   square

c.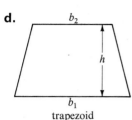
parallelogram

$\dfrac{b+b}{2} \cdot h$; let $b = 16$ yds. and $h = 4\tfrac{1}{4}$ yds.

d.

trapezoid

$\dfrac{b_1 + b_2}{2} \cdot h$; let $b_1 = 15.3$ in., $b_2 = 11.5$ in., and $h = 9.4$ in.

B 10. The nth triangular number: $\dfrac{n(n+1)}{2}$; let $n = 49$.

11. Degrees Centigrade to degrees Fahrenheit: $\tfrac{9}{5}C + 32$; let $C = 40°$.

12. Surface area of a cube-shaped block: $6(s)(s)$; let $s = 5.5$ in.

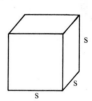

13. Lateral area of each box:

    a. $4lw$;
       let $l = 32$ in. and $w = 14$ in.

    b. $(s \cdot s) + (s \cdot s) + (s \cdot s) + (s \cdot s)$;
       let $s = 9$ cm.

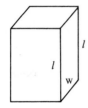 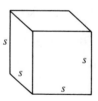

14. Sum of the degree measures of the interior angles of a polygon: $(n - 2)180$; let $n = 5$.

15. Water pressure (in lb. per square foot) exerted on a surface: hd; let $h = 31$ ft. and $d = 62.4$ lb. per cubic foot.

C 16. Degrees Fahrenheit to degrees Centigrade: $\tfrac{5}{9}(F - 32)$; let $F = 68°$.

17. Weight (in pounds) of water in a box-shaped tank: $62.4(l)(w)(h)$; let $l = 10$ ft., $w = 8$ ft., and $h = 3\tfrac{1}{2}$ ft.

18. Area of a cross section of a pipe: $\pi(R - r)(R + r)$; let $R = 23$ cm., $r = 16$ cm., $\pi = \tfrac{22}{7}$.

4–4 Factors and Coefficients in Algebraic Expressions

You will recall that numbers which are multiplied are called factors. When the number **12** is written as **3 · 4** we have **factored** 12. We might also **factor** 12 as **2 · 6** or as **1 · 12**. Until now we have considered whole numbers as factors of a whole number. In the case of an algebraic expression we may have factors which contain both variables and numbers. For example, some different ways of factoring **4m** include: **1 · 4m, 2 · 2m,** and **4 · m.** So we see that 4m has the factors 1, 2, 4, *m*, 2*m*, and 4*m*.

Some different ways of factoring the expression $\frac{2rs}{3}$ include:

$$\frac{2}{3}(rs); \quad \frac{1}{3}(2rs); \quad 2\left(\frac{rs}{3}\right); \quad \text{and} \quad \frac{2r}{3}(s).$$

In general, if an algebraic expression represents a product, the entire numerical factor is called the **coefficient** of the expression. Thus the coefficient of **15x** is **15**; the coefficient of $\frac{1}{3}t$ is $\frac{1}{3}$. For the expression **y**, the coefficient is understood to be **1**, since *y* and **1***y* have the same meaning. However, any factor of a product may be considered to be the **coefficient** of the product of the remaining factors. For example:

About the expression **3mn** we may say: **3** is the **coefficient** of *mn*;
3m is the **coefficient** of *n*;
3n is the **coefficient** of *m*.
About the expression $\frac{1}{5}rs$ we may say: $\frac{1}{5}$ is the **coefficient** of *rs*;
$\frac{1}{5}r$ is the **coefficient** of *s*;
$\frac{1}{5}s$ is the **coefficient** of *r*.

The algebraic expression **4k + 7t** consists of two parts, **4k** and **7t**, with a **+** sign between them. We say 4k and 7t are **terms** of the expression. The expression **3m + 2r − 6b** contains **three** terms. The expression **mn + $\frac{r}{3}$ − 2(xy)** also contains three terms. That is, parts of an algebraic expression that are separated by plus or minus symbols are **terms** of the expression.

ORAL EXERCISES

Name the coefficient of *k* in each expression.

1. 3*k*
2. 16*k*
3. *k*
4. $\frac{1}{10}k$
5. *mnk*
6. *krs*
7. *k*(3 + *a*)
8. (*a* + *b*)*k*
9. 0.25*k*

Variables, Expressions, and Sentences

Tell how many terms are in each expression. Give the value of each expression, if $x = 3$, $n = 6$, and $a = 12$.

SAMPLE. $\dfrac{a}{n} + x$ *What you say:* two terms; $\dfrac{12}{6} + 3 = 2 + 3 = 5$

10. $x + n$
11. $4n$
12. $2x + a$
13. $x + a + n$
14. $a - n$
15. $n - x$
16. $a - \dfrac{n}{x}$
17. $\dfrac{n}{x} + a$
18. $\tfrac{1}{2}n + \dfrac{a}{x}$
19. $nx + a$
20. $\dfrac{xn}{a}$
21. $\dfrac{x}{n} + \dfrac{a}{n}$

WRITTEN EXERCISES

Replace each question mark with the factor required to name the given product.

SAMPLE. Product: $6a$; $6(?)$, $3a(?)$, $a(?)$

Solution: $6(a)$, $3a(2)$, $a(6)$

A 1. Product: $12m$; $3(?)$, $2m(?)$, $(?)m$
2. Product: $15b$; $5(?)$, $15(?)$, $3b(?)$
3. Product: $\tfrac{1}{4}z$; $(?)z$, $\tfrac{1}{2}(?)$, $\tfrac{1}{4}(?)$
4. Product: $5pq$; $5p(?)$, $qp(?)$, $(?)q$
5. Product: $1.5ab$; $1.5b(?)$, $1.5(?)$, $(?)a$
6. Product: $\tfrac{3}{5}k$; $\tfrac{3}{5}(?)$, $\tfrac{1}{5}k(?)$, $k(?)$
7. Product: abc; $(?)bc$, $ac(?)$, $c(?)$
8. Product: $\dfrac{xy}{5}$; $\tfrac{1}{5}(?)$, $x(?)$, $y(?)$, $\dfrac{x}{5}(?)$
9. Product: $\dfrac{3r}{4}$; $\tfrac{3}{4}(?)$, $3r(?)$, $3(?)$, $r(?)$

B 10. Product: $\dfrac{mnq}{3}$; $\dfrac{m}{3}(?)$, $\tfrac{1}{3}(?)$, $(?)nq$, $(?)q$
11. Product: $ab(n + 2)$; $ab(?)$, $a(?)$, $b(?)$, $(n + 2)(?)$
12. Product: $\dfrac{2xy}{3}$; $\tfrac{2}{3}(?)$, $\dfrac{2x}{3}(?)$, $\dfrac{2y}{3}(?)$, $\dfrac{y}{3}(?)$
13. Product: $\dfrac{x}{2}(a + b)$; $\dfrac{x}{2}(?)$, $(a + b)(?)$, $\tfrac{1}{2}(?)(a + b)$
14. Product: $\dfrac{3ab}{5}$; $\tfrac{3}{5}(?)$, $\dfrac{3a}{5}(?)$, $3(?)$, $(?)3ab$
15. Product: $8(x + y)$; $(?)4(x + y)$; $(?)(x + y)$; $(?)(x + y)(2)$

Simplify each expression.

SAMPLE 1. $\frac{1}{4}(xw)(z)$ Solution: $\frac{xwz}{4}$

SAMPLE 2. $(3a)(5cd)$ Solution: $15acd$

16. $(3)(4)(r)$
17. $(3)(x)(2)(y)$
18. $(5)(3a)(b)$
19. $(3)(m)(z)(7)$
20. $\left(\frac{a}{5}\right)(3)(b)$
21. $(0.5)(r)(s)3$

22. $4(ab)(3c)$
23. $(t)(2n)\frac{1}{3}$
24. $\frac{5}{8}(a)(bc)$
25. $4\left(\frac{\pi}{5}\right)(d)$
26. $\frac{1}{2}(r)(3t)$
27. $(2)(3)(4y)(z)$

Write at least three different factored forms for each of the given expressions.

SAMPLE. $6xy$ Solution: $6(xy)$; $6x(y)$; $(3)(2)xy$; $(2x)(3y)$

|C|

28. $15a$
29. $7qr$
30. $12z$
31. $14mn$
32. $4cd$

33. $\frac{3}{4}t$
34. $0.75k$
35. $\frac{2m}{9}$
36. $3acm$
37. $\frac{abc}{2}$

38. $6(a+3)$
39. $\frac{2mk}{3}$
40. $4rst$
41. $\frac{7xy}{8}$
42. $\frac{3}{5}rq$

4-5 Algebraic Expressions and Generalized Forms

You already know that the set of even numbers is {0, 2, 4, 6, 8, 10, 12, ...} and that each even number has the number **2** as a factor. Do you agree that **2** is a factor of each of the following even numbers?

$$12 = 2(6) \qquad 52 = 2(26) \qquad 438 = 2(219)$$

We can say that a number is *even* if it can be written in the generalized form $2n$, where n is a whole number. If the set of whole numbers is used as the replacement set for *n*, then the result is the set of even numbers, as shown in the following table.

Whole Number (*n*)	0	1	2	3	4	5	6
Even Number (2*n*)	2·0 or 0	2·1 or 2	2·2 or 4	2·3 or 6	2·4 or 8	2·5 or 10	2·6 or 12

According to the table, the even number **6** is 2 more than the previous even number **4**; the even number **8** is 2 more than the previous even number **6**; and so on. We can say that every even number *e* has as its **successor** the even number *e* + 2.

Do you suppose that there is a generalized form for **odd** numbers? If **1** is added to the **even** number **6** the result is the **odd** number **7**; if **1** is added to the **even** number **14** the result is the **odd** number **15**; and so on. Thus, the generalized form for odd numbers would seem to be 2*n* + 1, where *n* is a whole number. Do you agree that if we use the set of whole numbers as the replacement set for *n* in 2*n* + 1, the result is the set of odd numbers?

Whole Number (*n*)	0	1	2	3	4	...
Odd Number (2*n* + 1)	0 + 1 or 1	2 + 1 or 3	4 + 1 or 5	6 + 1 or 7	8 + 1 or 9	...

The table of odd numbers above shows that the odd number **7** is 2 more than the previous odd number **5**; the odd number **9** is 2 more than the previous odd number **7**; and so on. We can say that every odd number *c* has as its **successor** the odd number *c* + 2.

If you were to "count by threes" you would name the numbers 3, 6, 9, 12, 15, 18, ... These are called **multiples of three**. The set of multiples of 7 is written {7, 14, 21, 28, 35, ...} and as you probably noticed the first multiple is **1 · 7** or **7 · 1**; the second multiple is 2 · 7, or 7 · 2; the third multiple is 3 · 7, or 7 · 3; and so on. Thus, the generalized form for the set of multiples of a number *a* is:

$$\{(1 \cdot a), (2 \cdot a), (3 \cdot a), (4 \cdot a), (5 \cdot a), (6 \cdot a), \ldots\}$$
or
$$\{(a \cdot 1), (a \cdot 2), (a \cdot 3), (a \cdot 4), (a \cdot 5), (a \cdot 6), \ldots\}$$

We can write the multiples of the number 9, for example, by replacing each *a* in the generalized form with 9.

$$\{(1 \cdot 9), (2 \cdot 9), (3 \cdot 9), (4 \cdot 9), (5 \cdot 9), (6 \cdot 9), \ldots\}$$

or

$$\{9, 18, 27, 36, 45, 54, \ldots\}$$

ORAL EXERCISES

Tell which of the following represent even numbers and which represent odd numbers.

1. 38
2. 145
3. 140
4. 251
5. $2 \cdot 7$
6. $(2 \cdot 15) + 1$
7. $2 \cdot 8$
8. $(2 \cdot 8) + 1$
9. $2 \cdot 19$
10. $1 + (2 \cdot 19)$
11. $(2 \cdot 7) + 2$
12. $9 + 2$
13. $11 + 2$
14. $(16 + 2) + 1$
15. $2 \cdot (2 \cdot 5)$

Tell whether each numeral in Column 1 represents an odd or an even number, and match it with the statement in Column 2 that explains it.

SAMPLE. $37 + 2$ *Solution:* Odd number; C

COLUMN 1

16. $2 \cdot 23$
17. $(2 \cdot 7) + 1$
18. $2 \cdot (3 + 5)$
19. $1 + (2 \cdot 6)$
20. $10 + 2$
21. $2 + 19$
22. $37 + 2$
23. $149 \cdot 2$

COLUMN 2

A. The product of 2 and any whole number is an even number.

B. The product of 2 and any whole number, increased by 1, is an odd number.

C. The sum of 2 and any odd number is an odd number.

D. The sum of 2 and any even number is an even number.

WRITTEN EXERCISES

Tell whether each statement is true or false.

A 1. Any even number, increased by 2, is an even number.
2. Any even number, increased by 1, is an odd number.

Variables, Expressions, and Sentences

3. If the product of 2 and any whole number is increased by 1, the result is an even number.
4. Any odd number, increased by 1, is an even number.
5. Any odd number, increased by 2, is an odd number.
6. In the set of odd numbers, the successor of a number is 2 more than the number.
7. In the set of even numbers, the successor of a number is 1 more than the number.

Show that each of the following represents a multiple of 5, 7, or 9 by writing it in one of these forms: $n \cdot 5$, $n \cdot 7$, $n \cdot 9$, or $5 \cdot n$, $7 \cdot n$, $9 \cdot n$, where n is a whole number.

SAMPLE 1. $(2 + 4)7$ Solution: $6 \cdot 7$; multiple of 7

SAMPLE 2. $\frac{10}{2} \cdot 4$ Solution: $5 \cdot 4$; multiple of 5

8. $(3 + 5)9$
9. $5 \cdot \frac{9}{3}$
10. $7(8 \div 4)$
11. $(3 \cdot 3)13$
12. $9 + 9 + 9 + 9$
13. $4 \cdot (9 + 0)$
14. $5(2 \cdot 2)$
15. $(6 + 1)7$
16. $(2 + 4)5$
17. $9(2 \cdot 3)$
18. $\frac{14}{2} \cdot (2 + 9)$
19. $\frac{10}{2} \cdot \frac{15}{5}$

Show that each number is even or is odd by writing it in the form $2n$ or $2n + 1$, where n is a whole number.

SAMPLE 1. 35   Solution: $34 + 1 = (2 \cdot 17) + 1$; odd

SAMPLE 2. $\frac{5 + 27}{2}$ Solution: $\frac{32}{2} = 16 = 2 \cdot 8$; even

20. 125
21. 62
22. 51
23. $(3 + 16)$
24. 15^2
25. $3 \cdot 42$

B 26.
27. $(3)(5)(2)$
28. $(3 \cdot 5)(2) + 1$
29. $\frac{3 + 6 + 21}{2}$
30. $10^2 + 3$
31. $\frac{5^3}{25}$

In each expression, x represents any even number and y represents any odd number. Tell which expressions represent even numbers and which represent odd numbers.

32. $x + 2$
33. $x + y$
34. $(y + 2) + 1$
35. $2x + y$
36. $x + 2y$
37. $2y + 1$
38. $2x + 1$
39. $1 + (x + 2)$
40. $2 \cdot 3x$

C 41. $2x + 2y$
42. $(2x + 2y) + 1$
43. $2(y + 1)$
44. $2(x + 1)$
45. $2(x)(x)$
46. $2(x)(y)$

Using Exponents

4–6 Expressions Containing Exponents

If an expression representing a product contains a repeated factor, the expression may also be written in **exponential notation**. For example, $5 \cdot 5 \cdot 5 \cdot 5$ may be written 5^4. The **4** is called an **exponent**, and the **5** is called the **base**. Written above and to the right, as shown here, the 4 means that **5** is used as a factor **four** times.

Exponents can also be used in writing variable expressions. For example, if the side of a square is indicated by the variable s, then the area of the square, which is $s \cdot s$, can be written s^2. The symbol s^2 is read "s squared," or "the square of s," or "s to the second power."

Suppose that each edge of a cube has a measure indicated by the variable e. The volume of the cube can be expressed by the algebraic expression $e \cdot e \cdot e$, where e is used as a factor **three** times. Using exponential notation this is written e^3, which is read "e cubed," or "the cube of e," or "e to the third power." In the symbol e^3, the variable e is called the *base*. The small raised numeral 3 is the **exponent** and tells how many times the **base** is used as a **factor**.

exponent ───┐
            e^3
base ───────┘

Do you suppose we can use exponential notation to write an algebraic expression like $4 \cdot m \cdot m \cdot m$ in simpler form? Since m is used as a factor **three** times, we can write the expression like this:

$$4 \cdot m \cdot m \cdot m = 4(m \cdot m \cdot m)$$
$$= 4(m^3)$$
$$= 4m^3$$

In the expression $(xy)(xy)(xy)(xy)$, since xy is used as a factor **four** times, we can use the exponent **4** to write

$$(xy)(xy)(xy)(xy) = (xy)^4$$

Study these examples carefully.
 (a) $18 \cdot t \cdot t \cdot t = 18(t \cdot t \cdot t) = 18t^3$
 (b) $(m)^5 = m^5$

(c) $(3 \cdot 4)^2 = (3 \cdot 4)(3 \cdot 4)$
$= (12)(12) = \mathbf{144}$
(d) $a \cdot x \cdot x \cdot x \cdot x = a(x \cdot x \cdot x \cdot x)$
$= a \cdot x^4 = \mathbf{ax^4}$
(e) $ab \cdot ab \cdot ab = \mathbf{(ab)^3}$
(f) $(2 + x)(2 + x)(2 + x) = \mathbf{(2 + x)^3}$

ORAL EXERCISES

Give the meaning of each expression in words.

SAMPLE 1. x^6 *What you say:* x to the sixth power
SAMPLE 2. $3y^2$ *What you say:* 3 times the square of y.
SAMPLE 3. $(a + b)^3$ *What you say:* the quantity a plus b, to the third power.

1. m^3
2. z^6
3. $(ab)^2$
4. ab^2
5. a^2b
6. $3t^5$
7. $2n^7$
8. $(mn)^3$
9. $(3 + a)^2$
10. $(k + 7)^{10}$
11. $5 - c^3$
12. $4 \cdot 5^2$
13. $(4 \cdot 5)^2$
14. $4^2 \cdot 5$
15. $(x + y)^3$
16. $x^3 + y^3$
17. $x + y^3$
18. $m^3 + n^2$

WRITTEN EXERCISES

Write each of the following as a product in which each use of an exponent is replaced by repeating a factor.

SAMPLE 1. $7k^3$ *Solution:* $7 \cdot k \cdot k \cdot k$
SAMPLE 2. $(5^2)(x^3)$ *Solution:* $5 \cdot 5 \cdot x \cdot x \cdot x$

1. y^4
2. m^6
3. $(rt)^2$
4. $3z^5$
5. $(xy)^3$
6. $(3 + a)^2$
7. $13k^4$
8. xy^3
9. $\dfrac{ab^2}{c^2}$
10. $5m^3n$
11. $7r^2s^3$
12. $(b - h)^3$
13. $\dfrac{a^3b}{m^3}$
14. $3(a + b)^3$
15. $(2^3)(d^2)$
16. $x^2y^2z^2$
17. $(xyz)^2$
18. $\dfrac{3 + b^3}{k^2}$

Chapter 4

Use exponential notation to write each expression in a simpler form.

SAMPLE 1. $3 \cdot k \cdot k \cdot k \cdot k$ Solution: $3k^4$
SAMPLE 2. $(w + x)(w + x)$ Solution: $(w + x)^2$
SAMPLE 3. $x \cdot y \cdot y \cdot x \cdot x$ Solution: $x^3 y^2$

19. $(ab)(ab)(ab)$
20. $5 \cdot x \cdot x \cdot x$
21. $(kt)(kt)$
22. $a \cdot a \cdot b \cdot b \cdot b$
23. $m \cdot m \cdot m \cdot x \cdot x \cdot x$
24. $(a + t)(a + t) \cdot 3$

25. $c \cdot d \cdot c \cdot d \cdot c$
26. $6 \left(\dfrac{n \cdot n \cdot n}{r \cdot r \cdot r \cdot r} \right)$
27. $\dfrac{3 \cdot 3 \cdot 3 \cdot g \cdot g}{m \cdot m}$
28. $v \cdot w \cdot w \cdot v \cdot w$
29. $(b - h)(b - h)(b - h)$
30. $s \cdot s \cdot s \cdot (g + m)(g + m)$

B 31. ten times the square of r
32. mn to the fourth power
33. five times the cube of w
34. the quantity m plus n, to the fifth power
35. z used as a factor six times
36. one-fifth the third power of a

Determine the value of each expression for the value assigned to each variable.

SAMPLE. $3x^2$; $x = 5$ Solution: $3 \cdot 5^2 = 3 \cdot 5 \cdot 5$
     $= 75$

37. $5y^3$; $y = 2$
38. $(3 + b)^3$; $b = 5$
39. $7x^4$; $x = 10$
40. $(9 - k)^4$; $k = 8$
41. $\dfrac{5 \cdot a^3}{45}$; $a = 3$

42. $6m^3$; $m = \frac{1}{2}$
43. $(a + b)^2$; $a = 4$ and $b = 5$
44. $m^2 n^3$; $m = 4$ and $n = 3$
45. $(5x)^3$; $x = 3$
46. $\dfrac{x^2 + y^2}{(x + y)^2}$; $x = 5$ and $y = 6$

C 47. $3x^2 + 2x$; $x = 7$
48. $2m^2 + 4m - 5$; $m = 1$
49. $5w^3 + w^2 + 2w$; $w = 2$
50. $2a^2 + 3ab + b^2$; $a = 2$ and $b = 10$
51. $r^2 - s^2 + 3r$; $r = 4$ and $s = 3$

Write at least four different factored forms for each of the following expressions.

SAMPLE. $3x^3y^2$ Solution: $3x^3(y^2)$; $3x(x^2)(y^2)$; $3y^2(x^3)$; $3x(x^2)(y)(y)$

52. $5c^2d^2$
53. $2xy^2$
54. ac^2d^3
55. $\dfrac{x^2y^2}{2}$
56. t^3k^4
57. $\dfrac{mnr^2}{3}$
58. $\dfrac{qr^4}{7}$
59. $6(a+2)^3$
60. $\dfrac{ab^2c^3}{10}$
61. $\dfrac{3m^2n^3}{2}$
62. $a^2b(n+3)^2$
63. $2.5\pi r^2$

Tell whether or not both expressions in each exercise name the same product for the given value of the variable.

SAMPLE. $3 \cdot x$ and x^3; let $x = 5$.

Solution: $3 \cdot 5 = 15$ and $5^3 = 5 \cdot 5 \cdot 5 = 125$
    $3 \cdot x$ and x^3 do not name the same product when $x = 5$.

64. r^5 and $5r$; let $r = 2$.
65. $2m$ and m^2; let $m = 3$.
66. $2k^4$ and k^8; let $k = 1$.
67. $3a^2$ and $(3a)^2$; let $a = 4$.
68. $a^2 + b^2$ and $(a+b)^2$; let $a = 3$ and $b = 5$.
69. $x^2 - y^2$ and $(x-y)^2$; let $x = 10$ and $y = 6$.
70. m^2n^2 and $(mn)^2$; let $m = 3$ and $n = 4$.

PROBLEMS

Find the value of each expression when replacements of variables are made as suggested.

  1. Volume of a cube: e^3.
    Let $e = 7$ inches.

2. Area of a circle: πr^2.
    Let $\pi = 3.14$ and $r = 10$ centimeters.

3. Area of a square: s^2.
   Let $s = 12\frac{1}{2}$ inches.

4. Surface area of a cube: $6e^2$.
   Let $e = 9$ meters.

B 5. Volume of a cone whose radius and height are equal: $\dfrac{\pi r^3}{3}$. Let $\pi = 3.1416$ and $r = 10$ inches.

6. Area of the shaded region in the figure at the right: $r^2(4 - \pi)$. Let $\pi = 3.14$ and $r = 3.6$ inches.

7. Volume of a sphere: $\dfrac{4\pi}{3}(r^3)$. Let $\pi = 3.14$ and $r = 5$ centimeters.

8. Volume of the figure shown at the right: $E^3 + e^3$. Let $E = 6$ inches and $e = 2\frac{1}{2}$ inches.

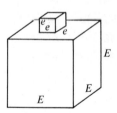

4–7 Exponents and Nomographs

Earlier you learned about using **nomographs** for addition and multiplication. The nomograph shown here can help you solve problems dealing with exponents. You know that $3^2 = 9$. If this statement is written in general form, using a for the base, n for the exponent, and b for the product, we have:

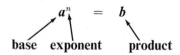

Notice which scales of the nomograph in Figure 4–1 are labeled a, n, and b, respectively.

Variables, Expressions, and Sentences

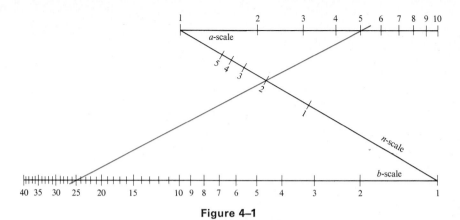

Figure 4–1

To use the nomograph to solve the problem $5^2 = ?$, we have drawn a line that intersects the graph of **5** on the **a-scale** and the graph of **2** on the **n-scale**. Since the line intersects the **b-scale** at **25**, we see that $5^2 = 25$. Check with the nomograph for yourself to see that $3^2 = 9$, as we noted at the beginning of this section.

Sometimes we shall want to simplify an expression like $3^2 \cdot 3^3$. Notice that the base **3** is the same in each factor. Writing each as a repeated factor, we have:

$$3^2 \cdot 3^3 = ?$$
$$(3 \cdot 3)(3 \cdot 3 \cdot 3) = ?$$
$$3 \cdot 3 \cdot 3 \cdot 3 \cdot 3 = 3^5$$
$$3^2 \cdot 3^3 = 3^{2+3} = 3^5$$

The base 3 remains unchanged and the exponents of the factors are added. Study the methods shown below for simplifying $a^3 \cdot a^4$ and $b(b) + (b \cdot b \cdot b)$.

$$a^3 \cdot a^4 = ? \qquad\qquad b(b) + (b \cdot b \cdot b) = ?$$
$$(a \cdot a \cdot a) \cdot (a \cdot a \cdot a \cdot a) = ? \qquad (b \cdot b) + (b \cdot b \cdot b) = b^2 + b^3$$
$$a \cdot a \cdot a \cdot a \cdot a \cdot a \cdot a = a^7$$

ORAL EXERCISES

Tell how to express each of the following in exponential notation, and show how it is written.

SAMPLE. $(4 \cdot 4)(4 \cdot 4 \cdot 4)$

What you say: four to the fifth power   *What you write:* 4^5

114   *Chapter 4*

1. $(2 \cdot 2 \cdot 2)$
2. $(5 \cdot 5 \cdot 5 \cdot 5 \cdot 5 \cdot 5)$
3. $(3 \cdot 3 \cdot 3 \cdot 3)$
4. $(3 \cdot 3 \cdot 3)(3 \cdot 3 \cdot 3 \cdot 3)$
5. $(6 \cdot 6)(6 \cdot 6 \cdot 6)$
6. $(10 \cdot 10 \cdot 10 \cdot 10)10$
7. $35(35 \cdot 35 \cdot 35)$
8. $r \cdot r \cdot r \cdot r \cdot r$
9. $(x \cdot x \cdot x) + x$
10. $(y \cdot y \cdot y \cdot y)(y \cdot y)$
11. $(a \cdot a \cdot a \cdot a) + (a \cdot a)$
12. $t(t \cdot t \cdot t \cdot t \cdot t)$
13. $(\frac{1}{2} \cdot \frac{1}{2} \cdot \frac{1}{2} \cdot \frac{1}{2})$
14. $\left(\frac{a}{3} \cdot \frac{a}{3}\right)\left(\frac{a}{3} \cdot \frac{a}{3} \cdot \frac{a}{3}\right)$
15. $(x + y)(x + y)(x + y)$
16. $(rs)(rs)(rs)rs$

WRITTEN EXERCISES

Write, in the form $a^n = b$, each number statement suggested by a solid line drawn across the nomograph.

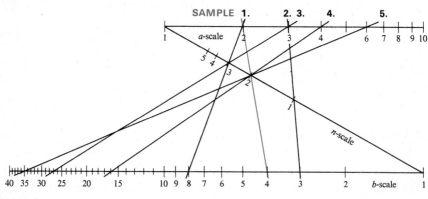

SAMPLE. (See line marked **SAMPLE**.)   Solution: $2^2 = 4$

A **1.–5.** (See nomograph above)

Use exponential notation to simplify each of the following expressions.

SAMPLE 1. $(3 \cdot 3 \cdot 3 \cdot 3)(5 \cdot 5)$ Solution: $3^4 \cdot 5^2$
SAMPLE 2. $(t \cdot t \cdot t \cdot t)(t \cdot t) + (r \cdot r)$ Solution: $t^6 + r^2$

6. $(5 \cdot 5 \cdot 5)(5 \cdot 5)$
7. $(3 \cdot 3) + (7 \cdot 7 \cdot 7 \cdot 7)$
8. $(mn)(mn)(mn)(mn)$
9. $(b \cdot b \cdot b) + (c \cdot c)$
10. $(rs \cdot rs)(rs \cdot rs \cdot rs)$
11. $(2 \cdot 2 \cdot 2 \cdot 2) + 2$
12. $(2 \cdot 2 \cdot 2)(10 \cdot 10 \cdot 10)$
13. $7 \cdot (a \cdot a \cdot a \cdot a \cdot a)$
14. $(4 \cdot 4 \cdot 4)(9 \cdot 9 \cdot 9 \cdot 9) + (5 \cdot 5)$
15. $(x + y)(x + y)(x + y)$
16. $(k \cdot k \cdot k)(n \cdot n \cdot n \cdot n)$
17. $\left(\frac{a}{5} \cdot \frac{a}{5} \cdot \frac{a}{5}\right) - \left(\frac{a}{5} \cdot \frac{a}{5}\right)$

Write each of the following in repeated-factor form, and then simplify by using exponential notation.

SAMPLE. $x^3 \cdot x^4$ Solution: $x^3 \cdot x^4 = (x \cdot x \cdot x)(x \cdot x \cdot x \cdot x)$
$= x \cdot x \cdot x \cdot x \cdot x \cdot x \cdot x = x^7$

18. $10^2 \cdot 10^3$
19. $6^5 \cdot 6^2$
20. $12^3 \cdot 12^6$
21. $n^4 \cdot n^2$
22. $b^3 \cdot b^2 \cdot b^2$
23. $2^3 \cdot 2^3 \cdot 2^2$
24. $a \cdot a \cdot a^2 \cdot a$

25. $(xyz)^4 \cdot (xyz)^2$
26. $(mn)^2 \cdot (mn)^3 \cdot (mn)$
27. $(3+4)^2 \cdot (3+4)^2$
28. $(3b)^3 \cdot (3b)^3$
29. $\left(\dfrac{ab}{3}\right) \cdot \left(\dfrac{ab}{3}\right)^2 \cdot \left(\dfrac{ab}{3}\right)^4$
30. $k(k^2)(x^2)$
31. $(mn^2)(rs^2)$

Show that each statement is true.

SAMPLE. $(3 \cdot 3 \cdot 3) + (3 \cdot 3) \neq 3^5$

Solution: $(3 \cdot 3 \cdot 3) + (3 \cdot 3) \neq 3^5$

| 27 + 9 | $3 \cdot 3 \cdot 3 \cdot 3 \cdot 3$ |
| 36 | 243 |

$36 \neq 243$

B
32. $(4 \cdot 4 \cdot 4) = (4 \cdot 4)(4)$
33. $(2 \cdot 2)(2 \cdot 2) = (4)(4)$
34. $3^3 \cdot 3^2 \neq 3^6$
35. $2^2 \cdot 2^4 \cdot 2^3 = 2^9$
36. $5^2 + 5^2 < 5^4$
37. $2^5 > 5^2$
38. $(4 \cdot 4 \cdot 4) = (8 \cdot 8)$
39. $10^7 \neq 10^2 + 10^2 + 10^3$
40. $10^3 \neq 10^5 - 10^2$
41. $10^3 + 10^3 \neq 10^6$

CHAPTER SUMMARY

Inventory of Structure and Concepts

1. **Grouping symbols**, such as parentheses, brackets, and fraction bars, are used to indicate the order in which operations are to be carried out in simplifying expressions. Some expressions contain two or more grouping symbols, one within another. In this case, the numeral in the innermost grouping symbol is usually simplified first, then the numeral in the next innermost symbol is simplified, and so on.

Chapter 4

2. Letters used to represent numbers in expressions and sentences are called **variables**. The specified set from which replacements for a variable are taken is called the **replacement set**. The **value** of an expression is determined by replacing each variable with a member of its specified replacement set.

3. Usually the numerical part of an algebraic expression that represents a product is called the **coefficient** of the expression. In general, however, **any** factor of a number expressed as a product may be considered to be the **coefficient** of the product of the remaining factors.

4. The parts of an algebraic expression that are separated by plus or minus symbols are **terms** of the expression.

5. Any number that can be expressed in the general form $2n$, where n is a whole number, is an **even** number; any number that can be expressed in the general form $2n + 1$, where n is a whole number, is an **odd** number.
The **successor** of any even number e is $e + 2$; the **successor** of any odd number r is $r + 2$.
The set of **multiples** of a number n is expressed by the general form $\{(1 \cdot n), (2 \cdot n), (3 \cdot n), (4 \cdot n), \ldots\}$.

6. **Exponential notation** is a short method for representing a product that contains a repeated factor. For the number a^n, which is in exponential notation, a is the **base** and n is the **exponent**.

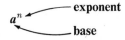

Vocabulary and Spelling

Pronounce, spell, and give the meaning of each of the following words and expressions.

parentheses (*p. 92*)
brackets (*p. 92*)
fraction bar (*p. 92*)
grouping symbols (*p. 92*)
variable (*p. 96*)
replacement set (*p. 96*)
variable expression (*p. 96*)
algebraic expression (*p. 96*)
value (of an expression) (*p. 96*)
factored (number) (*p. 102*)

coefficient (*p. 102*)
terms (of an expression) (*p. 102*)
even number (*p. 104*)
odd number (*p. 105*)
generalized form (*p. 105*)
multiple (*p. 105*)
exponential notation (*p. 108*)
base (*p. 108*)
exponent (*p. 108*)

Variables, Expressions, and Sentences

Chapter Test

Simplify each expression.

1. $10 - (5 - 2)$
2. $\dfrac{12 \times 2}{2 \times (3 + 1)}$
3. $(2 \times 3 \times 5) + (6 + 10 + 9)$
4. $\dfrac{22 + 6}{2 \times 7} + \dfrac{30 \div 5}{2} + 8$

Find the number named by each expression.

5. $5(3 + 7 + 9)$
6. $4[5 + (3 + 6)]$
7. $5[7 + (4 + 8)] + 12$
8. $(\tfrac{5}{8})(\tfrac{1}{2})5$

Find the value of each expression if $m = 4$, $n = \tfrac{1}{2}$, and $r = 5$.

9. $6m + n$
10. $\dfrac{m + r}{2}$
11. $\dfrac{3m + 6n}{r}$
12. $mn + nr$
13. $(m \cdot m) + 2r$
14. $\dfrac{m(m + r)}{r(2m + r)}$

Write an algebraic expression for each of the following.

15. The product of 3, k, and b
16. t less than the sum of 12 and x
17. The sum of p and q increased by the product of a and b

Use the algebraic expression $2l + 2w$ to find the perimeter of each rectangle whose dimensions are given below.

18. $l = 14$ inches
    $w = 9$ inches
19. $l = 23$ centimeters
    $w = 15$ centimeters
20. $l = 8\tfrac{1}{2}$ yards
    $w = 3\tfrac{1}{4}$ yards
21. $l = 10.3$ miles
    $w = 5.6$ miles

Replace each question mark with the factor required to name the given product.

22. Product: $18x$; $(?)x$, $6(?)$, $(?)(2x)$, $(\tfrac{1}{2})(?)$
23. Product: $\dfrac{2n}{5}$; $(?)n$, $\tfrac{1}{5}(?)$, $2(?)$, $\dfrac{n}{5}(?)$

Simplify each expression.

24. $(2)(5n)(k)$
25. $\tfrac{1}{5}(x)(3y)$

Write each even number in the form $2n$ and each odd number in the form $2n + 1$, where n is a whole number.

26. 33   **27.** 26   **28.** 105   **29.** 54

Write each of the following as a product of repeated factors.

30. 5^3 **31.** $(ab)^4$ **32.** $(x + w)^3$

Simplify each expression by using exponential notation.

33. $10 \cdot v \cdot v \cdot v$ **35.** $n \cdot n \cdot n \cdot m \cdot m$
34. $(3 \cdot 3) + (9 \cdot 9 \cdot 9)$ **36.** $(k \cdot k \cdot k \cdot k) + k$

Chapter Review

4–1, 2 Expressions and Punctuation

1. The grouping or punctuation symbols of mathematics include __?__, __?__, and __?__.
2. Simplify each of the following.

   a. $(5 \times 9) \div 3$ **b.** $3[6 + 2(3 + 5)]$

4–3 Expressions and Variables

3. In mathematical expressions, symbols such as letters are called __?__ if they represent numbers.
4. The specified set of numbers that may be used to replace a variable is called the __?__ set.
5. An expression that contains one or more variables is called a(n) __?__ expression.

Find all the possible values for each expression if the replacement set for k is $\{3, 5, 7, 9\}$.

6. $k + 12$ **7.** $\dfrac{k + 5}{2}$ **8.** $2(3 + k)$

Write an algebraic expression for each of the following.

9. The sum of 2 and 7, increased by 3
10. The product of 7 and 4, decreased by 10
11. The sum of r and s, divided by 5

Variables, Expressions, and Sentences 119

4–4 Factors and Coefficients in Algebraic Expressions

12. When numbers are multiplied the result is called the __?__.
13. Numbers that are multiplied are called __?__.
14. The numerical part of an algebraic expression representing a product is called the __?__.
15. The parts of an algebraic expression that are separated by + or − signs are called __?__ of the expression.

4–5 Algebraic Expressions and Generalized Forms

16. The set of even numbers is written {__?__, __?__, __?__, __?__, ...}.
17. The set of odd numbers is written {__?__, __?__, __?__, __?__, ...}.
18. The generalized form for an even number is __?__.
19. The generalized form for an odd number is __?__.
20. In the set of even numbers the successor of any even number t is __?__.
21. In the set of odd numbers the successor of any odd number r is __?__.
22. The set of multiples of 6 is written {__?__, __?__, __?__, __?__, ...}.

4–6 Expressions Containing Exponents

23. Using exponential notation, the expression $3 \cdot 3 \cdot 3 \cdot 3$ can be written as __?__.
24. In the expression t^2 the small raised numeral "2" is called the __?__; t is called the __?__.

Write each of the following by using exponential notation.

25. $7 \cdot 7 \cdot 7 \cdot 7 \cdot 7$
26. $x \cdot x \cdot x \cdot x$
27. $(ab)(ab)(ab)$
28. $(3 + t)(3 + t)(3 + t)$

4–7 Exponents and Nomographs

Use exponential notation to write the numeral described by each of the following.

29. Seven to the third power
30. Ten to the fourth power
31. Twelve squared
32. Eight cubed

Simplify each of the following.

33. $(5 \cdot 5)(5 \cdot 5 \cdot 5)$
34. $4^5 \cdot 4^2$
35. $k^3 \cdot k^5$
36. $(xy)^2 \cdot (xy)^3$

Review of Skills

Multiply.

1. $5 \cdot 8$
2. $9 \cdot 6$
3. $7 \cdot 7$
4. $4 \cdot 7$
5. $8 \cdot 7$
6. $9 \cdot 8$
7. $12 \cdot 6$
8. $10 \cdot 8$
9. $6 \cdot 8 \cdot 3$
10. $5 \cdot 7 \cdot 6$
11. $9 \cdot 5 \cdot 18$
12. $11 \cdot 12 \cdot 5$

Divide.

13. $40 \div 8$
14. $28 \div 4$
15. $72 \div 8$
16. $72 \div 6$
17. $18 \div 9$
18. $42 \div 6$
19. $56 \div 7$
20. $32 \div 4$
21. $10 \div \frac{1}{2}$
22. $12 \div \frac{1}{5}$
23. $12 \div \frac{2}{3}$
24. $75 \div \frac{3}{5}$

Add.

25. 362
    765
    357

26. $253.27
    56.21
    71.24

27. 74.1
    69.5
    35.8

28. 0.125
    0.003
    0.638
    2.790

29. $146 + 395 + 204$
30. $37.2 + 84.7 + 32.5$
31. $\$8.37 + \$12.49 + \$17.08$
32. $6.07 + 3.25 + 0.69$

Write each number as the product of two whole-number factors in as many ways as you can.

SAMPLE. 16   *Solution:* $1 \cdot 16; \ 2 \cdot 8; \ 4 \cdot 4$

33. 9
34. 6
35. 4
36. 12
37. 15
38. 18
39. 30
40. 24
41. 20
42. 22
43. 28
44. 36

Complete each of the following to make a true statement.

45. $\frac{1}{4} + \frac{3}{4} = ?$
46. $\frac{3}{4} + ? = 1$
47. $1\frac{1}{2} + ? = 3\frac{1}{2}$
48. $4\frac{1}{5} + \frac{3}{5} = ?$
49. $6\frac{1}{2} + 1\frac{1}{4} = ?$
50. $\frac{5}{10} + 3\frac{4}{10} = ?$
51. $\frac{3}{8} + 1\frac{1}{2} = ?$
52. $4 - 2\frac{1}{3} = ?$
53. $6\frac{1}{2} - 1\frac{1}{4} = ?$

Simplify each of the following, if $a = 2$, $b = 3$, and $c = 4$.

SAMPLE. $\frac{24}{b}$ *Solution:* $\frac{24}{3} = 8$

Variables, Expressions, and Sentences   **121**

54. $\dfrac{30}{a}$ **59.** $b + \frac{1}{2}$ **64.** a^3

55. $4b$ **60.** $8a$ **65.** $b \cdot b \cdot b$

56. $1 \cdot 3c$ **61.** $\frac{1}{3}b$ **66.** $2a + 2b$

57. b^2 **62.** $\frac{2}{3}b$ **67.** $\dfrac{a \cdot c}{2}$

58. $\dfrac{c}{2}$ **63.** $5c$ **68.** $\dfrac{54}{b}$

■ ■

**CHECK POINT
FOR EXPERTS**

Exponents and Scientific Notation

As we have seen, exponential notation is very useful in expressing, in a simple way, a product consisting of repeating factors. For example:

$$8 \cdot 8 \cdot 8 \cdot 8 \cdot 8 \cdot 8 = 8^6$$

Study the pattern below:

$$3 \cdot 3 \cdot 3 \cdot 3 = 3^4$$
$$3 \cdot 3 \cdot 3 = 3^3$$
$$3 \cdot 3 = 3^2$$
$$3 = 3^1$$

Do you see why $3 = 3^1$? In a similar manner $y^1 = y$ and $k^1 = k$. We read k^1 as "k to the first power."

Often a large number can be expressed as the product of two numbers of which one is a power of **10**. Remember that:

$$10 = 10 = 10^1$$
$$100 = 10 \cdot 10 = 10^2$$
$$1{,}000 = 10 \cdot 10 \cdot 10 = 10^3$$
$$10{,}000 = 10 \cdot 10 \cdot 10 \cdot 10 = 10^4$$
$$100{,}000 = 10 \cdot 10 \cdot 10 \cdot 10 \cdot 10 = 10^5$$
$$1{,}000{,}000 = 10 \cdot 10 \cdot 10 \cdot 10 \cdot 10 \cdot 10 = 10^6$$
$$10{,}000{,}000 = 10 \cdot 10 \cdot 10 \cdot 10 \cdot 10 \cdot 10 \cdot 10 = 10^7 \text{ etc.}$$

For example, here are two ways of writing **3800** as a product involving a power of 10:

$$3800 = 38 \times 100 \qquad 3800 = 3.8 \times 1000$$
$$ = 38 \times 10^2 \qquad = 3.8 \times 10^3$$

A number is said to be in **scientific notation** when it is written as the product of a number between 1 and 10 and a power of 10 expressed in exponential notation. Which of the methods shown above expresses **3800** in **scientific notation**? To write **21,500** in scientific notation, we note that **2.15** is a number between 1 and 10.

$$21,500 = 2.15 \times 10,000$$
$$ = 2.15 \times 10^4$$

Questions

1. Express each of the following powers of 10 in exponential notation.
   - a. one million
   - b. 10,000,000
   - c. one thousand
   - d. 100,000,000
   - e. one hundred
   - f. ten thousand
   - g. one billion
   - h. ten billion
   - i. 1,000,000

2. What power of 10 should replace n to make each statement true?
   - a. $9500 = 95 \times n$
   - b. $8340 = 83.4 \times n$
   - c. $10,000 = 1 \times n$
   - d. $3870 = 3.87 \times n$
   - e. $86,000 = 8.6 \times n$
   - f. $4,650,000 = 4.65 \times n$

3. Express each of the following as a regular decimal numeral.
   - a. 48.6×10^2
   - b. 3.245×10^2
   - c. 439.25×10^3
   - d. 8.7×10^6
   - e. 0.365×10^5
   - f. 1.007×10^1

4. Write each of the following in scientific notation.
   - a. 2300
   - b. 15,400
   - c. 580,000
   - d. 963,000
   - e. 385
   - f. 45
   - g. 18,600,000
   - h. 3,500,000
   - i. 125,000,000

THE HUMAN ELEMENT

Blaise Pascal

A young French boy, the son of a government tax collector, was required to help his father do much of the endless, dull, arithmetic calculations associated with the book work of collecting taxes. The boy was Blaise Pascal (1623–1662) who was destined to become a famous mathematician and scientist.

Although Pascal's father did not believe that anyone below the age of sixteen should begin the study of geometry, he allowed his son to begin earlier when he discovered that the boy had already proved a number of theorems for himself, without the help of any book or teacher. By the time he was sixteen, Pascal had already published a pamphlet containing results of his own original work, and was on friendly terms with some of France's greatest mathematicians.

When the young Pascal became bored with the calculating tasks assigned to him by his tax-collector father, he invented, at the age of nineteen, one of the earliest mechanical calculating machines that could carry out addition and subtraction. A picture of the machine is shown on page 90 of this book.

While he was still in his twenties, Pascal studied the array of numbers that has come to be known as *Pascal's Triangle*.

```
            1
          1   1
        1   2   1
      1   3   3   1
    1   4   6   4   1
```

In each row after the first, there is a **1** on each end, and each of the other numbers in the row is the sum of the two numbers above it. For example, the middle number in the fifth row is **6**, which is **3 + 3**. The number pattern can be extended indefinitely.

Pascal found the array helpful in calculating the likelihood, or *probability*, that certain events would occur. Although his work with probability arose from the calculating of odds in gambling, he laid the foundation for a branch of mathematics that is used for many purposes today. It forms the basis, for example, for calculating insurance risks, as well as for the study of heredity.

Ready for an early rocket flight . . .

Off to the first moonwalk . . .

Open Sentences

The first person to work out the mathematics of rocket action, the late Dr. Robert H. Goddard, is shown with his rocket model of March, 1926. This is the model which gave the first flight of any rocket with liquid propellants, and indicated how powerful a jet could be. After the first flight, the location of the engine was changed to the base of the rocket, where all rockets now have their engines. It was Dr. Goddard's perseverance with rocket experiments that laid the groundwork for the space explorations of the 1960's. The culmination of these is shown as the Apollo 11, riding a pillar of flame, rises to clear its mobile launcher at the Kennedy Space Center on its historic flight to man's first landing on the moon.

Equations

5–1 Open Equations and Inequalities

Let us quickly review what we have learned about the number sentences called equations and inequalities. In Chapter 1 of this book we talked about numerical expressions like $3 + 4$ and 12×4 and saw that number sentences are made up of expressions and relation symbols such as $=, \neq, <,$ and $>$. Number statements that use the $=$ symbol are called **equations**; those that use the $\neq, <,$ and $>$ symbols are called **inequalities**.

$$\text{Equation: } \underbrace{\text{expression}}_{4 \cdot \frac{1}{2}} \; \underbrace{\text{relation symbol}}_{=} \; \underbrace{\text{expression}}_{17 - 15}$$

$$\text{Inequality: } \underbrace{\text{expression}}_{2 + \frac{3}{4}} \; \underbrace{\text{relation symbol}}_{>} \; \underbrace{\text{expression}}_{3 \cdot \frac{1}{2}}$$

The number sentences above are called **statements** because they contain no variables and so we can tell whether each is **true** or **false**

Chapter 5

Suppose that we were told to write a number sentence for this word sentence: "Some number, increased by five, is the same as eight." Do you agree that we cannot write a number **statement** for this number sentence because we are not told the meaning of **some number**? But by using a variable, x in this case, as a symbol for "some number," we can write the equation

$$\underbrace{x + 5}_{\text{Some number increased by five}} \underbrace{=}_{\text{is the same as}} \underbrace{8}_{\text{eight}}$$

A number sentence that contains one or more variables is called an **open sentence**. In algebra we frequently refer to the expression on the left side of the relation symbol as the **left member** of the sentence; the other expression is referred to as the **right member**. The following are open sentences;

Equation: $\underbrace{\overbrace{3x + 4}^{\text{expression}}}_{\text{left member}} \quad \overbrace{=}^{\text{relation symbol}} \quad \underbrace{\overbrace{38 \div 2}^{\text{expression}}}_{\text{right member}}$

Inequality: $\underbrace{\overbrace{n + 7}^{\text{expression}}}_{\text{left member}} \quad \overbrace{<}^{\text{relation symbol}} \quad \underbrace{\overbrace{12 \cdot 1}^{\text{expression}}}_{\text{right member}}$

An open sentence is neither true nor false until the variable is replaced by a member of a designated replacement set.

ORAL EXERCISES

Tell which of the following sentences are *open* and which are *not open*.

1. $\frac{10}{x} + 2 = 8$
2. $3 + 7 < 10 + 1$
3. $8 + x > 15$
4. $k + 3k = \frac{1}{5}$
5. $3 + \frac{2}{3} = \frac{11}{3}$
6. $3x + 5 = 20$
7. $2 + 7 \neq 4x + 2$
8. $3 + 9 < t + 2$
9. $5 \div 2 > 3 \cdot \frac{1}{2}$
10. $7 > 10 \cdot 0$

Tell which sentences are true, false, or neither.

SAMPLE 1: $3 + 8 > 15$ *What you say:* False.

SAMPLE 2: $6 \div x = 10$ *What you say:* Neither.

11. $2 + 7 < 7 + 2$
12. $8 + 0 = 8$
13. $4.3 + 1.6 > 4.9$
14. $1.5 + x \neq 15$
15. $3(4 + 2) \neq 12$
16. $x + y = 3 + 19$
17. $3.04 > 3.004$
18. $\frac{1}{100} = .01$

WRITTEN EXERCISES

Match each number sentence in Column 1 with the correct item in Column 2.

COLUMN 1

1. $n + 3 > 5$
2. $2y \neq 23$
3. $8 = 5t - 3.8$
4. $6 + 3w = 10.5$
5. $k - 3^5 = 7$
6. $10 < 7b$
7. $10 \cdot r^3 = 11$
8. $y^4 < 20$

COLUMN 2

A. 8 is equal to five times a number decreased by 3.8.
B. 10 is less than seven times a number.
C. y to the fourth power is less than 20.
D. Some number, increased by 3, is more than 5.
E. Some number, decreased by 3^5, is equal to 7.
F. The product of 10 and r^3 is equal to 11.
G. Twice some number is not the same as 23.
H. 6 increased by three times a number is the same as 10.5.

Write an open sentence for each of the following.

SAMPLE: The sum of 8 and 4 is less than some number.

Solution: $8 + 4 < t$

9. 5 more than three times a number is less than 25.
10. The product of 10.5 and some number is the same as 2.1.
11. The difference of a number and $\frac{3}{4}$ is greater than $\frac{1}{2}$.
12. 8 less than some number is not the same as 14.
13. The product of 7 and the cube of some number is greater than m.
14. 15 less than some number is less than 32.

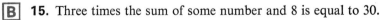
15. Three times the sum of some number and 8 is equal to 30.
16. 9 more than three times a number is equal to 24.
17. Some number to the fourth power, decreased by 2, is equal to 25.
18. Some number that is a repeated factor four times, increased by 7, is the same as 62.3.

Write a word statement for each open sentence.

SAMPLE: $6 + k = 13$ Solution: k more than 6 is the same as 13.

19. $m + 3 = 10$
20. $5 + r > 14$
21. $31 < k - 25$
22. $3b + 2 \neq 9$
23. $w \cdot w \cdot w > 7 + 2$
24. $5 + 4m = 17$
25. $x^2 + 4 = 18$
26. $10 + r^5 < 81$

5–2 Solving Equations

We have said that the open sentence $n + 7 = 10$ is neither true nor false until the variable is replaced by a member of a specified replacement set. To **solve** an open equation means to find among the numbers of the replacement set those that make the statement **true**. Each number that makes the statement true is called a **solution** or **root**; the set of all the numbers that make the statement true is the **solution set** or **truth set** of the open sentence.

Suppose $\{1, 2, 3, 4, 5\}$ is the replacement set for the variable n in $n + 7 = 10$. We can make a **truth table** as shown here.

n-replacement	$n + 7 = 10$	True/False
1	$1 + 7 = 10$	False
2	$2 + 7 = 10$	False
3	$3 + 7 = 10$	True
4	$4 + 7 = 10$	False
5	$5 + 7 = 10$	False

Do you agree that a true statement results when n is replaced by **3**?

Thus, 3 is a **solution**. Since 3 is the only member of the replacement set that makes the statement true, the truth set or **solution set** for $n + 7 = 10$ is $\{3\}$. The **graph** of the solution set is:

Suppose that we are to solve the open sentence $4x + 3 = 27$ and that the replacement set for the variable x is $\{$the whole numbers$\}$ Since it is not possible to try all of the whole numbers as replacements for x, we use a slightly different strategy to find the solution set. Study this solution carefully.

Given the equation $4x + 3 = 27$,
we recall that $? + 3 = 27$
is equivalent to $? = 27 - 3$,
which is true only when $? = 24$.
So we conclude that $4x + 3 = 27$
is equivalent to $4x = 24$,
which is true only when $x = 6$.

A test of 6 as a replacement for x in the sentence $4x + 3 = 27$ shows that **6** is a member of the **solution** set. So the solution set is either $\{6\}$ or a larger set that contains 6. Do you agree that it contains only the member 6? Thus we say that the solution set of the equation $4x + 3 = 27$ is $\{6\}$.

The variable m appears twice in the equation $3 + m = 13 - m$. When solving an equation where the same variable appears more than once, that variable must be replaced by the same number each time it appears. Thus, if the replacement set for m is $\{1, 3, 5, 7\}$ we have:

m-replacement	$3 + m = 13 - m$	True/False
1	$3 + 1 = 13 - 1$	False
3	$3 + 3 = 13 - 3$	False
5	$3 + 5 = 13 - 5$	True
7	$3 + 7 = 13 - 7$	False

Therefore, the **solution** set is $\{5\}$.

ORAL EXERCISES

Name the solution set for each open equation. The replacement set for each variable is $\{2, 4, 6\}$.

SAMPLE 1: $s + 5 = 9$ *What you say:* $\{4\}$

SAMPLE 2: $8 + r = 10.5$ *What you say:* $\emptyset$

1. $12 + k = 18$
2. $m - 2 = 0$
3. $3 \cdot h = 4 + 1$
4. $n \cdot 0 = 0$
5. $4k + 1.3 = 17.3$
6. $x \div 3 = 3$
7. $3 + 7 = y + 8$
8. $n^2 + 3 = 39$
9. $12 \div 3 = s^2$
10. $1.3m + 2 = 18$

WRITTEN EXERCISES

Make a truth table and find the solution set for each open sentence. Use the specified replacement set for each variable.

SAMPLE: $3(2 + k) = 12$; $\{1, 2, 3\}$

Solution:

k-replacement	$3(2 + k) = 12$	True/False
1	$3(2 + 1) = 12$	False
2	$3(2 + 2) = 12$	True
3	$3(2 + 3) = 12$	False

Therefore, the solution set is $\{2\}$.

A
1. $x + 7 = 8 + 4$; $\{1, 3, 5\}$
2. $16 - 3n = 4$; $\{0, 2, 4, 6\}$
3. $3.5m = 3.5$; $\{0, 1\}$
4. $2(r + 1) = 15$; $\{6, 7, 8\}$
5. $x + 3x = 2$; $\{\frac{1}{2}, \frac{1}{3}, \frac{1}{4}\}$
6. $k^2 + 2 = 27$; $\{3, 5, 7\}$
7. $\frac{s}{2} = 3 + 4$; $\{10, 12, 14, 16\}$
8. $5(w - 4) = 5$; $\{1, 2, 4, 6\}$
9. $15 = (3 + t)3$; $\{1, 2, 3, 4\}$
10. $2n + 4 = n + 7$; $\{0, 1, 2, 3\}$
11. $m + 0 = m$; $\{$whole numbers$\}$
12. $3r^2 - 4 = 10r + 11$; $\{1, 3, 5, 7\}$

Find the solution set for each open sentence if the replacement set for each variable is $\{$the numbers of arithmetic$\}$.

SAMPLE: $3 + 3m = 27$

Solution: Since $3 + ? = 27$
is equivalent to $? = 27 - 3$,
which is true only when $? = 24$,
we conclude that $3 + 3m = 27$
is equivalent to $3m = 24$,
which is true only when $m = 8$.

Check:

$3 + 3m$	$= 27$
$3 + 3(8)$	27
$3 + 24$	27
27	27

The solution set is $\{8\}$.

13. $2v + 6 = 20$
14. $18 - 2 = 4w + 4$
15. $4x + 2 = 38$
16. $\frac{k}{4} + 3 = 8$
17. $4.5 = m + 3.0$
18. $3\frac{1}{4} = 1\frac{1}{4} + y$
19. $a + 3 = 4.6$
20. $19 + \frac{t}{3} = 24$

21. $\frac{3}{4} + k = 1.0$
22. $3(n + 4) = 21$
23. $\frac{1}{4}b + 3 = 15$
24. $k - 7 = 6$

Name the solution set for each open sentence. Use the specified replacement set.

SAMPLE: $3n + 2 = n + 14$; $\{2, 4, 6\}$

Solution:

n-replacement	$3n + 2 = n + 14$	True/False
2	$3(2) + 2 = 2 + 14$	False
4	$3(4) + 2 = 4 + 14$	False
6	$3(6) + 2 = 6 + 14$	True

Solution set: $\{6\}$.

B **25.** $m + 12 = 2 + 6m$; $\{0, 1, 2, 3\}$
26. $2a + 3a = 3 \cdot 10$; $\{2, 4, 6, 8\}$
27. $4d - 16 = d + 4$; $\{1, 2, 3, 4\}$
28. $\frac{6k}{3} + 4 = 10 + k$; $\{6, 8, 10, 12\}$
29. $g = 3g - 20$; $\{6, 8, 10, 12\}$
30. $3m - m = 8 + m$; $\{5, 6, 7, 8\}$
31. $z + z + z = 3(z)$; $\{0, 2, 4, 6\}$
32. $2r + 0 = r \cdot r$; $\{0, 1, 2, 3\}$

Find the solution set for each open sentence and graph it. Use the specified replacement set.

SAMPLE: $3m + m = 4m$; $\{2, 3, 4\}$

Solution: $\{2, 3, 4\}$

C **33.** $2n \cdot 1 = n + n$; $\{1, 3, 5, 7\}$
34. $2k \cdot 1 = k \cdot k$; $\{0, 1, 2, 3, 4\}$
35. $0 + (r \cdot r) = 0 + (r + r)$; $\{0, \frac{1}{2}, 1, 1\frac{1}{2}, 2\}$
36. $2m + 2 = \frac{m}{2} + 6$; $\{1, 2, 3, 4, 5\}$
37. $4z + 2 = 4z + 3$; {the whole numbers}
38. $3 + (x + \frac{1}{2}) = 18 \cdot \frac{1}{2}$; {the numbers of arithmetic}

5–3 Using Variables and Writing Expressions

When solving a problem you must first read it carefully to find all the information that is given. Next, you must express this information by using the symbolism of algebra.

EXAMPLE. Consider the books shown on the shelf in the illustration. The shelf is 14 inches long and each book is m inches wide. How can we express the length of the unused part of the shelf?

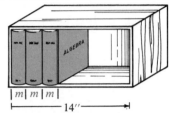

Solution: Let m = width, in inches, of each book.
Then $3m$ = width, in inches, of three books.
Therefore, $14 - 3m$ is the length of the unused part of the shelf.

ORAL EXERCISES

Write an algebraic expression for each of the following.

1. Concrete blocks of uniform size are stacked as shown here. The height in inches of each block is t, and its width in inches is b.
   a. What is the height in inches of the stack?
   b. What is the width in inches of the stack?

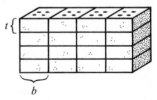

2. Each larger bag of flour shown here weighs k pounds and each smaller bag weighs t pounds.
   a. What is the total weight of flour in the pile of larger bags?
   b. What is the total weight of the flour in the pile of smaller bags?
   c. How much heavier is the flour in the pile of larger bags than the flour in the pile of smaller bags?

3. Tom has saved s dollars, and Sue has saved half as much as Tom. How much has Sue saved?

4. The shelves of a bookcase are spaced as shown in the illustration. Assume that x and y stand for numbers of inches.
What is the height in inches of the bookcase?

5. Fran has a collection of x records. Pete has 2 more than twice as many as Fran. How many records does Pete have?

6. One basketball player is 74 inches tall. A smaller player is only y inches tall. How many inches taller is the first player?

7. If the width in inches of a folding chair is h, what is the length in inches of a row of eight chairs?

WRITTEN EXERCISES

Write an algebraic expression to answer each question.

SAMPLE: Given the number x, what is 3 more than twice that number?

Solution: $2x + 3$

1. Given the number t, what is 5 times that number?
2. Given the number m, what is one-half that number?
3. Given the number y, what is 4 less than that number?
4. Given the number k, what is 2 more than twice that number?
5. Given the number $3x$, what is 8 more than that number?
6. Given the number $5r$, what is 3 more than one-half that number?

Write an algebraic expression for each of the following.

7. 3 more than d
8. 5 more than $3k$
9. 2 less than n
10. $2m$ more than 7
11. $3t$ less than 10
12. a less than h
13. 5 less than $2y$
14. one-half of $5z$
15. 5 times the sum of a and b
16. 3 more than one-half z
17. $3a$ more than xy
18. 10 less than 3 times b
19. 5 more than 7 times r
20. one-third the product of 2 and s

134   Chapter 5

Write the algebraic expression indicated in each exercise.

B **21.** The first line segment pictured here is three times as long as the other line segment. If the first segment has a length of t inches, give the length of the other segment in terms of t.

22. The figure below shows four equal squares. The length of the side of each square is m inches. Give the overall length of the figure in terms of m.

23. Suppose h represents some odd number. Give an expression that names the next larger odd number.

24. Suppose e represents some even number. Give an expression that names the next larger even number.

25. A bicycle shop sold d bicycles during its first week of operation. The second week the number of bicycle sales was 5 more than twice the number sold the first week. Give an expression for the second week's sales.

26. The sum of two numbers is 19. If s is the smaller number give an expression for the larger number.

27. The length of a rectangular swimming pool is 8 feet more than twice its width. Use the variable y to stand for the width, and give an expression for the length of the pool.

28. A rectangle is 4 feet longer than it is wide. Use the variable y to represent the length. Then give an expression for the width in terms of y.

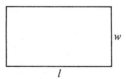

Problem Solving

5–4 Writing Equations and Solving Problems

You have learned how to solve several different kinds of equations that are useful in solving problems. When you are faced with a problem, you must first analyze it, then write an equation that serves as a model of the problem, and finally solve the equation.

EXAMPLE. A rain gauge, used to measure rainfall, collects 8.4 inches of rain water during the months of March and April. The rainfall in April was three times the rainfall in March. How many inches of rain fell in March? in April?

Solution: Since we do not know how much rain fell in March or April, a variable must be used.

Let k = **the number of inches of rainfall in March**

We know that the rainfall in April was three times as great as in March.

Thus, $3k$ = **the number of inches of rainfall in April**.
Since   **March rainfall + April rainfall = 8.4,**
we can write $1k$ + $3k$ = 8.4.
                                    $4k$ = 8.4
                                      k = **2.1**

Therefore, the rainfall in March was **2.1** inches and the rainfall in April was **6.3** inches. We can check our work by verifying that the combined rainfall was **8.4** inches.

Check: 2.1 inches (March)
         6.3 inches (April)
         8.4 inches

ORAL EXERCISES

Use the information given to answer the questions.

SAMPLE: Truck driver: travels $2x$ miles on Tuesday; x miles on Wednesday; 660 miles total.
How far on Tuesday? on Wednesday?
Hint: Tuesday miles + Wednesday miles = 660
              $2x$ + x = 660

What you say: $3x$ = 660
                                  x = 220

He drove **440** miles on Tuesday and **220** miles on Wednesday.

1. Mr. Sander's budget: spent d dollars on clothes; $4d$ dollars on food; $5d$ dollars on rent; $180 total spent. How much for clothes? for food? for rent?
   Hint: clothes + food + rent = 180
   $$d + 4d + 5d = 180$$

2. Rectangle: $10m$ inches long; $8m$ inches wide; perimeter is 72 inches. What is the length? the width?
   Hint: 2 · length + 2 · width = 72
   $$2 \cdot (10m) + 2 \cdot (8m) = 72$$

3. Two consecutive numbers: first number is n; second number is $n + 1$; sum of the numbers is 29. What is the first number? the second number?
   Hint: first number + second number = 29
   $$n + n + 1 = 29$$

4. Baseball team record: played 45 games; won 4 times as many games as lost. How many games won? games lost?
   Hint: games won + games lost = 45
   $$4h + h = 45$$

5. Two consecutive even numbers: first number is w; second number is $w + 2$; sum of the numbers is 42. What are the two numbers?
   Hint: first number + second number = 42
   $$w + w + 2 = 42$$

6. Rectangle: s inches wide; twice as long as it is wide; perimeter is 54 inches. What is the width? the length?
   Hint: 2 · width + 2 · length = 54
   $$2(s) + 2(2s) = 54$$

7. Two consecutive odd numbers: first number is t; second number is $t + 2$; sum of the numbers is 80. What are the two numbers?
   Hint: first number + second number = 80
   $$t + t + 2 = 80$$

PROBLEMS

Write an equation for each problem. Then solve it to answer the question.

SAMPLE: The number of fiction books in the library is five times the number of reference books. There are 1200 reference books. How many fiction books are in the library?

Solution: Let *x* stand for the number of fiction books.
Then, $x = 5(1200)$
$x = 6000$
There are 6000 books in the library.

A
1. Ken worked part time after school for a weekly salary. After six weeks he had earned $78. What was his weekly salary?
2. The high school football team won twice as many games as it lost. They won six games. How many did they lose?
3. Mr. Blair earns $75 a month more than Mr. Cory. Mr. Blair's salary is $700 a month. How much does Mr. Cory earn per month?
4. Tom's score on a mathematics test was 6 points less than Phil's. If Tom's score was 81 points, what was Phil's score?
5. One number is $2\frac{1}{2}$ times a second number. If the second number is 18, what is the other number?
6. Karen's temperature is 2.3 degrees above the normal temperature of 98.6 degrees. What is Karen's temperature?

B
7. The length of a rectangle is five times the width. If the perimeter is 180 feet, find the length and width.
8. The sum of two consecutive numbers is 45. What are the two numbers?
9. A ball team played 19 games. The team won three more games than they lost. Find the number of games won and the number of games lost.
10. The sum of two consecutive even numbers is 62. What are the two numbers?
11. The length of a field is 10 yards more than the width. The perimeter is 180 yards. Find the length and width.
12. Sam is 3 years older than Bill. The sum of their ages is 33 years. How old is each boy?
13. A large bag of laundry weighs 15 pounds more than a smaller bag of laundry. Together they weigh 61 pounds. How much does each bag of laundry weigh?
14. A mathematics class of 30 students contains 6 more boys than girls. How many boys are in the class? How many girls?
15. Mr. Hayes drove a certain number of miles on Friday and half as many miles on Saturday. Altogether he drove 225 miles. How far did he drive on each day?
16. Beth's age is 2 years more than twice her brother Bill's age. If Bill is 10 years old, how old is Beth?
17. The sum of three consecutive numbers is 30. What are the numbers?
18. The width of a rectangle is 3 inches less than its length. If the perimeter of the rectangle is 22 inches, find the length and width.

5-5 Using Formulas to Solve Problems

Have you wondered why the number sentences that we call equations are important in the study of algebra? One reason is that equations, often appearing as formulas, help us solve many kinds of practical problems. A **formula** is a general rule used to solve a particular kind of problem. For example, suppose Mr. Thomas wants to fence off a rectangular-shaped portion of his yard. The width is to be 42 feet and the length is to be 60 feet. How much fencing is needed?

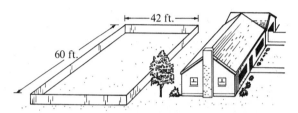

The question "How much fencing is needed?" is another way of asking for the perimeter of the portion of the yard that is to be fenced. We know how to find the perimeter of a rectangle.

$$\text{Perimeter} = (2 \times \text{length}) + (2 \times \text{width})$$
$$P = 2l + 2w$$
$$P = (2 \cdot 60) + (2 \cdot 42)$$
$$= 120 + 84 = 204$$

204 feet of fencing is needed.

ORAL EXERCISES

Match each formula in Column 1 with the correct item in Column 2.

COLUMN 1

1. $A = s^2$
2. $P = 2l + 2w$
3. $A = \pi r^2$
4. $C = \pi d$
5. $A = lw$
6. $P = 4s$
7. $A = \dfrac{bh}{2}$

COLUMN 2

A. Perimeter of a square
B. Area of a circle
C. Area of a square
D. Area of a triangle
E. Perimeter of a rectangle
F. Circumference of a circle
G. Area of a rectangle

State the formula you would use to solve each problem.

SAMPLE: Area = ?

What you say: $A = \pi r^2$

8. Perimeter = ?
   Area = ?

10. Area = ?

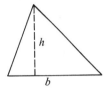

9. Perimeter = ?
   Area = ?

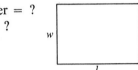

11. Circumference = ?

WRITTEN EXERCISES

Tell which formula is required to solve each problem. Then complete the solution.

SAMPLE: Find the area of a circle if the radius is 10 inches. Use $\pi = 3.1416$.

Solution: $A = \pi r^2$
$A = 3.1416 \cdot 10 \cdot 10$
$A = 314.16$ (square inches)

A
1. Find the circumference of a circle if the radius is 14 inches. Use $\pi = \frac{22}{7}$.
2. Find the area of a square if the length of a side is 15 inches.
3. Find the perimeter of a rectangle if the length is 13 inches and the width is 25 inches.
4. The height and base of a triangle are 17 inches and 12 inches, respectively. Find the area of the triangle.
5. The length and width of a rectangle are 3.8 inches and 6.2 inches. Find the area of the rectangle.
6. Find the perimeter of a square if the length of a side is 18 centimeters.
7. The perimeter of a square is 56 inches. What is the length of a side?
8. The area of a square is 64 square inches. What is the length of a side?

B   9. The perimeter of a rectangle is 48 inches and the width is 10 inches. Find the length of the rectangle.

10. Find the area of a square if the perimeter is 48 yards.

11. The diameter of a circle is 21 inches. Find the circumference of the circle. Use $\pi = \frac{22}{7}$.

12. The diameter of a circle is 14 inches. Find the area of the circle. Use $\pi = \frac{22}{7}$.

C   13. Find the distance around each figure. Use $\pi = \frac{22}{7}$.

a. 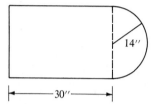   b.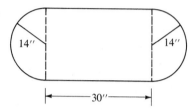

14. Find the area of the shaded part of each figure.

a. 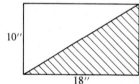   b.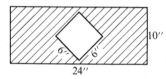

c.   d. 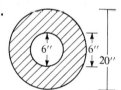   e.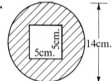

Use $\pi = 3.1416$.   Use $\pi = 3.1416$.   Use $\pi = \frac{22}{7}$.

Inequalities

5–6   Solving Inequalities

You are already familiar with open sentences that are inequalities. Suppose we want to solve the inequality $3 + 2t < 15$ when the replacement set for t is $\{2, 4, 6\}$. The following truth table will help us organize the work.

t-replacement	$3 + 2t < 15$	True/False
2	$3 + (2 \cdot 2) < 15$	True
4	$3 + (2 \cdot 4) < 15$	True
6	$3 + (2 \cdot 6) < 15$	False

Therefore, the solution set is $\{2, 4\}$.

Now let us use the set of whole numbers as the replacement set for solving the inequality $2x + 3 > 5$.

x-replacement	$2x + 3 > 5$	True/False
0	$0 + 3 > 5$	False
1	$2 + 3 > 5$	False
2	$4 + 3 > 5$	True
3	$6 + 3 > 5$	True
4	$8 + 3 > 5$	True
⋮	⋮	⋮

Do you agree that the solution set is $\{2, 3, 4, 5, \ldots\}$?

What is the solution set for the open sentence $2k > 4$ if the replacement set is the set of arithmetic numbers? Any replacement for k that is greater than 2 will result in a true statement. Therefore, the solution set is {arithmetic numbers greater than 2} and the graph of the solution set is as follows:

ORAL EXERCISES

State the question that is suggested by each inequality if the replacement set is {the numbers of arithmetic}.

SAMPLE 1: $k < 3$ *What you say:* Which numbers of arithmetic are less than 3?

SAMPLE 2: $3r > 7$ *What you say:* Which numbers of arithmetic, when multiplied by 3, are greater than 7?

1. $t > 5$
2. $n > 3.5$
3. $s < 14$
4. $2a < 5$
5. $3d > 10$
6. $m + 3 > 7$
7. $k + 2 < 5$
8. $3r + 2 > 17$

Tell whether each resulting statement is true or false when these replacements are used for the variables: $s = 4$, $t = 7$, and $k = 1$.

SAMPLE: $3t + 2 < 15$ *What you say:* $21 + 2 < 15$
$23 < 15$; false

9. $2s < 12$
10. $k + 3 > 3$
11. $t + 1 > 8$
12. $2 + 3t > 16$
13. $\dfrac{s}{2} < 1$
14. $\dfrac{t}{3} < 2$
15. $\dfrac{1}{3} > \dfrac{k}{5}$
16. $k + 3 > k + 2$
17. $3s + 4 < 4s$
18. $10 > 3s - 4$
19. $2k > 3 \cdot 5$
20. $5t \neq 40$
21. $1 + s < 5$
22. $3s \neq s \cdot s \cdot s$
23. $k > k + 0$
24. $t \cdot t > 2t$
25. $s < s \cdot 1$
26. $k + k > k \cdot k$

WRITTEN EXERCISES

For each member of the replacement set, show whether a true or a false statement results when the variable in the sentence is replaced by the number.

SAMPLE: $s + 3 > 8$; $\{5, 5.5, 6\}$

Solution: $5 + 3 > 8$ $5.5 + 3 > 8$ $6 + 3 > 8$
         $8 > 8$; false $8.5 > 8$; true $9 > 8$; true

A 1. $w + 2 > 8$; $\{5, 7, 9\}$
2. $n - 3 < 6$; $\{6, 8, 10\}$
3. $2z + 5 > 20$; $\{5, 10, 15\}$
4. $10 < 6 + r$; $\{0, 1, 2, 3\}$
5. $8 < s - 3$; $\{4, 6, 8\}$
6. $7 > 2r + 1$; $\{0, 1, 2, 3\}$
7. $5 < 3 + 2n$; $\{1, 2, 3, 4\}$
8. $24 < 5a - 6$; $\{2, 4, 6, 8\}$
9. $3n < n + 4$; $\{0, 1, 2, 3\}$
10. $\dfrac{b}{4} > 13 - b$; $\{8, 10, 12\}$

11. $12 > 4 + 2k$; $\{3, 6, 9\}$
12. $5t > 3$; $\{0, \frac{1}{2}, 1, \frac{3}{2}\}$
13. $3.2 > 6s$; $\{0.4, 0.5, 0.6\}$
14. $4.2 > 2s + 0.5$; $\{0, 2, 4, 6\}$
15. $m + 2m < m^3$; $\{0, 1, 2, 3\}$
16. $\frac{3c}{4} < 5$; $\{3, 5, 7, 9\}$

Find each solution set if the replacement set is {the numbers of arithmetic}. Graph each solution set on a number line.

SAMPLE 1: $n + 2 > 5$

Solution: {the numbers of arithmetic greater than 3}

SAMPLE 2: $6 > 3k$

Solution: {the numbers of arithmetic less than 2}

17. $b > 4$
18. $t > 3\frac{1}{2}$
19. $a < 3$
20. $5 > 2r$
21. $8 < 4c$
22. $m < 7$
23. $3 > y$
24. $2z > 4$
25. $3t > 9$

B 26. $x - 1 < 5$
27. $2n + 1 > 6$
28. $\frac{s}{2} > 3$
29. $a + 2 < 3$
30. $9 > 2n$
31. $\frac{k}{3} < 4$
32. $5n < 13$
33. $q + 3 < 4.5$
34. $\frac{p}{5} > 1$

5–7 Inequalities and the Relationship Symbols ≥ and ≤

The symbol ≥ means "is greater than or equal to" and the symbol ≤ means "is less than or equal to." The following examples illustrate the use of these symbols in inequalities when the replacement set is {the numbers of arithmetic}.

EXAMPLE 1. In the case of $x + 2 \geq 5$ if x is replaced by 3 or any number **greater than 3**, a true statement results. The graph of the solution set is:

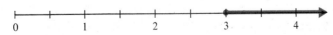

Notice that there is a solid dot at the point that is the graph of 3, showing that the number 3 is in the set.

EXAMPLE 2. In the case of $x + 1 \leq 3$ if x is replaced by **2** or any number **less than 2**, a true statement results. The graph of the solution set is as follows:

The solid dot at the point that is the graph of **2** means that the number 2 is included in the set.

ORAL EXERCISES

State the meaning of each inequality in words. The replacement set is {the numbers of arithmetic}.

SAMPLE: $2n \geq 5$ *What you say:* Which numbers of arithmetic, when multiplied by 2, are greater than or equal to 5?

1. $d \geq 12$
2. $q \geq 5$
3. $t \leq 3.5$
4. $z \leq 6$
5. $3r \geq 10$
6. $5n \leq 1.2$
7. $m + 1 \geq 5$
8. $n - 1 \leq 9$
9. $2w + 2 \geq 4$

Tell whether each resulting statement is true or false when these replacements are used for the variables: $a = 3$, $c = 2$, and $m = 5$.

10. $a \leq 3.1$
11. $m \geq 5$
12. $c \leq 3.9$
13. $c \geq 2$
14. $m \leq 4.8$
15. $a + 1 \geq 2$
16. $3m \leq 25$
17. $2m + 1 \geq 9$
18. $4a - 8 \leq 4.1$

WRITTEN EXERCISES

For each member of the replacement set, show whether a true or a false statement results when the variable is replaced by the number.

SAMPLE: $k - 3 \leq 7$; {8, 10, 12}

Solution: $8 - 3 \leq 7$ $10 - 3 \leq 7$ $12 - 3 \leq 7$
             $5 \leq 7$; true $7 \leq 7$; true $9 \leq 7$; false

A
1. $m \leq 8$; {4, 6, 8}
2. $f \leq 5$; {4, 5, 6, 7}
3. $a \geq 7$; {3, 4, 5, 6}
4. $n + 2 \leq 8$; {4, 6, 8}
5. $t - 3 \leq 5$; {8, 10, 12}
6. $3 + r \geq 9$; {0, 5, 10}

7. $10 - s \leq 10$; $\{0, 1, 2\}$
8. $5 + z \neq 7$; $\{0, 1, 2, 3\}$
9. $3t \geq 12$; $\{0, 2, 4, 6\}$
10. $6w \geq 14$; $\{1, 2, 3, 4\}$
11. $9 \leq 3d$; $\{0, 1, 2, 3\}$
12. $6 \leq 10k$; $\{0, 2, 4, 6\}$
13. $n + 1 \geq 12$; $\{5, 10, 15\}$
14. $4k + 0 \geq 10$; $\{0, 1, 2\}$

B
15. $3t + 2 \leq 1$; $\{4, 6, 89\}$
16. $4b - 8 \geq 0$; $\{2, 4, 6\}$
17. $\frac{m}{3} \geq 1$; $\{1, 2, 3, \pi\}$
18. $23 \geq 10q + 3$; $\{0, 2, 4\}$
19. $3s + s \geq 3.9$; $\{0, \frac{1}{2}, 1\}$
20. $15 \geq 5e + 1$; $\{0, 1, 2\}$
21. $\frac{y}{4} + 1 \leq 4$; $\{4, 6, 8, 9\}$
22. $\frac{4k}{5} \geq 1$; $\{0, \frac{1}{2}, 1, \frac{3}{2}\}$

Match each inequality in Column 1 with the graph of its solution set in Column 2. The replacement set is {the numbers of arithmetic}.

COLUMN 1

23. $6m \geq 12$

24. $3a \leq 6$

25. $2k \geq 5$

26. $10 \leq 3m$

27. $3t \neq 6$

COLUMN 2

A.

B. (graph 0 to 4, shaded from 2 right)

C. (graph 0 to 4, shaded from 0 right with open at 2)

D. (graph 0 to 3, shaded from about 2.5 right)

E. (graph 0 to 4, shaded from 0 to 1)

Describe the set of numbers represented by each number line graph.

SAMPLE: (graph showing shading from $1\frac{1}{3}$ right)

Solution: {the arithmetic numbers greater than or equal to $1\frac{1}{3}$}

28. (graph 0 to 5)

29.

30.

31.

32.

33.

Describe the solution set for each inequality. The replacement set is {the numbers of arithmetic}.

SAMPLE: $3r \geq 7$

Solution: {the numbers of arithmetic greater than or equal to $2\frac{1}{3}$}

C
34. $n \geq 12$
35. $t \leq \frac{1}{2}$
36. $3n \geq 5$
37. $6 \leq 4k$
38. $\frac{s}{2} \leq 3$
39. $\frac{x}{3} \geq 1$

40. $r + 2 > 6$
41. $3 + m \geq 5.2$
42. $z + 1 \neq 4$
43. $t - 2 \leq 7$
44. $3v + 2 \leq 9$
45. $\frac{n+1}{2} < 3$

Graph the solution set of each inequality on a number line. The replacement set is {the numbers of arithmetic}.

46. $2r \leq 5$
47. $2k + 1 \geq 10$
48. $3t + 2 < 9$
49. $\frac{a}{2} \geq 2$

50. $3x \neq 15 - 7$
51. $11 \leq 3 + n$
52. $5 \geq 3y + 1$
53. $3 < \frac{x}{2}$

5–8 Writing Inequalities and Solving Problems

You have learned how to write inequalities and how to determine the solution set from a specified replacement set. Now let us consider problems for which we can write and solve inequalities.

EXAMPLE. The sum of 13 and some number is greater than 21. What numbers can satisfy these conditions?

First we must write an inequality. Do you agree that the following inequality fits the problem?

$$13 + n > 21$$

If we agree that the replacement set is {the numbers of arithmetic}, any number greater than **8** can replace the variable and give a true statement. Thus the solution set is {**the numbers of arithmetic > 8**}.

ORAL EXERCISES

Name the solution set for the open sentence suggested by each problem, if the replacement set is {the numbers of arithmetic}.

SAMPLE: If 5 is subtracted from a certain number, the remainder is equal to or greater than 10. What numbers meet this condition?
Hint: $n - 5 \geq 10$

Solution: Since $15 - 5 = 10$, we see that if n is any number greater than 15, $n - 5 > 10$. Thus the solution set is

{the numbers of arithmetic ≥ 15}.

1. The longer of two line segments is 14 inches long. What could be the length of the shorter one if their combined length is less than 23 inches?
Hint: $14 + n < 23$

2. The sum of 19 and a certain number is equal to or less than 35. What numbers can meet this condition?
Hint: $19 + r \leq 35$

3. The length of a rectangle is 28 inches. What might the width be if the sum of the length and width is to be equal to or less than 54 inches?
Hint: $28 + t \leq 54$

4. A large water storage tank has a maximum capacity of 4500 gallons. If it contains 2384 gallons of water, how many additional gallons of water might be pumped into the tank?
Hint: $2384 + p \leq 4500$

5. The perimeter of a square is greater than 32 inches. What numbers might represent the length in inches of one side of the square?
Hint: $4s > 32$

6. The flying range of an aircraft is limited to 480 miles by the size of the fuel tanks. After flying 261 miles how many additional miles might the aircraft fly without refueling?
Hint: $261 + m \leq 480$

7. The length of a rectangle is 32 feet. What might the width be if the perimeter is not more than 124 feet?
Hint: $(2 \cdot 32) + (2 \cdot w) \leq 124$

8. The lengths of two sides of the pictured triangle are 45 cm. and 82 cm. What might be the length of the third side if the perimeter is at least 200 cm.?
Hint: $45 + 82 + w \geq 200$ and $w < 45 + 82$

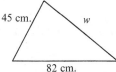

PROBLEMS

Write an inequality for each problem. Then write the solution set if the replacement set is {the numbers of arithmetic}.

SAMPLE. A rectangle is 7 feet wide. If its perimeter is at least 34 feet, what numbers might describe the length in feet of the rectangle?

Solution: Let n stand for the number of feet in the length.

Then
$$2n + 2(7) \geq 34$$
$$2n \geq 20$$
$$n \geq 10$$

The solution set is {the numbers of arithmetic ≥ 10}.

A 1. Mr. Kay traveled 800 miles by airplane and some additional miles by car. Altogether he traveled less than 1200 miles. What numbers indicate the number of miles he might have traveled by car?

2. The sum of 34 and some number is equal to or greater than 65. What numbers can meet these conditions?

3. A given line segment is 20 inches long. What numbers might describe the length in inches of a second line segment if their combined length is 38 inches or less?

4. The perimeter of a square is equal to or less than 68 inches. What numbers might describe the length of a side of the square in inches?

5. The number 3.4 is equal to or less than 2.6 increased by some number. What numbers meet these conditions?

6. A container holds a maximum of 250 cubic centimeters of liquid. Suppose 88.4 cubic centimeters of water is put into the container. What numbers describe the additional number of cubic centimeters of water that can be added?

B 7. The total weight of a truck may not be more than 15,000 pounds on a certain road. If an empty truck weighs 9680 pounds, what numbers describe the load in pounds the truck may carry on this road.

8. The length of a rectangle is 8 inches. The area of the rectangle is less than 36 square inches. What numbers might represent the width in inches?

9. The area of a square is at least 100 square feet. What numbers might represent the length of a side in feet?

10. The lengths of two sides of a triangle are 2.8 inches and 6.4 inches, respectively, and the perimeter is less than 16.9 inches. What numbers might represent the length of the third side in inches?

CHAPTER SUMMARY

Inventory of Structure and Concepts

1. Number statements that use the = relation symbol are **equations**; those that use the symbols ≠, <, >, ≤, and ≥ are **inequalities**.
2. Symbols, such as letters, that are used to represent numbers are called **variables**. A number sentence that contains at least one variable is an **open** sentence.
3. Solving an open sentence requires finding, from a specified **replacement set**, those numbers that can replace the variable and result in a true statement. Each such number is a **root** or **solution**; all such numbers make up the **solution set** or **truth set**.

Vocabulary and Spelling

variable (*p. 125*)
open sentence (*p. 126*)
left member (*p. 126*)
right member (*p. 126*)
replacement set (*p. 128*)
solution (*p. 128*)
root (*p. 128*)

solution set (*p. 128*)
truth set (*p. 128*)
truth table (*p. 128*)
formula (*p. 138*)
perimeter (*p. 138*)
circumference (*p. 138*)
area (*p. 138*)

Chapter Test

Write an open sentence for each of the following. Do not determine the solution set.

1. If 5 is decreased by some number the difference is 3.
2. If some number is multiplied by 15 the resulting product is 75.
3. If some number is increased by 5, the result is less than 15.

Find the solution set for each open sentence if the replacement set is {0, 1, 2, 3, 4, ... , 9, 10}. Give your answer in roster form.

4. $k + 12 = 2k - 3$
5. $2y - 3 < 5$
6. $2x - 3 = 11$
7. $\frac{x}{3} \geq 2$

Graph the solution set for each of the following on a number line. Assume each replacement set to be {the numbers of arithmetic}.

8. $2x + 3 < 12$

9. $3x - 5 \neq 4$

10. $\dfrac{k}{3} + 1 \geq 2$

11. $2(x + 1) = 12$

Write an algebraic expression for each of Questions 12–14.

12. One-half a given number increased by 3
13. 15 decreased by twice a given number
14. If $2k + 1$ represents an odd number, the next larger odd number
15. The area of a circle is equal to π times the square of the radius. Assuming $\pi = 3.14$, determine the area of a circle whose diameter is 10 feet.
16. If the area of a rectangle is 40 square feet and its length is 8 feet, what is its perimeter?

Write an open sentence for each problem. Then solve it to answer the question.

17. Mr. Stamm is twice as old as his son John. If the sum of their ages is 63 years, how old is John?
18. If the sum of two consecutive whole numbers is less than 11, what values could the smaller of the two whole numbers assume?

Chapter Review

5–1 Open Equations and Inequalities

Write an open sentence for each of the following.

1. Some number increased by 5 is equal to 24.
2. Two less than some number is greater than 3.
3. If the square of some number is increased by 2, the result is equal to 15.
4. If 15 is decreased by two times a number, the result is not equal to 7.

5–2 Solving Equations

Find the solution set for each equation if the replacement set is {the whole numbers}.

5. $n + 5 = 12 - 5$

6. $2x - 3 = 15$

7. $\dfrac{k}{4} = 15$
8. $2(y + 3) = 14$
9. $m + 12 = 2m - 6$
10. $15 - 3 = 4z$

11. Find the root of the equation $22 = 30 - x$.
12. Make a graph of the solution set of $x^2 - 2x = 0$ if the replacement set is $\{0, 1, 2, 3, 4, 5\}$.

5–3 Using Variables and Writing Expressions

Write an algebraic expression for each of the following.

13. 5 less than twice the number n
14. 5 more than one-half x
15. $6y$ increased by 15
16. If $2k$ represents an even number, the next larger even number
17. The cost of John's car, which was $10 more than twice the cost (y) of Steve's car

5–4 Writing Equations and Solving Problems

Write an equation for each problem. Then solve it to answer the question.

18. There are 270 boys in a certain school and 50 more boys than girls. How many girls are there?
19. A baseball team scored 17 runs in 3 games. They scored 5 runs in the first game and twice as many runs in the second game as in the third game. How many runs did they score in the third game?
20. A team scored 65 points in a basketball game, making 10 more field goals than foul shots. A field goal is worth 2 points and a foul shot is worth 1 point. How many field goals did the team make?
21. The length of a rectangle is 10 feet more than its width. If its perimeter is 180 feet, what is its width?

5–5 Using Formulas to Solve Problems

22. Find the perimeter of a square if the length of one side is 12 inches.
23. The diameter of a circle is 15 inches. Find its circumference. Let $\pi = 3.14$.

152 *Chapter 5*

24. The area of a rectangle is 56 square feet. If its length is 8 feet, what is its width?

25. Find the area of the figure to the right. Let $\pi = 3.14$.

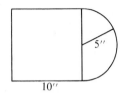

5-6 Solving Inequalities

Find the solution set for each of the following if the replacement set is {the whole numbers less than 10}. Give your answer in roster form.

26. $n - 3 < 6$

27. $8 - y > 4$

28. $12 < x + 4$

29. $4 < n - 6$

Graph the solution set for each of the following if the replacement set is {the numbers of arithmetic}.

30. $t - 5.5 < 12.5$

31. $4x - 2 < 14$

32. $\dfrac{k}{6} > \dfrac{1}{2}$

33. $3 > 3b$

5-7 Inequalities and the Relationship Symbols $\geq$ and $\leq$

Graph the solution set for each of the following, if the replacement set is {the numbers of arithmetic}.

34. $x \leq 4$

35. $k \neq 2$

36. $t - 3 \geq 1$

37. $\dfrac{y}{3} + 4 \leq 6$

5-8 Writing Inequalities and Solving Problems

Write an inequality for each problem. *Do not* solve the problem.

38. What numbers meet the condition that the sum of 12 and the number is less than or equal to 15?

39. The label on a jar of jelly indicates a minimum net weight of 13 ounces. What might the jelly weigh and still meet the requirements of the label?

40. If a rocket has a maximum range of 500 miles what might be the target distance that this rocket could reach?

Review of Skills

Indicate whether each statement is true or false.

1. $2 + 5 = 7$, so $7 = 2 + 5$.
2. $5 \cdot 8 = 40$, so $40 = 5 \cdot 8$.
3. $18 - 3 = 15$, so $15 = 18 - 3$.
4. $\frac{36}{12} = 3$, so $3 = \frac{36}{12}$.
5. $2 + 8 = 10$ and $10 = 2 \cdot 5$, so $2 + 8 = 2 \cdot 5$.
6. $3 \cdot 6 = 18$ and $18 = 2 \cdot 9$, so $3 \cdot 6 = 2 \cdot 9$.
7. $\frac{20}{5} = 4$ and $4 = 10 - 6$, so $\frac{20}{5} = 10 - 6$.
8. $16 = 10 + 6$ and $10 = 12 - 2$, so $16 = (12 - 2) + 6$.
9. $6 \cdot 9 = 54$ and $9 = 5 + 4$, so $6 \cdot (5 + 4) = 54$.
10. $5 \cdot (10 + 2) = 5 \cdot 10 + 5 \cdot 2$
11. $\frac{15 + 3}{3} = \frac{15}{3} + \frac{3}{3}$
12. $3 + (2 + 1) = (3 + 2) + 1$
13. $13 - (3 + 2) = (13 - 3) + 2$
14. $5 \cdot (2 \cdot 4) = (5 \cdot 2) \cdot 4$
15. $28 \cdot (8 \div 2) = (28 \cdot 8) \div 2$
16. $37 + 15 = 15 + 37$
17. $21 \cdot 4 = 4 \cdot 21$
18. $2 \cdot (3 + 5) = (3 + 5) \cdot 2$
19. $\frac{364}{2} = \frac{300}{2} + \frac{60}{2} + \frac{4}{2}$
20. $3 \cdot (6 - 1) = 3 \cdot 6 - 3 \cdot 1$
21. $\frac{24 - 18}{6} = \frac{24}{6} - \frac{18}{6}$
22. $15 + (8 - 2) = (15 + 8) - 2$
23. $18 - (5 - 3) = (18 - 5) - 3$
24. $24 \div (2 \cdot 3) = (24 \div 2) \cdot 3$
25. $18 \div (9 \div 3) = (18 \div 9) \div 3$
26. $18 - 6 = 6 - 18$
27. $27 \div 9 = 9 \div 27$
28. $(4 + 7) \cdot 3 = 4 \cdot 3 + 7 \cdot 3$
29. $4 \cdot (432) = 4 \cdot 400 + 4 \cdot 30 + 4 \cdot 2$

For each open sentence, state its solution set.

30. $10 = 2n$
31. $n = \frac{10}{2}$
32. $0 = 2n$
33. $n = \frac{0}{2}$
34. $0 \cdot n = 5$
35. $0 = \frac{5}{n}$
36. $0 \cdot n = 0$
37. $0 = \frac{0}{n}$

Simplify each expression.

38. $54 + 0$
39. $\frac{54}{1}$
40. $0 \cdot 0$
41. $54 + 1$
42. $\frac{0}{4}$
43. $\frac{13}{13}$
44. $54 \cdot 0$
45. $54 \cdot 1$
46. $1 - \frac{4}{4}$
47. $(2^{10} + 1) \cdot 0$
48. $3 \cdot \frac{3}{3}$
49. $54 - 1$

CHECK POINT FOR EXPERTS

Special Sets of Numbers

You are already familiar with the set of whole numbers, {**0, 1, 2, 3,** ...}. The whole numbers are sometimes called **cardinal** numbers, while the numbers used to indicate order, such as first, second, third, and so on, are called **ordinal** numbers.

Consider the set of even whole numbers. The first of these is **0**, the second is **2**, the third is **4**, and so on. Look carefully and find how each even whole number is related to its corresponding ordinal number.

Even numbers: 0,   2,   4,   6,   8,   10, ...
Ordinal numbers: 1st, 2nd, 3rd, 4th, 5th, 6th, ...

Did you discover that (**2 × ordinal number**) − **2** names the corresponding **even number**? Or you may have said that (**even number ÷ 2**) **+ 1** names the corresponding **ordinal number**.

Now let us see if there is a pattern that holds true for the odd whole numbers and the corresponding ordinal numbers.

Odd numbers: 1,   3,   5,   7,   9,   11, ...
Ordinal numbers: 1st, 2nd, 3rd, 4th, 5th, 6th, ...

Do you agree that (**2 × ordinal number**) − **1** names the corresponding odd number and (**odd number + 1**) **÷ 2** names the corresponding **ordinal number**?

The numbers that we call **square numbers**, {**1, 4, 9, 16, 25,** ...}, can be represented by square arrays of objects. Observe the pattern below, in which the first array is **1 × 1**, the second array is **2 × 2**, the third is **3 × 3**, and so on.

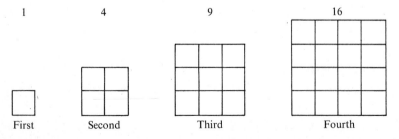

As you have probably discovered, (**ordinal number**)² names the corresponding **square number**.

Look at these three-dimensional arrays that represent **cubic numbers**. Do you see the pattern? The first array is **1 × 1 × 1**, the second is **2 × 2 × 2**, the third is **3 × 3 × 3**, and so on.

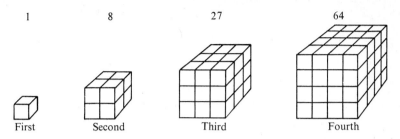

1	8	27	64
First	Second	Third	Fourth

The set of cubic numbers is {**1, 8, 27, 64, 125, ...**}. Do you see that (**ordinal number**)³ names the corresponding cubic number?

Questions

1. What are the first ten members of the set of square numbers?
2. What is the ninth number in the set of even numbers? the fifteenth? the 319th?
3. What is the eighth number in the set of odd numbers? the twelfth? the eighty-seventh?
4. What ordinal number corresponds with 146 in the set of even numbers? What ordinal number corresponds with 304 in the set of even numbers?
5. What are the first ten members of the set of cubic numbers?
6. What is the corresponding cubic number for each of these ordinal numbers?
   - a. third
   - b. twelfth
   - c. twentieth
   - d. fortieth
   - e. tenth
   - f. one hundredth
7. Which is greater:
   - a. The sum of the first ten odd numbers or the sum of the first ten even numbers?
   - b. The sum of the first one hundred odd numbers or the sum of the first one hundred even numbers?
8. Study this number pattern. Then answer the questions.

$$1 = 1 \text{ or } 1^2$$
$$1 + 3 = 4 \text{ or } 2^2$$
$$1 + 3 + 5 = 9 \text{ or } 3^2$$
$$1 + 3 + 5 + 7 = 16 \text{ or } 4^2$$

What is the sum of the first six odd numbers? the first ten odd numbers? the first forty odd numbers?

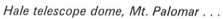

Mayan observatory . . .

Hale telescope dome, Mt. Palomar . . .

Operations, Axioms, and Equations

That mathematics and astronomy were highly developed in ancient Mexico is indicated by the remarkable accuracy of their calendars. At Chichén Itzá in Mexico the circular tower, or caracol, is believed to have been part of a Mayan observatory, and may also have been used for religious services. Built about 1000 A.D., it has a small observation chamber near the top, with square openings in the wall leading to the outside. These openings fix important astronomical lines of sight. The circular tower bears a striking resemblance to the Palomar Observatory in California, which houses the 200-inch Hale Reflector, the largest telescope in the world. This telescope collects 360,000 times as much light as does the human eye, and can photograph objects at a distance of 4 billion light-years.

Basic Properties for Equality and Existence

6–1 Basic Operations and Axioms

When you begin to learn a new game for the very first time you must accept the rules of the game and apply them consistently. Although you may never have thought of mathematics as a kind of game with rules to follow, such is very much the case. We accept and use many ideas whose truth we must assume.

The study of algebra is concerned with the nature of operations, with rules for performing these operations, and with the ideas we assume to be true about numbers and operations. The rules and assumptions of mathematics are called **axioms** or **postulates**. We shall consider first three fundamental axioms related to the idea of equality.

(1) You would probably agree that the following statements are true:

$$3 = 3; \quad 38\tfrac{1}{2} = 38\tfrac{1}{2}; \quad 3845 = 3845.$$

157

158 *Chapter 6*

One of the basic axioms in mathematics is that any number is equal to itself. We assume the truth of this idea and call it the **reflexive property of equality**. We can state this axiom in a generalized way:

For any number r, $r = r$.

(2) Would you agree that the following statements are true?

$$3 + 7 = 10, \text{ so } 10 = 3 + 7.$$
$$8 \times 3 = 20 + 4, \text{ so } 20 + 4 = 8 \times 3.$$

This assumption is also an axiom and is called the **symmetric property of equality**. In general we say:

For any numbers r and s,

if $r = s$, then $s = r$.

(3) Study carefully the following statement and tell whether or not you agree with the conclusion.

Since $5 + 3 = 8$ and $8 = 2 \cdot 2 \cdot 2$, we conclude that $5 + 3 = 2 \cdot 2 \cdot 2$.

In general, we state the axiom called the **transitive property of equality** as follows:

For any numbers r, s, t,

if $r = s$ and $s = t$, then $r = t$.

ORAL EXERCISES

Tell which property of equality is illustrated in each exercise.

SAMPLE 1: $2 + 3 = 5$ and $5 = 4 + 1$, so $2 + 3 = 4 + 1$.

What you say: The transitive property of equality.

SAMPLE 2: $3 \cdot 3 = 9$, so $9 = 3 \cdot 3$.

What you say: The symmetric property of equality.

1. $4 \cdot 5 = 20$, so $20 = 4 \cdot 5$.

2. $3^3 = 27$, so $27 = 3^3$.
3. $6 + 6 = 12$ and $12 = 3 \cdot 4$, so $6 + 6 = 3 \cdot 4$.
4. $3 = 12 \div 4$, so $12 \div 4 = 3$.
5. $10 + r = 10 + r$
6. $15 - k = 15 - k$
7. $15 = 5 \cdot 3$ and $5 \cdot 3 = 10 + 5$, so $15 = 10 + 5$.
8. If $a = b$, then $b = a$.
9. If $x + y = a + b$, then $a + b = x + y$.
10. If $A = \pi r^2$, then $\pi r^2 = A$.
11. $\dfrac{a}{3} = \dfrac{a}{3}$
12. If $a + b = \frac{72}{4}$, and $\frac{72}{4} = 18$, then $a + b = 18$.
13. If $3x = 15 - 6$, and $15 - 6 = 9$, then $3x = 9$.
14. $\frac{9}{5}y = \frac{9}{5}y$
15. If $7a + 2 = 37$, and $37 = 35 + 2$, then $7a + 2 = 35 + 2$.

WRITTEN EXERCISES

Use each equation to illustrate the symmetric property of equality.

SAMPLE: $4 \cdot 9 = 6^2$ Solution: $4 \cdot 9 = 6^2$, so $6^2 = 4 \cdot 9$.

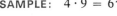

1. $10 + 8 = 18$
2. $m + 7 = 10 + 4$
3. $x^3 = x^2 \cdot x$
4. $\pi + t^2 = 5\pi$
5. $3(2 + a) = 3a + 6$
6. $10(m - 30) = (m - 30)10$
7. $\dfrac{10 + 6}{4} + k = 2^2 + k$
8. $a \cdot 0 = 0 \cdot b$

Use the transitive property of equality to complete each of the following.

SAMPLE: $40 = 5 \cdot 8$ and $5 \cdot 8 = 10 \cdot 4$, so __?__.
Solution: $40 = 5 \cdot 8$ and $5 \cdot 8 = 10 \cdot 4$, so $40 = 10 \cdot 4$.

9. $3 \cdot 8 = 4 \cdot 6$ and $4 \cdot 6 = 20 + 4$, so __?__.
10. $2 \cdot 3^2 = 12 + 6$ and $12 + 6 = 18$, so __?__.
11. $4 \cdot 12 = 40 + 8$ and $40 + 8 = 3 \cdot 4^2$, so __?__.

B 12. __?__ and $7 = 14 \div 2$, so $6 + 1 = 14 \div 2$.
13. __?__ and $4 \cdot 1 = 2^2$, so $3 + 1 = 2^2$.
14. If $x = y$ and $y = w$, then __?__.
15. If $a + b = t$ and $t = rs$, then __?__.

160   Chapter 6

16. If __?__ and $g = k$, then $b = k$.
17. $3 + 5 = 8$ and __?__, so $3 + 5 = 2 \cdot 2^2$.
18. If $q = w$ and __?__, then $q = z$.

Tell which property of equality is used to reach the first conclusion, and which is used to reach the second conclusion.

C **19.** $15 - 3 = 3 \cdot 4$ and $2 \cdot 6 = 3 \cdot 4$
   (1) so $3 \cdot 4 = 2 \cdot 6$
   (2) and $15 - 3 = 2 \cdot 6$.
20. $3 \cdot 6 = 18$ and $17 + 1 = 18$
   (1) so $18 = 17 + 1$
   (2) and $3 \cdot 6 = 17 + 1$.
21. $2(4 + 3) = 2(7) = 14$ and $14 = 10 + 4$
   (1) so $2(4 + 3) = 14$
   (2) and $2(4 + 3) = 10 + 4$.
22. $8 \times \frac{1}{2} = 8 \div 2$, and $8 \div 2 = 0 + 4$, and $0 + 4 = 2^2$
   (1) so $8 \div 2 = 2^2$
   (2) and $8 \times \frac{1}{2} = 2^2$.

6–2 Existence and Uniqueness; the Closure Property

You may recall that earlier we said that the addition and multiplication operations could be thought of as a matching of a pair of numbers with some third number. This idea is illustrated here with whole numbers.

Addition

$(2, 3)$ $(0, 7)$ $(5, 4)$ $(9, 3)$
0, 1, 2, 3, 4, 5, 6, 7, 8, 9, 10, 11, 12, 13, 14, 15, ...
    $(1, 4)$ $(3, 2)$ $(2, 6)$ $(5, 3)$

Multiplication

As you might guess from the diagram, and from sampling other pairs of numbers, the following assumptions seem justified:
   (1) If you **add** two whole numbers the result is a **whole number**. Furthermore, there is **one and only one** whole number that is the correct result.
   (2) If you **multiply** two whole numbers the result is a **whole number**. Furthermore, there is **one and only one** whole number that is the correct result.

Each statement represents **existence** and **uniqueness** properties for the operation named.

Since the sum of any two whole numbers is also a whole number, we say that the set of whole numbers is **closed under addition**. Do you agree that the **product** of any two whole numbers is also a whole number? Study these examples.

$$5 \cdot 9 = 45 \qquad 308 \cdot 4 = 1232 \qquad 19 \cdot 15 = 285$$

The set of whole numbers is **closed under multiplication**.

In general, we describe the **closure property** as follows:

A given set of numbers is closed under an operation performed on its elements if each result is also an element of the given set.

Thus we see that the closure property depends on both the set of numbers used and the operation performed. For example, the set of even numbers is **closed** under **addition**, as suggested by these examples.

EXAMPLE 1. $8 + 14 = 22$ EXAMPLE 2. $36 + 52 = 88$

EXAMPLE 3. $470 + 294 = 764$

In each, the sum of two even numbers is an even number.

The following examples show that the set of whole numbers is **not closed** under **subtraction**, by showing some whole-number subtraction problems that do not have whole-number answers.

$$10 - 6 = 4 \qquad 9 - 14 = ? \qquad 35 - 162 = ?$$

You have learned that it is possible to name a number by using one of many different numerals. An indicated sum or product of numbers does not change regardless of the numerals used to represent the numbers.

Since $6 + 3 = 9$, we know $5(6 + 3) = 5(9)$.

Since $18 = 20 - 2$, we know $6(18) = 6(20 - 2)$.

The examples above illustrate the **substitution principle**:

For any numbers *m* and *n*, if *m* = *n*, then *m* and *n* may be substituted for each other.

162 *Chapter 6*

Observe how the substitution principle is used to simplify the following expression.

$$3(4 + 6 + 2)$$
$$3(4 + 8)\qquad \text{8 is substituted for } 6 + 2$$
$$3 \cdot (12)\qquad \text{12 is substituted for } 4 + 8$$
$$36\qquad \text{36 is substituted for } (3 \cdot 12)$$

ORAL EXERCISES

Describe the substitutions made in simplifying each expression.

SAMPLE: $3 \cdot 2 \cdot 5 \cdot 1$ *Solution:* $3 \cdot 2 \cdot 5 \cdot 1$
$6\cdot 5 \cdot 1$ $6\cdot 5 \cdot 1$ 6 is substituted for $3 \cdot 2$
$30\cdot 1$ $30\cdot 1$ 30 is substituted for $6 \cdot 5$
30 30 30 is substituted for $30 \cdot 1$

1. $10 + 3 + 7 + 2$
    $13 + 7 + 2$
    $13 + 9$
    22

2. $6 \cdot 2 \cdot 4 \cdot 3$
    $6 \cdot 2 \cdot 12$
    $12 \cdot 12$
    144

3. $(9 \cdot 2) + 3 \cdot 2$
    $(9 \cdot 2) + 6$
    $18 + 6$
    24

4. $2 \cdot 5 \cdot 3 \cdot 4$
    $2 \cdot 5 \cdot 12$
    $10 \cdot 12$
    120

5. $(4 + 5 + 8)6$
    $(9 + 8)6$
    $17 \cdot 6$
    102

6. $2 + 3(5 \cdot 6)$
    $2 + 3(30)$
    $2 + 90$
    92

Tell whether or not the specified set of numbers is closed under the indicated operation. If it is *not closed* give at least one example to support your answer.

SAMPLE 1: $\{0, 1, 2, 3, 4, \ldots\}$; division

What you say: Not closed; $2 \div 3 = \frac{2}{3}$, and $\frac{2}{3}$ is not a member of the specified set.

SAMPLE 2: $\{0, 2\}$; addition

What you say: Not closed; $2 + 2 = 4$, and **4** is not a member of $\{0, 2\}$.

7. $\{10, 20, 30, 40, 50\}$; addition
8. $\{0, 2, 4, 6, 8, \ldots\}$; multiplication
9. $\{3, 6, 9, 12, 15, \ldots\}$; multiplication
10. $\{0, 1\}$; addition
11. $\{100, 200, 300, 400, \ldots\}$; addition
12. $\{\frac{1}{2}, 1, \frac{1}{4}, 2, \frac{1}{8}, 4, \frac{1}{16}, \ldots\}$; division
13. $\{0, 2\}$; multiplication
14. $\{0, 2, 4\}$; subtraction
15. $\{0, 1, 2, 3, 4, \ldots\}$; subtraction
16. $\{0, 1\}$; multiplication

WRITTEN EXERCISES

Tell whether or not each set is closed for addition, subtraction, multiplication, and division.

1. $\{0, 2, 4, 6\}$
2. $\{1\}$
3. $\{5, 10, 15, 20, \ldots\}$
4. $\{10, 20, 30, 40, 50, \ldots\}$
5. $\{0\}$
6. $\{0, 10\}$
7. $\{1, 4, 9, 16, 25, 36, \ldots\}$
8. $\{1^3, 2^3, 3^3, 4^3, \ldots\}$
9. {numbers between 0 and 1}
10. {numbers less than 1}
11. {multiples of 6}
12. {the odd numbers}
13. {the prime numbers}
14. $\{1, 3, 6, 10, 15, 21, \ldots\}$

Simplify each expression by using the substitution principle in the manner indicated.

SAMPLE: $3 + 8 + 2 + 10$
$\underbrace{}\ ?\ + 2 + 10$
$\underbrace{}\ ?\ + 10$
$\underbrace{}\ ?$

Solution: $3 + 8 + 2 + 10$
$\underbrace{}\ 11\ + 2 + 10$
$\underbrace{}\ 13\ + 10$
$\underbrace{}\ 23$

15. $\frac{1}{2} \cdot \frac{1}{3} \cdot \frac{6}{7} \cdot \frac{1}{4}$
$?\ \cdot \frac{6}{7} \cdot \frac{1}{4}$
$?\ \cdot \frac{1}{4}$
$?$

16. $2^2 + 3^2 + 1^4 + 2^3$
$2^2 + \ ?\ + 2^3$
$2^2 + \ ?$
$?$

17. $2.3 + 7.1 + \underbrace{8.8 + 0.2}$
$\underbrace{2.3 + 7.1 + \quad ?}$
$\underbrace{2.3 + \quad ?}$
$?$

18. $\underbrace{8(50 - 2)}$
$\underbrace{8 \cdot ?}$
$?$

19. $\underbrace{\tfrac{1}{2} + \tfrac{3}{8}} + \tfrac{1}{8} + \tfrac{3}{4}$
$\underbrace{? \quad + \tfrac{1}{8}} + \tfrac{3}{4}$
$\underbrace{? \quad + \tfrac{3}{4}}$
$?$

20. $\underbrace{(15 \div 3)} + 6 + 3$
$\underbrace{? \quad + 6} + 3$
$\underbrace{? \quad + 3}$
$?$

Commutative, Associative, and Distributive Properties

6–3 Commutative and Associative Properties

The **commutative property** concerns itself with the **order** in which two numbers are added or multiplied. In either **addition** or **multiplication,** the order of the two numbers does not affect the result.

The commutative property of addition for numbers of arithmetic is illustrated by the statements

$$5 + 7 = 7 + 5; \quad 10.8 + b = b + 10.8.$$

We can show the commutative property on the number line, as illustrated in the number line interpretations of $1\tfrac{1}{2} + 3$ and $3 + 1\tfrac{1}{2}$, shown in Figure 6–1. Do you agree that in both cases we end up at the same point, $4\tfrac{1}{2}$?

Figure 6–1

The axiom for the **commutative property of addition** is stated as follows:

For every number **r**, and every number **s**,

$$r + s = s + r.$$

The following statements illustrate the **commutative property** of **multiplication**. Again, notice that the **order** of the numbers does not affect the result.

$$5 \cdot 10 = 10 \cdot 5; \quad 12 \cdot k = k \cdot 12.$$

Figure 6–2 shows interpretations of $3 \cdot 4$ and $4 \cdot 3$ on the number line. Do we end up at the same point, **12**, in both cases?

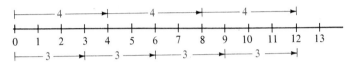

Figure 6–2

The axiom for the **commutative property of multiplication** is as follows:

For every number **r**, and every number **s**,

$$rs = sr.$$

You may have already tested out the idea of the commutative property with subtraction and division. The statements below show that the **subtraction** and **division** operations are **not commutative**.

$$10 - 5 \neq 5 - 10 \qquad 10 \div 5 \neq 5 \div 10$$

The **associative property** concerns itself with the **grouping** of three numbers to be added or multiplied. If you are to find the sum $6 + 9 + 12$, you may group the numbers in either way shown below, and get the same result.

$$
\begin{array}{cc}
(6 + 9) + 12 & 6 + (9 + 12) \\
\underbrace{}_{15} \; + 12 & 6 + \underbrace{}_{21} \\
\underbrace{}_{27} & \underbrace{}_{27}
\end{array}
$$

The axiom for the **associative property of addition** is stated as follows:

For every number **r**, every number **s**, and every number **t**,

$$r + (s + t) = (r + s) + t.$$

If you are to find the product $3 \cdot 5 \cdot 2$ you may group the numbers in either way shown, and get the same result.

$$3 \cdot \underbrace{(5 \cdot 2)} \qquad \underbrace{(3 \cdot 5)} \cdot 2$$
$$3 \cdot \underbrace{\quad 10 \quad} \qquad \underbrace{15 \quad} \cdot 2$$
$$\quad 30 \qquad\qquad\qquad 30$$

The axiom for the **associative property of multiplication** is stated as follows:

For every number *r*, every number *s*, and every number *t*,

$$r(st) = (rs)t.$$

•

ORAL EXERCISES

Tell whether each true statement illustrates the commutative or the associative property of addition or of multiplication.

SAMPLE 1: $4 + 10 = 10 + 4$

What you say: Commutative property of addition

SAMPLE 2: $4(3 \cdot 7) = (3 \cdot 7)4$

What you say: Commutative property of multiplication

1. $3 + 9 = 9 + 3$
2. $4 \cdot 18 = 18 \cdot 4$
3. $(3 \cdot 5)10 = 3(5 \cdot 10)$
4. $\frac{1}{2} + \frac{2}{3} = \frac{2}{3} + \frac{1}{2}$
5. $4 \cdot (19 \cdot 10) = (4 \cdot 19) \cdot 10$
6. $8 + (9 + 3) = (8 + 9) + 3$
7. $(\frac{3}{4} + \frac{1}{2}) + \frac{1}{2} = \frac{3}{4} + (\frac{1}{2} + \frac{1}{2})$
8. $(3 + 2) + 5 = 5 + (3 + 2)$
9. $\frac{1}{2} \cdot (3 + 4) = (3 + 4) \cdot \frac{1}{2}$
10. $(5 \cdot 7) + 3 = 3 + (5 \cdot 7)$
11. $(9 \cdot 4) \cdot 7 = 9 \cdot (4 \cdot 7)$
12. $(5 \cdot 6) \cdot 8 = 8 \cdot (5 \cdot 6)$
13. $(6 \cdot 3.8) \cdot 5 = 6 \cdot (3.8 \cdot 5)$
14. $1\frac{1}{8} + 2\frac{1}{2} = 2\frac{1}{2} + 1\frac{1}{8}$
15. $6 + (7 + 12) = (6 + 7) + 12$
16. $19 + (3 + 9) = (3 + 9) + 19$
17. $\frac{2}{3} \cdot 12 = 12 \cdot \frac{2}{3}$
18. $(.003 + .045) + 1 = .003 + (.045 + 1)$
19. $(5^2 \cdot 5^3) \cdot 5^4 = 5^2 \cdot (5^3 \cdot 5^4)$
20. $10\% + 30\% = 30\% + 10\%$
21. $(7 + \frac{1}{2}) + 3 = 7 + (\frac{1}{2} + 3)$
22. $7 + (\frac{1}{2} + 3) = 7 + (3 + \frac{1}{2})$
23. $7 + (3 + \frac{1}{2}) = (7 + 3) + \frac{1}{2}$
24. $(\frac{1}{3} \cdot 8) \cdot 6 = (8 \cdot \frac{1}{3}) \cdot 6$
25. $(8 \cdot \frac{1}{3}) \cdot 6 = 8 \cdot (\frac{1}{3} \cdot 6)$

WRITTEN EXERCISES

Show that each of the following sentences is true. Also name the property illustrated.

SAMPLE: $10 + (18 + 6) = (10 + 18) + 6$

Solution:
$$10 + (18 + 6) = (10 + 18) + 6$$

$10 + 24$	$28 + 6$
34	34

Associative property of addition

1. $10.8 + 6.6 = 6.6 + 10.8$
2. $(3 \cdot 9) \cdot 5 = 3 \cdot (9 \cdot 5)$
3. $(12 \cdot 6) \cdot 10 = 10 \cdot (12 \cdot 6)$
4. $15 \cdot (8 \cdot 7) = (15 \cdot 8) \cdot 7$
5. $(0.7 + 0.9) + 0.2 = 0.7 + (0.9 + 0.2)$
6. $\frac{3}{4} \cdot (\frac{4}{3} \cdot 10) = (\frac{3}{4} \cdot \frac{4}{3}) \cdot 10$
7. $2\frac{1}{2} + 3\frac{1}{8} = 3\frac{1}{8} + 2\frac{1}{2}$
8. $59 + (18 + 2) = (59 + 18) + 2$
9. $(\frac{1}{8} + \frac{1}{2}) + \frac{3}{4} = \frac{1}{8} + (\frac{1}{2} + \frac{3}{4})$
10. $(1\frac{1}{3} + 2\frac{1}{3}) + 5 = 5 + (1\frac{1}{3} + 2\frac{1}{3})$

Show that each statement is true when $a = 6$, $b = 1.6$, $c = 7$, and $d = 3.2$. Also name the property illustrated.

SAMPLE: $a + d = d + a$

Solution:
$$a + d = d + a$$

$6 + 3.2$	$3.2 + 6$
9.2	9.2

Commutative property of addition

11. $a \cdot c = c \cdot a$
12. $(a \cdot d) \cdot c = a \cdot (d \cdot c)$
13. $bc = cb$
14. $a \cdot d = a \cdot d$
15. $(b \cdot c)a = b(c \cdot a)$
16. $a \cdot \frac{d}{b} = \frac{d}{b} \cdot a$
17. $(cd)a = a(cd)$
18. $c + d = d + c$
19. $(bd)c = b(dc)$
20. $(bd)a = b(da)$
21. $(a + c)d = d(a + c)$
22. $b + (ac) = (ac) + b$
23. $b + (a + c) = (a + c) + b$
24. $\frac{b}{d} + c = c + \frac{b}{d}$
25. $(a + c) + d = a + (c + d)$
26. $a + (b + c) = a + (c + b)$

168 Chapter 6

Justify each lettered step in the problem by naming either a property of addition or multiplication, or the substitution principle.

SAMPLE:

$6 + (9 + 4) = 6 + (4 + 9)$
$= (6 + 4) + 9$
$ 10 + 9$
$= 19$

Solution:

Commutative property of addition
Associative property of addition
Substitution principle
Substitution principle

27. $12 + (15 + 6) = (15 + 6) + 12$
$= 15 + (6 + 12)$
$= 15 + 18$
$= 33$

28. $8 + (12 + 6) = 8 + (6 + 12)$
$= (8 + 6) + 12$
$= 14 + 12$
$= 26$

29. $10 \cdot (5 \cdot 7) = 10 \cdot (7 \cdot 5)$
$= (10 \cdot 7) \cdot 5$
$= 70 \cdot 5$
$= 350$

30. $4 \cdot (9 \cdot 5) = (9 \cdot 5) \cdot 4$
$= 9 \cdot (5 \cdot 4)$
$= 9 \cdot 20$
$= 180$

Replace each ? with $=$ or $\neq$ to make true statements. The replacement set for each variable is {the numbers of arithmetic}.

[C] **31.** $34 + 19 \; ? \; 19 + 43$
32. $a + b \; ? \; b + a$
33. $(m + n) + k \; ? \; m + (n + k)$
34. $(m \cdot n)k \; ? \; m(n \cdot k)$
35. $\frac{3 + 9}{6} + 5 \; ? \; 5 + \frac{3 + 9}{6}$
36. $(3 + 4) - 5 \; ? \; 5 - (3 + 4)$
37. $18 \div 6 \; ? \; 6 \div 18$

38. $(3 + 7) \cdot 6 \; ? \; (3 + 6) \cdot 7$
39. $xy \; ? \; yx$
40. $3 + (4 \cdot 5) \; ? \; 4 + (3 \cdot 5)$
41. $(a + b)c \; ? \; c(a + b)$
42. $20 - 6 \; ? \; 6 - 20$
43. $2^2 \div 4 \; ? \; 4 \div 2^2$
44. $(8 - 5) - 3 \; ? \; (8 - 4) - 4$

6–4 The Distributive Property

You have frequently used the formula $P = 2l + 2w$ to calculate the perimeter of a rectangle. In the case of the rectangle in Figure 6–3 the measure of the perimeter is 40 inches.

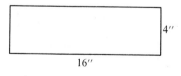

Figure 6–3

$P = 2l + 2w$
$ = (2 \cdot 4) + (2 \cdot 16)$
$ = 8 + 32$
$ = 40 \text{ (inches)}$

Do you suppose the formula $P = 2(l + w)$ will give the same result?

$$P = 2(l + w)$$
$$= 2(4 + 16)$$
$$= 2(20)$$
$$= 40 \text{ (inches)}$$

You have often used the **distributive property** in ordinary arithmetic, although you may not have been aware of it. Study these methods for completing the multiplication problem $34 \cdot 2$.

$$\begin{array}{r} 34 \\ \times 2 \\ \hline 68 \end{array} \qquad \begin{array}{r} 30 + 4 \\ \times 2 \\ \hline 60 + 8 = 68 \end{array} \qquad \begin{array}{rl} 2(30 + 4) &= (2 \cdot 30) + (2 \cdot 4) \\ &= 60 + 8 \\ &= 68 \end{array}$$

In each method, multiplication by 2 is **distributed** over **addition**. That is, each term of the sum $30 + 4$ is multiplied by 2. The axiom for the **distributive property** is stated as follows:

For every number *r*, every number *s*, and every number *t*,

$$r(s + t) = rs + rt.$$

As you have probably realized, the commutative property allows us to write both

$$r(s + t) = (s + t)r$$
and
$$rs + rt = sr + tr.$$

Hence we can use the substitution principle and state the axiom for the **distributive property** this way, also:

For every number *r*, every number *s*, and every number *t*,

$$(s + t)r = sr + tr.$$

ORAL EXERCISES

Tell how to complete each sentence to illustrate the distributive property.

SAMPLE: $8(2 + 7) = \underline{\ ?\ }$ *What you say:* $8(2 + 7) = 8 \cdot 2 + 8 \cdot 7$

1. $6(5 + 3) = \underline{\ ?\ }$ 2. $\underline{\ ?\ } = 5(8 + 2)$

3. $3(9 + 7) = \underline{\ ?\ }$
4. $10(6 + 8) = \underline{\ ?\ }$
5. $12(2 + 5) = \underline{\ ?\ }$
6. $(4 + 5)8 = \underline{\ ?\ }$
7. $(3 + 7)6 = \underline{\ ?\ }$
8. $\underline{\ ?\ } = 2 \cdot 5 + 2 \cdot 9$
9. $\underline{\ ?\ } = 8 \cdot 12 + 8 \cdot 6$

10. $\underline{\ ?\ } = 10(9 + 12)$
11. $\underline{\ ?\ } = (15 + 7)3$
12. $a(b + c) = \underline{\ ?\ }$
13. $(x + y)t = \underline{\ ?\ }$
14. $\underline{\ ?\ } = mk + mb$
15. $\underline{\ ?\ } = 6 \cdot 5 + 8 \cdot 5$
16. $7 \cdot 9 + 6 \cdot 9 = \underline{\ ?\ }$

WRITTEN EXERCISES

Show that each statement is true.

SAMPLE 1: $8(9 + 6) = 8 \cdot 9 + 8 \cdot 6$ Solution: $8(9 + 6) = 8 \cdot 9 + 8 \cdot 6$

$$\begin{array}{c|c} 8 \cdot 15 & 72 + 48 \\ 120 & 120 \end{array}$$

SAMPLE 2: $3(4 + \tfrac{1}{2}) = 3 \cdot 4 + 3 \cdot \tfrac{1}{2}$ Solution: $3(4 + \tfrac{1}{2}) = 3 \cdot 4 + 3 \cdot \tfrac{1}{2}$

$$\begin{array}{c|c} 3(\tfrac{9}{2}) & 12 + \tfrac{3}{2} \\ \tfrac{27}{2} & 12 + 1\tfrac{1}{2} \\ 13\tfrac{1}{2} & 13\tfrac{1}{2} \end{array}$$

1. $4 \cdot (8 + 6) = 4 \cdot 8 + 4 \cdot 6$
2. $8 \cdot 7 + 8 \cdot 2 = 8(7 + 2)$
3. $9 \cdot 5 + 6 \cdot 5 = (9 + 6)5$
4. $(10 + 3)6 = 10 \cdot 6 + 3 \cdot 6$
5. $7(4 + 0) = 7 \cdot 4 + 7 \cdot 0$
6. $1(9 + 8) = 1 \cdot 9 + 1 \cdot 8$

7. $10 \cdot 6 + 10 \cdot 8 = 10(6 + 8)$
8. $5(2 + \tfrac{1}{8}) = 5 \cdot 2 + 5 \cdot \tfrac{1}{8}$
9. $6(4 + \tfrac{1}{10}) = 6 \cdot 4 + 6 \cdot \tfrac{1}{10}$
10. $\tfrac{1}{4}(8 + 12) = \tfrac{1}{4} \cdot 8 + \tfrac{1}{4} \cdot 12$
11. $\tfrac{2}{3} \cdot 9 + \tfrac{2}{3} \cdot 6 = (9 + 6)\tfrac{2}{3}$
12. $6(3 + 0.4) = 6 \cdot 3 + 6 \cdot 0.4$

Apply the distributive property to complete each problem. Use the substitution principle as shown in the sample.

SAMPLE: $8 \cdot 125$ Solution: $8 \cdot 125 = 8(100 + 20 + 5)$
$= 8 \cdot 100 + 8 \cdot 20 + 8 \cdot 5$
$= 800 + 160 + 40$
$= 1000$

13. $6 \cdot 315$
14. $8 \cdot 421$
15. $3 \cdot 244$
16. $7 \cdot 361$

17. $231 \cdot 4$
18. $904 \cdot 8$
19. $306 \cdot 3$
20. $5 \cdot 1342$

21. $10 \cdot 236$
22. $2741 \cdot 6$
23. $8426 \cdot 2$
24. $3005 \cdot 4$

Operations, Axioms, and Equations

Replace each ? with = or ≠ to make true statements. The replacement set for each variable is {the numbers of arithmetic}.

SAMPLE: (80 + 3)4 ? 80 · 4 + 80 · 3

Solution: (80 + 3)4 ≠ 80 · 4 + 80 · 3

B
25. 5(90 + 6) ? 5 · 90 + 5 · 6
26. 3(5 + $\frac{1}{8}$) ? 3 · 5 + 3 · $\frac{1}{8}$
27. 10(3 + 9) ? 10 · 3 + 9 · 3
28. 6 · $7\frac{1}{10}$? 6 · 7 + 6 · $\frac{1}{10}$
29. $4\frac{1}{5}$ · 3 ? 4.3 + $\frac{1}{5}$ · 3
30. 15(7 + 0.8) ? (0.8 + 7)15
31. 10 · 7 + 9 · 7 ? (7 + 10)9
32. $x(3 + 2 + 5)$? $3x + 2x + 5x$
33. 18 · 9 + 18 · 5 ? 5(9 + 18)
34. $a(b + c)$? $ab + ac$

C
35. $(x + y)t$? $xt + xy$, and $t \neq x$
36. $(p + q)m$? $qm + pm$, and $p \neq m$
37. $ak + tk$? $k(a + t)$
38. $g(a + b + c)$? $ga + gb + gc$
39. $(m + t + r)z$? $zm + zt + zr$
40. $r(st + mn)$? $rst + rmn$

Justify each step by naming the substitution principle, the distributive property, or some other property.

41. 8 · 15 = 8(10 + 5)
    = 8 · 10 + 8 · 5
    = 80 + 40
    = 120

42. 16 · 7 = (10 + 6)7
    = 7(10 + 6)
    = 7 · 10 + 7 · 6
    = 70 + 42
    = 112

43. 12 · 8 = (10 + 2)8
    = 10 · 8 + 2 · 8
    = 80 + 16
    = 96

44. 5 · 423 = 5(400 + 20 + 3)
    = (400 + 20 + 3)5
    = (400)5 + (20)5 + (3)5
    = 2000 + 100 + 15
    = 2115

6–5 Other Applications of the Distributive Property

We have seen that the two basic operations, addition and multiplication, are related through the distributive property. Now we consider the idea of **multiplication distributed** over **subtraction**, as illustrated when $30 - 2$ is substituted for 28.

$$\begin{array}{r} 28 \\ \times\, 3 \\ \hline 84 \end{array} \qquad \begin{array}{r} 30 - 2 \\ \times\, 3 \\ \hline 90 - 6 = 84 \end{array} \qquad \begin{aligned} 3(30 - 2) &= (3 \cdot 30) - (3 \cdot 2) \\ &= 90 - 6 \\ &= 84 \end{aligned}$$

We state the axiom for the **distributive property of multiplication over subtraction** as:

For each **r**, each **s**, and each **t**,

$$r(s - t) = rs - rt.$$

Another useful application of the distributive property relates the operations of **division** and **addition**. The following problem illustrates the distributive property for division over addition.

$$\begin{array}{r}32\\4\overline{)128}\end{array} \qquad \begin{array}{r}25 + 5 + 2 = 32\\4\overline{)100 + 20 + 8}\end{array}$$

$$\frac{100 + 20 + 8}{4} = 25 + 5 + 2 = 32$$

Study the following method for solving the division problem $384 \div 12$, and note the use of the **distributive property**.

[Step 1]
$$\begin{array}{r}30\\10 + 2\overline{)300 + 80 + 4}\\300 + 60\\\hline 20 + 4\end{array}$$

[Step 2]
$$\begin{array}{r}30 + 2\\10 + 2\overline{)300 + 80 + 4}\\300 + 60\\\hline 20 + 4\\20 + 4\\\hline\end{array}$$

ORAL EXERCISES

Tell how to complete each sentence to illustrate the distributive property.

SAMPLE 1: $\dfrac{200 + 80 + 4}{4} = \underline{}$

What you say: $\dfrac{200 + 80 + 4}{4} = \dfrac{200}{4} + \dfrac{80}{4} + \dfrac{4}{4}$

SAMPLE 2: $6(20 - 1) = \underline{}$

What you say: $6(20 - 1) = 6 \cdot 20 - 6 \cdot 1$

1. $9(10 - 2) = \underline{}$
2. $6(30 - 1) = \underline{}$
3. $(15 - 3)5 = \underline{}$
4. $\dfrac{100 + 30 + 5}{5} = \underline{}$
5. $8(x - y) = \underline{}$
6. $(b - a)7 = \underline{}$

Operations, Axioms, and Equations   **173**

7. $\dfrac{300 + 60 + 3}{3} = \underline{\ ?\ }$ 11. $(3 - \tfrac{1}{8})4 = \underline{\ ?\ }$

8. $\dfrac{200 + 40 + 5}{5} = \underline{\ ?\ }$ 12. $\underline{\ ?\ } = \dfrac{300}{6} + \dfrac{60}{6} + \dfrac{6}{6}$

9. $\underline{\ ?\ } = 5 \cdot 12 - 5 \cdot 6$ 13. $6(5 - \tfrac{1}{4}) = \underline{\ ?\ }$

10. $\underline{\ ?\ } = 9 \cdot 30 - 9 \cdot 2$ 14. $4(9 - 0.2) = \underline{\ ?\ }$

WRITTEN EXERCISES

Show that each statement is true.

SAMPLE: $(20 - 2)3 = 3 \cdot 20 - 3 \cdot 2$

Solution: $(20 - 2)3 = 3 \cdot 20 - 3 \cdot 2$
$$\begin{array}{c|c} 18 \cdot 3 & 60 - 6 \\ 54 & 54 \end{array}$$

A
1. $7(10 - 2) = 7 \cdot 10 - 7 \cdot 2$
2. $9(30 - 1) = 9 \cdot 30 - 9 \cdot 1$
3. $(20 - 3)3 = 3 \cdot 20 - 3 \cdot 3$
4. $(6 - \tfrac{1}{5})4 = 4 \cdot 6 - 4 \cdot \tfrac{1}{5}$
5. $\dfrac{400 + 20 + 8}{4} = \dfrac{400}{4} + \dfrac{20}{4} + \dfrac{8}{4}$
6. $10(3 - \tfrac{1}{8}) = 10 \cdot 3 - 10 \cdot \tfrac{1}{8}$
7. $4(8 - 0.25) = 4 \cdot 8 - 4 \cdot 0.25$
8. $(6 - \tfrac{1}{3})2 = 2 \cdot 6 - 2 \cdot \tfrac{1}{3}$
9. $\dfrac{200 + 40 + 5}{5} = \dfrac{200}{5} + \dfrac{40}{5} + \dfrac{5}{5}$
10. $4(\tfrac{3}{4} - \tfrac{1}{8}) = 4 \cdot \tfrac{3}{4} - 4 \cdot \tfrac{1}{8}$

To complete each problem, use the **substitution principle** and one of the applications of the **distributive property** discussed in this section.

SAMPLE 1: $18 \cdot 3$ Solution: $18 \cdot 3 = (20 - 2)3$
$= 20 \cdot 3 - 2 \cdot 3$
$= 60 - 6 = 54$

SAMPLE 2: $448 \div 8$ Solution: $448 \div 8 = \dfrac{400 + 40 + 8}{8}$
$= \dfrac{400}{8} + \dfrac{40}{8} + \dfrac{8}{8}$
$= 50 + 5 + 1 = 56$

B
11. $4 \cdot 19$
12. $5 \cdot 28$
13. $48 \cdot 3$
14. $39 \cdot 6$
15. $96 \div 3$
16. $\dfrac{636}{3}$
17. $\dfrac{864}{2}$
18. $3 \cdot 99$
19. $230 \div 10$
20. $118 \cdot 4$
21. $3 \cdot 4\tfrac{7}{8}$
22. $5 \cdot 6\tfrac{9}{10}$
23. $4\tfrac{7}{8} \cdot 3$
24. $\dfrac{1488}{4}$
25. $5 \cdot 9.8$

Complete each division exercise in the manner indicated on Page 172.

[C] 26. $20 + 4 \overline{)200 + 80 + 8}$ 29. $10 + 2 \overline{)100 + 50 + 6}$
27. $10 + 4 \overline{)100 + 60 + 8}$ 30. $10 + 6 \overline{)100 + 70 + 6}$
28. $20 + 4 \overline{)500 + 70 + 6}$ 31. $30 + 2 \overline{)700 + 30 + 6}$
(Hint: rewrite the dividend as $400 + 160 + 16$.)

Other Number Properties

6–6 Properties of Zero and One

Suppose 0 is added to 5; the result is 5. Suppose 0 is added to $\frac{5}{8}$; the result is $\frac{5}{8}$. If you were to continue this process you would find that when zero and any number are added the **sum** is the other number. In the language of mathematics, zero is called the **additive identity element**. The following axiom states the **additive property of zero**.

For each number r, $r + 0 = 0 + r = r$.

Note that, while $5 - 0 = 5$ is **true**, $0 - 5 = 5$ is **false**.

To investigate the **multiplicative property** of **zero** let us consider these true statements.

$8 \cdot 0 = 0$ $\qquad$ $0 \cdot \frac{3}{4} = 0$ $\qquad$ $0 = 2\frac{3}{4} \cdot 0$ $\qquad$ $k \cdot 0 = 0$

Do you agree that if one of the factors of a product is **0**, the product must be also **0**?

For each number r, $r \cdot 0 = 0 \cdot r = 0$.

You will recall from previous work that division by **0** is not possible. You will understand this now if you recall that if there were a value for x such that $\frac{6}{0} = x$ were true, then the equivalent sentence $x \cdot 0 = 6$ would also be true. Do you agree that there is *no* replacement for x that makes $x \cdot 0 = 6$ a true statement? Hence the division of any **nonzero** number by **0** is said to be **undefined**.

Operations, Axioms, and Equations   175

The statement that $\frac{0}{0}$ is also undefined follows from a different reason. If $\frac{0}{0} = x$ were true for some value of x, an equivalent sentence would be $x \cdot 0 = 0$. Do you agree that this is true for *every* number of arithmetic that might replace x? This time we have *too many* answers, so we say that $\frac{0}{0}$ is also **undefined**.

The following examples illustrate the **multiplicative property of one**.

$$35 \cdot 1 = 35 \qquad 1 \cdot \frac{5}{8} = \frac{5}{8} \qquad m \cdot 1 = m$$

That is, if **1** and any number are multiplied, the **product** is the other number. For this reason, **1** is called the **multiplicative identity element**.

For each number r, $r \cdot 1 = 1 \cdot r = r$.

Note that, while $18 \div 1 = 18$ is **true**, $1 \div 18 = 18$ is **false**.

ORAL EXERCISES

Tell whether each numeral names the number **1** or the number **0**.

1. $0 + 0$
2. $1 \cdot 0$
3. $0 + 1$
4. $1 - 0$
5. $\frac{0}{1}$
6. $6 \cdot 0$
7. $0 \cdot 15$
8. $\frac{357}{357}$
9. $\frac{2+4}{2+4}$

Name the solution set for each open sentence. The replacement set for each variable is {the numbers of arithmetic}.

SAMPLE: $n \cdot 1 = \frac{3}{4}$ *What you say:* $\{\frac{3}{4}\}$

10. $1 \cdot 35 = k$
11. $\frac{4}{5} \cdot 1 = t$
12. $m = 0 + 12$
13. $r = 1 \cdot \frac{3}{4}$
14. $18 \cdot 0 = w$
15. $0 \cdot \frac{7}{8} = z$
16. $0 + \frac{6}{10} = x$
17. $b + 0 = 8$
18. $0 + f = \frac{1}{2}$
19. $c \cdot 1 = 26$
20. $m = 15 \cdot 1$
21. $\frac{1}{6} = t \cdot 1$
22. $1 \cdot y = \frac{2}{5}$
23. $\frac{d}{1} = 0.35$
24. $p - 0 = 1.6$
25. $0(3 + 5) = a$
26. $k \cdot 14 = 14$
27. $\frac{3}{3} \cdot 45 = x$

WRITTEN EXERCISES

Write the solution set for each open sentence. The replacement set for each variable is {the whole numbers}.

A
1. $m \cdot 1.50 = 1.50$
2. $\frac{3}{4} + r = \frac{3}{4}$
3. $65 \div s = 65$
4. $8.9 = 8.9 - n$
5. $0(5 + 7) = a$
6. $0.0375 + d = 0.0375$
7. $x(1 + 2.5) = 0$
8. $6.25 - y = 6.25$
9. $\frac{5}{5} \cdot b = 14$
10. $\frac{12}{3} \cdot \frac{100}{100} = w$
11. $4^2 \cdot 1 = t$
12. $0 \cdot 15^2 = z$
13. $0(3 + 4 + 5) = y$
14. $(10 \div 10) \cdot 7 = x$
15. $k \cdot \frac{4}{4} = (3 + 6)$
16. $(10 - 3)\frac{7}{7} = q$

B
17. $(3 + 9) + m = (3 + 9)$
18. $y - 0 = (10 + 14)$
19. $n + 0 = n$
20. $1 \cdot r = r$
21. $0 + n = n + 0$
22. $b + 0 \neq b$
23. $\frac{10}{10} \cdot r = 15^2$
24. $(5 + 0)(1)^2 = k$
25. $0 \div n = n$
26. $s - 0 = s$
27. $m \cdot 1 \neq m$
28. $0 - r = r$

C
29. $\frac{m}{m} + 0 = 1$
30. $\frac{5}{5} \cdot a \cdot a = 36$
31. $0 + \frac{x}{x} = 1$
32. $0 + \frac{r}{r} = 3$
33. $b^2 \cdot \frac{4}{4} = 25$
34. $y^2 \cdot 5^2 = 25$
35. $(x \cdot x \cdot x) + 1 = 2$
36. $\frac{b}{b} \cdot 1 = 5$
37. $(15 + 0)k^2 = 3$
38. $0 + \frac{x}{x} = 1 + 0$

6–7 Functions and Variables

You will recall that earlier we worked with a function machine and used function rules to write pairs of numbers. A set of number pairs is **a function** when **no two different pairs** have the **same first number**. For the machine in Figure 6–4, the $f(x)$ values [recall that $f(x)$ means "the value of the function for x"] are found by adding **5** to each value for x.

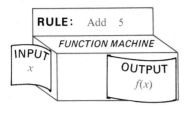

Figure 6-4

x	$f(x) = x + 5$	$(x, f(x))$
0	5	(0, 5)
1	6	(1, 6)
2	7	(2, 7)
3	8	(3, 8)
4	9	(4, 9)

The set of number pairs, taken from the table above, is the function **{(0, 5), (1, 6), (2, 7), (3, 8), (4, 9)}**

Suppose our function machine operates according to the rule **$3x + 2$**, as shown in Figure 6–5. This fact can be represented by the function equation $f(x) = 3x + 2$. Observe the accompanying table. How would you fill in the missing entries using the given replacements for x?

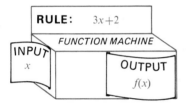

Figure 6-5

x	$f(x) = 3x + 2$	$(x, f(x))$
0	2	(0, 2)
2	8	(2, 8)
4	14	?
6	?	?

Do you see that the set of number pairs developed in the table is the function **{(0, 2), (2, 8), (4, 14), (6, 20)}** ?

Let us consider the function machine in Figure 6–6 which operates according to the rule "square." In this case let us use as the replacement set {the odd numbers}.

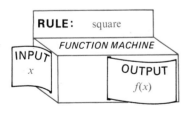

Figure 6-6

x	$f(x) = x^2$	$(x, f(x))$
1	1	(1, 1)
3	9	(3, 9)
5	25	(5, 25)
7	49	(7, 49)
⋮	⋮	⋮

Do you agree that the set of number pairs developed by this machine is a function? Do any two number pairs have the same first number? Thus the function is **{(1, 1), (3, 9), (5, 25), (7, 49), (9, 81), ...}**

ORAL EXERCISES

Use the function machine and tell how to fill in the accompanying table.

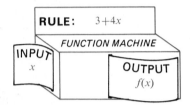

	x	$f(x) = 3 + 4x$	$(x, f(x))$
1.	0	3	?
2.	$\frac{1}{2}$	5	?
3.	1	?	?
4.	$1\frac{1}{2}$	?	?
5.	2	?	?
6.	$2\frac{1}{2}$	?	?

Tell how to complete each table according to the function rule.

Rule: $x(2 + 3)$

	x	$f(x)$	$(x, f(x))$
7.	0	0	?
8.	2	10	?
9.	4	20	?
10.	6	?	?
11.	8	?	?
12.	10	?	?

Rule: $3(n + 2)$

	n	$f(n)$	$(n, f(n))$
13.	1	?	?
14.	3	?	?
15.	5	21	?
16.	7	?	?
17.	9	?	?
18.	12	?	?

WRITTEN EXERCISES

Use the function machine and the function rule to complete each table for the given replacement set for the variable.

A

	m	$f(m) = m^2$	$(m, f(m))$
1.	0	0	(0, 0)
2.	1	1	(1, 1)
3.	2	4	(2, 4)
4.	3	9	?
5.	4	?	?
6.	5	?	?
7.	6	?	?
8.	7	?	?

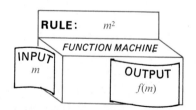

Replacement set: {0, 1, 2, 3, 4, 5, 6, 7}

Operations, Axioms, and Equations   179

	t	$f(t) = t \cdot \frac{1}{t}$	$(t, f(t))$
9.	5	1	(5, 1)
10.	10	1	(10, 1)
11.	15	1	?
12.	20	?	?
13.	25	?	?
14.	30	?	?
15.	35	?	?
16.	40	?	?

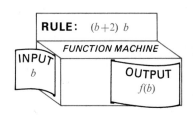

RULE: $t \cdot \frac{1}{t}$

Replacement set:
{5, 10, 15, 20, 25, 30, 35, 40}

	b	$f(b) = (b + 2)b$	$(b, f(b))$
17.	0	?	?
18.	2	?	?
19.	4	24	(4, 24)
20.	6	48	(6, 48)
21.	8	?	?
22.	10	?	?
23.	12	?	?
24.	14	?	?

RULE: $(b+2)b$

Replacement set:
{0, 2, 4, 6, 8, 10, 12, 14}

Use the function rule and the given replacement set to write the set of ordered pairs that represents the function. Graph the function on a lattice of the dimensions suggested.

SAMPLE: Rule: r^2
Replacement set: {0, 1, 2}
4 × 4 lattice

Solution: {(0, 0), (1, 1), (2, 4)}

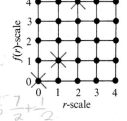

B 25. Rule: $6 - x^2$
Replacement set: {0, 1, 2}
6 × 6 lattice

26. Rule: $\frac{t}{2} + \frac{1}{2}$
Replacement set: {1, 3, 5, 7}
7 × 7 lattice

27. Rule: $s^2 + 2$
Replacement set: {0, 1, 2}
6 × 6 lattice

28. Rule: $(w \cdot w)\dfrac{1}{w}$
Replacement set: $\{1, 2, 3, 4, 5\}$
5×5 lattice

29. Rule: $0 + \dfrac{x}{x}$
Replacement set: $\{1, 2, 3, 4, 5\}$
5×5 lattice

30. Rule: $\dfrac{3}{3} \cdot \dfrac{z}{z}$
Replacement set: $\{1, 2, 3, 4, 5\}$
5×5 lattice

31. Rule: $m^2 - m$
Replacement set: $\{0, 1, 2, 3\}$
6×6 lattice

32. Rule: $\dfrac{1}{3} + \dfrac{k}{3}$
Replacement set: $\{2, 5, 8\}$
8×8 lattice

Use the given equation and replacement set to make a table of values of the variable, function of the variable, and the resulting number pairs.

SAMPLE: $2m + 2 = f(m)$ Solution:
$\{0, 2, 4, 6\}$

m	$f(m) = 2m + 2$	$(m, f(m))$
0	2	(0, 2)
2	6	(2, 6)
4	10	(4, 10)
6	14	(6, 14)

33. $r - 4 = f(r)$
$\{4, 5, 6, 7, 8\}$

34. $2d + 5 = f(d)$
$\{0, 1, 2, 3, 4, 5\}$

35. $17 - n^2 = f(n)$
$\{0, 1, 2, 3, 4\}$

36. $3x + \tfrac{1}{2} = f(x)$
$\{0, \tfrac{1}{2}, 1, \tfrac{3}{2}, 2, \tfrac{5}{2}\}$

37. $2a \cdot \tfrac{1}{3} = f(a)$
$\{0, 1, 2, 3, 4, 5\}$

38. $(m + 3)\tfrac{1}{2} = f(m)$
$\{1, 3, 5, 7, 9, 11\}$

Use the given equation as a function rule to complete the accompanying set of ordered pairs.

SAMPLE: $5x + 2 = f(x)$; $\{(1, 7), (2, 12), (3, ?), (4, ?), (5, ?)\}$

Solution: $\{(1, 7), (2, 12), (3, 17), (4, 22), (5, 27)\}$

[C] **39.** $(a^2 + 2) = f(a)$; $\{(0, 2), (1, 3), (2, 6), (3, ?), (4, ?), (5, ?)\}$
40. $(2m + m) = f(m)$; $\{(5, 15), (7, 21), (8, ?), (10, ?), (11, ?), (13, ?)\}$

41. $t^2 - t = f(t)$; $\{(0, 0), (2, 2), (4, ?), (6, ?), (8, ?), (10, ?)\}$
42. $10a - a^2 = f(a)$; $\{(0, ?), (1, ?), (2, ?), (3, ?), (4, ?), (5, ?), (6, ?)\}$
43. $(x + \pi) = f(x)$ for $\pi = 3.14$;
    $\{(0, 3.14), (1, 4.14), (2, ?), (3, ?), (4, ?), (5, ?)\}$
44. $(3 + m) - m = f(m)$;
    $\{(1, 3), (3, 3), (5, ?), (7, ?), (9, ?), (11, ?), (13, ?)\}$
45. $x + \dfrac{1}{x} = f(x)$; $\{(5, 5\tfrac{1}{5}), (6, 6\tfrac{1}{6}), (7, ?), (8, ?), (9, ?), (10, ?)\}$
46. $3(m^2 + 1) = f(m)$;
    $\{(0, 3), (2, 15), (4, ?), (6, ?), (8, ?), (10, ?), (12, ?)\}$

CHAPTER SUMMARY

Inventory of Structure and Concepts

1. The many fundamental ideas and assumptions that are simply accepted and used in mathematics are called **axioms** or **postulates**.

2. The **basic properties** of **equality** include the following:

   a. The **reflexive property of equality**: a number is equal to itself. In symbols, $t = t$.
   b. The **symmetric property of equality**: the left and right members of an equation can be interchanged without changing the meaning of the equation. In symbols, if $a = b$, then $b = a$.
   c. The **transitive property of equality**: when two expressions are equal to some third expression the first two expressions are equal to each other. In symbols, if $a = b$ and $b = c$, then $a = c$.

3. The **existence** and **uniqueness** properties deal with these assumptions for operations involving numbers of arithmetic:

   When the addition operation is performed on two numbers there is **exactly one number** that is equal to their **sum**.
   When the multiplication operation is performed on two numbers there is **exactly one number** that is equal to their **product**.

4. The **closure property** states that:

   A given set of numbers is **closed** under an operation performed on its elements if each result is also an element of the given set.

5. The following axioms describe properties of basic operations for the numbers of arithmetic.

   Addition:

   The **commutative property**: for every number r and every number s, $r + s = s + r$.

   The **associative property**: for every number r, every number s, and every number t, $r + (s + t) = (r + s) + t$.

   The **closure property**: for every arithmetic number r and every arithmetic number s, $r + s$ is an arithmetic number.

   The **additive property of zero**: for every number r, $0 + r = r + 0 = r$.

   Multiplication:

   The **commutative property**: for every number r and every number s, $rs = sr$.

   The **associative property**: for every number r, every number s, and every number t, $r(st) = (rs)t$.

   The **closure property**: for every arithmetic number r and every arithmetic number s, rs is an arithmetic number.

   The **multiplicative property of one**: for every number r, $1 \cdot r = r \cdot 1 = r$.

   Multiplication and Addition:

   The **distributive property** relates multiplication and addition; for each r, each s, and each t, $r(s + t) = rs + rt$.

6. The **substitution principle** states that if two different numerals represent the same number, either one may be substituted for the other.

7. Division by 0 is not possible.

Vocabulary and Spelling

axiom (*p. 157*)
postulate (*p. 157*)
properties of equality (*p. 158*)
reflexive property (*p. 158*)
symmetric property (*p. 158*)
transitive property (*p. 158*)
existence (*p. 161*)
uniqueness (*p. 161*)
closure property (*p. 161*)
substitution principle (*p. 161*)

commutative property (*p. 164*)
associative property (*p. 165*)
distributive property (*p. 169*)
additive identity element (*p. 174*)
additive property of zero (*p. 174*)
multiplicative identity element (*p. 174*)
multiplicative property of zero (*p. 174*)
multiplicative property of one (*p. 175*)
function (*p. 176*)

Chapter Test

Match each of the statements or equations in Column 1 with the appropriate property listed in Column 2. You may use a property more than once.

COLUMN 1

1. $3 + 2 = 2 + 3$
2. If $2 + n = 1$ then $1 = 2 + n$.
3. $3(x + y) = 3x + 3y$
4. $2(xy) = (2x) \cdot y$
5. $0 + n = n$
6. $15 = 15$
7. $13 + (5 + 2) = 13 + (2 + 5)$
8. $2n + 5n = (2 + 5) \cdot n$
9. $0 \cdot n = 0$
10. If $a = b$ and $b = c$, then $a = c$.
11. An even number + an even number = an even number.
12. $5(3 + 2) = (3 + 2) \cdot 5$
13. If $a = c + d$ and $c = 5$, then $a = 5 + d$.
14. $1(x + y) = x + y$
15. $3(12 - 9) = 3 \cdot 12 - 3 \cdot 9$
16. $13 + (5 + 2) = (13 + 5) + 2$
17. $\frac{3}{4} \cdot \frac{5}{5} = \frac{3}{4}$
18. $a(x + y) = (x + y)a$
19. $\pi + (3 - 3) = \pi$
20. If $5\frac{2}{3} = y$, then $y = 5\frac{2}{3}$.

COLUMN 2

A. Reflexive property of equality
B. Symmetric property of equality
C. Transitive property of equality
D. Closure under addition
E. Closure under multiplication
F. Substitution principle
G. Commutative property of multiplication
H. Associative property of multiplication
I. Commutative property of addition
J. Associative property of addition
K. Distributive property
L. Multiplicative property of one
M. Additive property of zero
N. Multiplicative property of zero

Using the given equation and replacement set, make a table showing values of the variable, values of the function of the variable, and the resulting number pairs.

21. $12 - x^2 = f(x)$; $\{0, 1, 2, 3\}$

Use the given equation as a function rule to complete the accompanying set of ordered pairs.

22. $f(t) = \frac{t}{2}(10 - t)$; $\{(0, ?), (2, ?), (4, ?), (10, ?)\}$

Chapter Review

6–1 Basic Operations and Axioms

In each of the following problems state which property of equality is used.

1. $a + b = a + b$
2. If $n + 2 = 5$, then $5 = n + 2$.
3. If $a = x - y$ and $x - y = b$, then $a = b$.
4. If $12 \cdot (7 - 2) = x$, then $x = 12 \cdot (7 - 2)$.
5. If $\frac{y}{3} = 4a$ and $12 = 4a$, then $4a = 12$.
6. If $15 = 3n$ and $3n = 2x$, then $15 = 2x$.

6–2 The Existence and Uniqueness Property

7. Because $2 + 7$ always equals 9 we can say the sum of 2 and 7 is __?__.

Given the following sets:

$A = \{0\}$ $B = \{1\}$ $C = \{0, 1\}$ $D = \{4, 8, 12, \ldots\}$
$E = \{\text{Odd numbers}\}$ $F = \{\text{Prime numbers}\}$ $G = \{\frac{1}{10}, \frac{1}{20}, \frac{1}{30}, \ldots\}$
$H = \{1, 2\}$ $I = \{\frac{0}{4}, \frac{1}{4}, \frac{2}{4}, \frac{3}{4}, \frac{4}{4}, \frac{5}{4}, \ldots\}$

8. Which of the above sets are closed under addition?
9. Which of the above sets are closed under subtraction?
10. Which of the above sets are closed under multiplication?
11. Which of the above sets are closed under division?
12. If $a = b + c$, and $b = xy$, then $a = xy + c$ is an example of the __?__ principle.

6–3 Commutative and Associative Properties

Name the property that is illustrated by each of the following examples.

13. $a + b = b + a$
14. $b(x + y) = b(y + x)$
15. $a(bc) = (ab)c$
16. $a(xy) = (xy)a$
17. $(13 + a) + 2 = 13 + (a + 2)$
18. $5 \cdot 3 + 7 = 3 \cdot 5 + 7$

Which of the following statements are true?

19. $(3\frac{3}{4} + \frac{1}{2}) + \frac{5}{8} = 3\frac{3}{4} + (\frac{1}{2} + \frac{5}{8})$
20. $2^3 = 3^2$
21. $2^5 \cdot 3^4 = 3^4 \cdot 2^5$
22. $4 + (8 \cdot 2) = (4 + 8) \cdot 2$

6–4 The Distributive Property

Complete the following by applying the distributive property.

23. $(10 + 18) \cdot 7 = ?$
24. $ab + ac = ?$
25. $2x + 2y = ?$
26. $3n + 5n = ?$
27. $(a + b) \cdot 5 = ?$
28. $2 \cdot 3 \cdot 4 + 2 \cdot 3 \cdot 5 = ?$
29. Is addition distributive over multiplication? Give an example to justify your response.

6–5 Other Applications of the Distributive Property

Find the product or quotient of each by applying the distributive property.

SAMPLE: $98 \cdot 4$ Solution: $98 \cdot 4 = (100 - 2) \cdot 4$
$= 100 \cdot 4 - 2 \cdot 4 = 400 - 8 = 392$

30. $53 \cdot 3$
32. $6 \cdot 49$
34. $6 \cdot 4\frac{1}{2}$
36. $8\frac{1}{4} \cdot 8$
31. $\dfrac{40 + 16}{8}$
33. $\dfrac{70 + 35}{7}$
35. $\frac{17}{4} + \frac{7}{4}$
37. $\frac{15}{9} + \frac{3}{9}$

6–6 Properties of Zero and One

Write the solution set for each open sentence. The replacement set for each variable is {the whole numbers}.

38. $63.5 - n = 63.5$
41. $54 \cdot n = 0$
44. $(63.5)n = 63.5$
39. $n + 0 = n$
42. $0 \cdot n = n$
45. $\frac{54}{54}(n + 1) = 1$
40. $\dfrac{n}{n} \cdot 14 = 14$
43. $15 \cdot n = 3$
46. $\dfrac{0}{n} = 0$

6–7 Functions and Variables

47. Using the function rule $2x - 1$ and the replacement set $\{1, 2, 3\}$, graph the function on a 5 × 5 lattice.

Using the given equation and replacement set, make a table showing values of the variable, values of the function of the variable, and the resulting number pairs.

48. $f(n) = n^2 - 2n;$ $\{0, 2, 3, 4, 5\}$

Use the given equation as a function rule to complete the accompanying set of ordered pairs.

49. $(x + 1)^2 = f(x);$ $\{(0, ?), (1, ?), (2, ?), (10, ?)\}$

Review of Skills

For each of the following, find a whole number value of n which will make the sentence true.

1. $n = 5 + 11$
2. $15 = n + 3$
3. $14 - n = 3$
4. $42 = 7n$
5. $3n + 4 = 13$
6. $2n + n = 12$
7. $3 + n = 0$
8. $30 - n = 0$
9. $\dfrac{n}{4} + 5 = 10$
10. $n = 15 - 9$
11. $n + 12 = 17$
12. $17 = 25 - n$
13. $32 - 2n = 8$
14. $3n - n = 12$
15. $3n = 0$
16. $3n = 12 - n$
17. $\dfrac{n}{4} = 9$
18. $\dfrac{24}{n} = 3$
19. $n = 4 \cdot 5$
20. $20 + n = 28$
21. $3n = 12$
22. $10 - 5n = 0$
23. $2n = n + 7$
24. $3(n - 1) = 0$
25. $6n - 2 = n + 8$
26. $\dfrac{n}{4} = 0$
27. $\dfrac{n + 2}{5} = 3$

For each of the following, name the property (commutative, distributive, associative, or multiplicative identity) that is illustrated.

28. $2 + 5 = 5 + 2$
29. $12 \cdot 5 = 5 \cdot 12$
30. $n + 5 = 5 + n$
31. $15 + (3 - 2)n = 15 + 3n - 2n$
32. $5(12 - 3) = 5 \cdot 12 - 5 \cdot 3$
33. $(2 + 5) + 4 = 2 + (5 + 4)$
34. $3n + 2n = (3 + 2)n$
35. $3(5n) = (3 \cdot 5)n$

■ ■

**CHECK POINT
FOR EXPERTS**

Properties of Operations

The familiar tables on page 187 show some of the basic addition and multiplication facts for whole numbers. By studying such tables, we can discover some of the properties of these and other operations. For example, the table can tell us at a glance whether or not an operation is commutative, and whether or not it has an identity element.

+	0	1	2	3	4	5
0	0	1	2	3	4	5
1	1	2	3	4	5	6
2	2	3	4	5	6	7
3	3	4	5	6	7	8
4	4	5	6	7	8	9
5	5	6	7	8	9	10

×	0	1	2	3	4	5
0	0	0	0	0	0	0
1	0	1	2	3	4	5
2	0	2	4	6	8	10
3	0	3	6	9	12	15
4	0	4	8	12	16	20
5	0	5	10	15	20	25

A diagonal has been drawn across each of the tables from the upper left-hand corner to the lower right-hand corner. Notice that if the table were to be cut out and folded along the diagonal, corresponding entries on either side of the diagonal would come together. Thus we say that the table is **symmetric about the diagonal**. The operation shown in such a table is **commutative** only if the table entries are **symmetric about the diagonal**.

In the addition table, observe that the **first row** of entries is exactly like the row of column headings, and the **first column** of entries is exactly like the row headings. Since **0** is the table heading for both the **first row** of entries and the **first column** of entries, we see that **0** is the **identity element for addition**.

In the multiplication table, which column has entries that duplicate the row headings? Which row has entries that duplicate the column headings? What number is the table heading for the **second column** of entries? for the **second row** of entries? What is the **identity element for multiplication**?

Questions

1. For the operation * shown in the table:

    a. Are the table entries symmetric about the diagonal?
    b. Is the * operation commutative?
    c. Does the * operation have an identity element? If so, what is it?

*	0	1	2	3
0	0	0	0	0
1	0	1	2	3
2	0	2	4	6
3	0	3	6	9

2. For the operation ▽ shown in the table:

   a. Is the ▽ operation commutative?
   b. Does the ▽ operation have an identity element? If so, what is it?
   c. Find the following:
      (1) (a ▽ b) ▽ c and a ▽ (b ▽ c)
      (2) (b ▽ c) ▽ b and b ▽ (c ▽ b)
   d. Do you think that the ▽ operation is associative?

▽	a	b	c
a	a	b	c
b	b	a	b
c	c	b	a

3. For the ⊡ operation shown in the table:

   a. Are the table entries symmetric about the diagonal?
   b. Is the ⊡ operation commutative?
   c. Does the ⊡ operation have an identity element? If so, what is it?

⊡	①	⊕	⊗	⊖
①	①	①	①	①
⊕	①	⊕	①	⊕
⊗	①	①	⊗	①
⊖	①	⊕	①	⊖

4. For the △ operation shown in the table:

   a. Are the table entires symmetric about the diagonal?
   b. Which statements are true and which are false?
      (1) f △ g is the same as g △ f.
      (2) h △ g is the same as g △ h.
      (3) i △ f is the same as f △ i.
   c. Is the △ operation commutative?
   d. Does the △ operation have an identity element? If so, what is it?

△	f	g	h	i
f	d	i	f	g
g	h	d	i	f
h	g	h	d	i
i	f	g	h	d

THE HUMAN ELEMENT

Leonardo of Pisa:

FIBONACCI

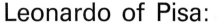

After the end of the Golden Age of Greece, few new mathematical discoveries were made in Europe for almost a thousand years. European mathematics began to reawaken about the year 1201 A.D., when the book *Liber Abaci* was published in Italy.

The author of *Liber Abaci* was Leonardo of Pisa (1170–1250), a wealthy Italian merchant. In this book, Leonardo introduced to Europeans the mathematical works he discovered among the Moslem and Hindu people whom he visited in his travels as a merchant. The numerals we use today are called Hindu-Arabic numerals, and Leonardo was one of those who helped bring them to Europe. Hindu-Arabic numerals are far more efficient than the Roman numerals that were widely used both in commerce and in mathematical studies by Europeans before that time. One of the greatest advantages of the Hindu-Arabic system is that it contains a symbol for zero.

Leonardo published many of his works under the name "Fibonacci," which is thought by some writers to be a family name meaning "son of Bonaccio." The number sequence below is called the *Fibonacci Sequence* because he used it in a famous problem.

$$1, \quad 1, \quad \underbrace{2,}_{1+1} \quad \underbrace{3,}_{1+2} \quad \underbrace{5,}_{2+3} \quad \underbrace{8,}_{3+5} \quad \underbrace{13,}_{5+8} \quad \underbrace{21,}_{8+13} \quad \underbrace{34}_{13+21}$$

Do you see that, beginning with the third term, 2, each term of the sequence is the sum of the two preceding terms? Of course the sequence can be extended indefinitely. The numbers in the sequence are called *Fibonacci numbers*.

Telephone exchange about 1888 . . .

Computerized switchboard . . .

Equations and Problem Solving

The picture of a telephone exchange of 1888 shows a section of the New York City central office, about twelve years after the telephone was invented by Alexander Graham Bell. Switchboards, plugs, and wires were manipulated by hand, and calls could be made only to places that were relatively nearby. Over the years, constant research and development have increased the efficiency of the telephone as a means of communication. Simple pushbutton consoles are all that are required for the operation of a modern computerized switchboard. Although most calls today are dialled directly, the equipment shown is used to help the customer get faster service on person-to-person calls, calls from coin telephones, and others that require operator assistance.

The Equality Properties

7–1 Combining Similar Terms

As we extend our knowledge of algebra it will often be necessary to write expressions such as $10x + 6x$ in a simpler form. By applying the distributive property we can show that $10x + 6x$ may be written simply as $16x$.

$$10x + 6x = (10 + 6)x \quad \text{Distributive property}$$
$$= 16x \quad \text{Substitution principle}$$

Terms such as $5b$ and $8b$ are called **similar terms** or **like terms** because each has the same variable factor, b. For the same reason, $3xy$ and $10xy$ are similar terms and we can show that $3xy + 10xy$ may be written as $13xy$.

$$3xy + 10xy = (3 + 10)xy \quad \text{Distributive property}$$
$$= 13xy \quad \text{Substitution principle}$$

The terms 8r and 3k are called **unlike terms** because each has a different variable factor. Such an expression as $8r + 3k$ cannot be simplified further.

As you may have already guessed, an expression such as $10t - 3t$ may be written simply as $7t$. Again we use the distributive property to show why this is so.

$$10t - 3t = (10 - 3)t \quad \text{Distributive property}$$
$$= 7t \quad \text{Substitution principle}$$

ORAL EXERCISES

Tell how to simplify each expression.

SAMPLE: $2k + 5k$ *What you say:* $7k$

1. $6n + 3n$
2. $12a + 7a$
3. $8b - 5b$
4. $25s + 10s$
5. $34m - 12m$
6. $\frac{1}{2}t + 4t$
7. $3n + 2n + 4n$
8. $16k - 7k$
9. $3x - 3x$
10. $y + y$
11. $c + c + c$
12. $n(3 + 6)$
13. $2xy + 10xy$
14. $24rs + 8rs$
15. $15cd - 10cd$

Name the properties that justify the steps in each exercise.

16. $9d + 6d = (9 + 6)d$
    $= 15d$
17. $15kt + 4kt = (15 + 4)kt$
    $= 19kt$
18. $3a + 5a + 9a = (3 + 5 + 9)a$
    $= 17a$
19. $10cd - 3cd = (10 - 3)cd$
    $= 7cd$

WRITTEN EXERCISES

Combine similar terms to simplify each expression.

SAMPLE 1: $3y + 4y + 2c$ *Solution:* $7y + 2c$

SAMPLE 2: $5(3a + 5) - a$ *Solution:* $15a + 25 - a = 14a + 25$

1. $8k + 5k$
2. $9s + 3s + 2s$
3. $12y - 7y$
4. $4x + 7x + x$
5. $3n + 10n + 5$
6. $6y + 5y - 2y$
7. $3xy + 5xy + 8xy$
8. $8rs + 6rs + 5rs$
9. $3(5t + 2) + t$
10. $6(2h + 4) + 2h$
11. $8mn + 3b + 2mn + b$
12. $5w + (2w + 3)7$

13. $5y + 8y - 2y$
14. $38ab - 18ab$
15. $3(m + 2) - 6$
16. $(7b + 3b) - 10b$

17. $2(2w + 3) + 3w$
18. $3d + 2(4 + d)$
19. $4h + 3k + 2h + 5k$
20. $8c + 9 + 3c - 4$

B 21. $18z + (5z - 2z)$
22. $(10p - 4p) - 3p$
23. $3t + 2r + t + r$
24. $5q + 8q + 2r + 3$
25. $24k + 12 - 4k + 3$
26. $8y + y + 5 + 10 + 6$

27. $15 + w + 6y - y + 12$
28. $19 + 5b + b - 3a + 6$
29. $5cd + 4c - 2cd + 8c$
30. $9rs + 6s - 4rs + 7s$
31. $6m + 7n - m + n + 2m$
32. $8xy + 3x + 7xy - 2x$

SAMPLE: $5(c + d) + 6(c + 2d)$

Solution: $5(c + d) + 6(c + 2d) = 5c + 5d + 6c + 12d$
$= 11c + 17d$

33. $4(r + t) + 7(r + t)$
34. $7(x + y) + 2(x + 3y)$
35. $2(a + 3) + 4(a - 1)$
36. $8(k + 2) + 3(2k - 1)$
37. $5(2 + m) + 9(3 + m)$

38. $3(4d + 2) + 2(5d - 2)$
39. $9(3c + d) + 4(4c - d)$
40. $(6f + 2)5 + 3(2f - 3)$
41. $4(2a + c + 1) + 2(c + 1)$
42. $5(s + 1) + 8(3r + 2s + 2)$

C 43. $8[3a + 3(5 + x)] + 20$
44. $4[5k + 4(2m + 3k + 2)]$
45. $25 + 3[5a + 4(2a - 2)]$

7–2 Addition and Subtraction Properties of Equality

Suppose a scale such as that shown in Figure 7–1 has equal weights placed on each side. The scale is balanced.

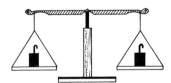

Figure 7–1

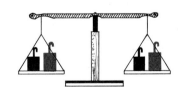

Figure 7–2

Now let us assume that equal weights are added to each side as in Figure 7–2. Do you agree that the scale is still balanced? This illustration brings out the very important **addition property of equality**.

Study the following ideas carefully.

This statement is true:	$10 = 10$
Now add 6 to each number	$10 + 6 = 10 + 6$
Now we see that	$16 = 16$

We say that if two numbers are equal and the same number is added to both, the sums are equal.

For each r, each s, and each t,

if $r = s$, then $r + t = s + t$.

In a similar manner we can show that there is a **subtraction property of equality** if we assume that the indicated subtraction is possible for the set of numbers being used.

This statement is true:	$8 = 8$
Now subtract 3 from each number	$8 - 3 = 8 - 3$
We now see that	$5 = 5$

For each r, each s, and each t such that $r > t$,

if $r = s$, then $r - t = s - t$.

The addition and subtraction properties of equality are very useful in solving equations. Study the following method for finding the replacement for x in $x - 3 = 12$. Notice that our strategy is to change the equation so that the variable stands alone as one member of the equation. This process is called **transformation**.

$x - 3 = 12$	Given equation
$x - 3 + 3 = 12 + 3$	Add 3 to each member
$x = 15$	Substitution principle

In the solution above, notice that we have used the additive property of zero in substituting x for $x - 3 + 3$, since $x - 3 + 3 = x + 0$.

We can check the accuracy of our work by replacing x with 15 in the original equation $x - 3 = 12$.

Check: $x - 3 = 12$

$15 - 3$	12
12	12

Since both the left member and the right member of the equation name the number 12 we know that 15 is the correct replacement for x.

Therefore, the solution set is $\{15\}$.

Study the following solution of the equation $y + 5 = 17$. Again the strategy will be to **transform** the equation so that the variable stands alone as one member of the equation.

$y + 5 = 17$	Given equation
$y + 5 - 5 = 17 - 5$	Subtract 5 from each member
$y = 12$	Substitution principle

The solution set is $\{12\}$, and we again check our accuracy as follows:

Check: $y + 5 = 17$
$$\begin{array}{c|c} 12 + 5 & 17 \\ 17 & 17 \end{array}$$

ORAL EXERCISES

Give the simplest name for each expression.

SAMPLE 1: $8 - 5 + 5$ *What you say:* 8
SAMPLE 2: $t + 7 - 7$ *What you say:* t

1. $x + 2 - 2$
2. $t + 6 - 6$
3. $10 + 3 - 3$
4. $12 - 5 + 5$
5. $a + \frac{1}{2} - \frac{1}{2}$
6. $b + \frac{3}{5} - \frac{3}{5}$
7. $m + 0.3 - 0.3$
8. $y + 0.03 - 0.03$
9. $k - \frac{7}{8} + \frac{7}{8}$
10. $13 + \frac{1}{4} - \frac{1}{4}$
11. $\frac{7}{8} + \frac{5}{2} - \frac{5}{2}$
12. $d - \frac{8}{5} + \frac{8}{5}$
13. $1.75 - 0.23 + 0.23$
14. $2\frac{1}{8} - 0.003 + 0.003$

Tell what number should be added to each expression or subtracted from each expression so that the variable will stand alone.

SAMPLE: $k + 8$ *What you say:* subtract 8

15. $w + 10$
16. $c + 3$
17. $n - 15$
18. $k - \frac{3}{4}$
19. $a + \frac{1}{5}$
20. $m + 0.3$
21. $b - 0.15$
22. $d - 2.5$
23. $c + 0.03$

WRITTEN EXERCISES

Copy and complete the solution of each equation.

SAMPLE:
$$p - 8 = 5$$
$$p - 8 + 8 = 5 + 8$$
$$p = ?$$

Solution:
$$p - 8 = 5$$
$$p - 8 + 8 = 5 + 8$$
$$p = 13$$

Check:
$$\begin{array}{c|c} p - 8 & = 5 \\ \hline 13 - 8 & 5 \\ 5 & 5 \end{array}$$

The solution set is $\{13\}$.

A 1. $$k + 2 = 5$$
$$k + 2 - 2 = 5 - 2$$
$$k = ?$$

2. $$8 + a = 19$$
$$a + 8 - 8 = 19 - 8$$
$$a = ?$$

3. $$h + \tfrac{2}{3} = 10$$
$$h + \tfrac{2}{3} - \tfrac{2}{3} = 10 - \tfrac{2}{3}$$
$$h = ?$$

4. $$2\tfrac{3}{4} = m - \tfrac{1}{4}$$
$$2\tfrac{3}{4} + \tfrac{1}{4} = m - \tfrac{1}{4} + \tfrac{1}{4}$$
$$? = m$$

5. $$k - \tfrac{3}{8} = 4$$
$$k - \tfrac{3}{8} + \tfrac{3}{8} = 4 + \tfrac{3}{8}$$
$$k = ?$$

6. $$d + \tfrac{1}{5} = \tfrac{4}{5}$$
$$d + \tfrac{1}{5} - \tfrac{1}{5} = \tfrac{4}{5} - \tfrac{1}{5}$$
$$d = ?$$

7. $$14 = b + 7$$
$$14 - 7 = b + 7 - 7$$
$$? = b$$

8. $$25 = r - 10$$
$$25 + 10 = r - 10 + 10$$
$$? = r$$

9. $$d + 0.34 = 0.77$$
$$d + 0.34 - 0.34 = 0.77 - 0.34$$
$$d = ?$$

10. $$a + 0.25 = 1.075$$
$$a + 0.25 - 0.25 = 1.075 - 0.25$$
$$a = ?$$

Solve each equation. Begin by stating what number should be added to each member or subtracted from each member.

SAMPLE: $y + 3 = 10$

Solution: Subtract 3 from each member.

$$y + 3 - 3 = 10 - 3$$
$$y = 7$$

Check:
$$\begin{array}{c|c} y + 3 & = 10 \\ \hline 7 + 3 & 10 \\ 10 & 10 \end{array}$$

The solution set is $\{7\}$.

11. $y + 2 = 8$
12. $n + 5 = 6$
13. $16 + k = 21$
14. $c + \tfrac{1}{5} = \tfrac{4}{5}$

15. $x - 1 = 7$
16. $y - 6 = 9$
17. $k + 7 = 13$
18. $m + \frac{2}{3} = \frac{4}{3}$
19. $10 + b = 15$

20. $a - 19 = 30$
21. $7 = x + 4$
22. $15 = w - 3$
23. $r - 0.4 = 2.3$
24. $t + 0.03 = 3.44$

B 25. $\frac{3}{8} = n - \frac{5}{8}$
26. $\frac{9}{10} = v + \frac{3}{10}$
27. $\frac{7}{3} = a - \frac{2}{3}$

28. $\frac{2}{5} = w + \frac{1}{5}$
29. $\frac{10}{3} = \frac{2}{3} + d$
30. $\frac{6}{5} = \frac{2}{5} + r$

Solve each equation. Check your answer.

SAMPLE: $x + 12 = 35$

Solution: $x + 12 = 35$
$x + 12 - 12 = 35 - 12$
$x = 23$

Check: $x + 12 = 35$
$23 + 12 \mid 35$
$35 \mid 35$

The solution set is $\{23\}$.

31. $t + 45 = 68$
32. $a + 24 = 53$
33. $26 = b + 26$
34. $38 = d + 16$
35. $14 + k = 38$
36. $36 + k = 55$
37. $n - 28 = 62$
38. $r - 45 = 81$

39. $c - 25 = 0$
40. $h + 2.4 = 8.7$
41. $t - 3.9 = 1.4$
42. $k - 13 = 0$
43. $m + 3.4 = 12.1$
44. $s - 5.25 = 0.04$
45. $h - 0.78 = 9.2$
46. $p - 0.37 = 4.4$

C 47. $\frac{2}{5} + r = \frac{8}{5}$
48. $\frac{2}{7} + b = \frac{10}{7}$
49. $\frac{3}{5} = t - \frac{4}{5}$
50. $\frac{9}{10} = k - \frac{11}{10}$

51. $x - 273 = 0$
52. $w - 0 = 35$
53. $0.13 + y = 1.45$
54. $a - 0.71 = 0.30$

7–3 The Division Property of Equality

Perhaps you noticed, in the equations which we have solved by using the addition and subtraction properties of equality, that the coefficient of each variable was **1**. Thus we were able to determine the solution set as soon as we had added to or subtracted from each member of the equation.

Now let us consider the equation $4x - 3 = 17$. If we add 3 to each member, we have the equation $4x = 20$.

We need next to find a transformation that will give us an equation in which the coefficient of the variable x is 1. Remember that "$4x$" means "x multiplied by 4." What operation is the inverse of multiplying by 4? Do you agree that it is dividing by 4? So we have

$$4x - 3 = 17$$
$$4x = 20$$
$$\frac{4x}{4} = \frac{20}{4}$$
$$x = 5$$

Check:
$$4x - 3 = 17$$
$$4(5) - 3 \mid 17$$
$$20 - 3 \mid 17$$
$$17 \mid 17$$

The solution set is $\{5\}$.

Thus we state the **division property of equality**:

For each r, each s, and each t that is not zero,

if $r = s$, then $\dfrac{r}{t} = \dfrac{s}{t}$.

Suppose that we are to solve the equation $3n + 4 = 19$. In this case we transform the equation twice: first by using the **subtraction property of equality**, and then by using the **division property of equality**. Study this solution carefully.

$3n + 4 = 19$	Given equation
$3n + 4 - 4 = 19 - 4$	Subtract 4 from each member
$3n = 15$	Substitution principle
$\dfrac{3n}{3} = \dfrac{15}{3}$	Divide each member by 3
$n = 5$	Substitution principle

The solution set is $\{5\}$.

ORAL EXERCISES

Name the solution set for each equation. Begin by telling the number by which each member is to be divided.

SAMPLE: $5t = 35$ *What you say:* Divide each member by 5.
The solution set is $\{7\}$.

Equations and Problem Solving   199

1. $4r = 24$
2. $8q = 32$
3. $45 = 5m$
4. $56 = 7w$
5. $15m = 75$
6. $38 \cdot s = 380$
7. $t \cdot 3 = 39$
8. $k \cdot 7 = 63$
9. $6b = 3$
10. $10r = 5$
11. $3a = 2$
12. $5w = 4$
13. $7n = 10$
14. $15z = 3$
15. $30 = 6 \cdot c$
16. $72 = m \cdot 6$
17. $20 = 5 \cdot w$
18. $3 = k \cdot 7$

WRITTEN EXERCISES

Solve each equation. Check your answer.

SAMPLE: $2n - 3 = 9$

Solution:
$$2n - 3 = 9$$
$$2n - 3 + 3 = 9 + 3$$
$$2n = 12$$
$$\frac{2n}{2} = \frac{12}{2}$$
$$n = 6$$

Check:
$$2n - 3 = 9$$
$2(6) - 3$	9
$12 - 3$	9
9	9

The solution set is $\{6\}$.

A
1. $15t = 180$
2. $25k = 650$
3. $14b = 35$
4. $17m = 30$
5. $125y = 1000$
6. $350a = 1050$
7. $1750 = 25b$
8. $690 = 23t$
9. $20c = 8$
10. $18e = 270$
11. $3z = .105$
12. $4h = 1.24$
13. $3.5 = 5 \cdot p$
14. $9 \cdot d = 81$
15. $2c = 3.5$
16. $12t = 1.5$
17. $5x = \frac{1}{2}$
18. $3y = \frac{1}{3}$
19. $4b + 2 = 14$
20. $10n + 4 = 84$
21. $5a - 7 = 28$
22. $9x - 10 = 80$
23. $4c = (10 + 14)$
24. $8y = (1.5 + 10.5)$
25. $1.5y = 45$
26. $1.2w = 6$

B
27. $25 = 6t - 17$
28. $58 = 7m + 2$
29. $31 = 5 + 8a$
30. $15 = 12 + 2b$
31. $22 = 4s + 22$
32. $(12 + 14) = 3k + 5$
33. $0 = 7m - 14$
34. $0.9y - 3 = 42$
35. $1.6x - 4 = 28$
36. $0 = 12b - 3$

[C] 37. $3n - \frac{1}{3} = \frac{8}{3}$
38. $2b + 0.05 = 0.17$
39. $7x + 0.3 = 3.8$

40. $0.081 = 0.009 + 0.12c$
41. $2.03 = 0.03 + 6t$
42. $(0.34 + 0.17) = 6d - 0.03$

7–4 The Multiplication Property of Equality

Let us next investigate the solution of the equation $\frac{t}{4} = 7$. You should recall that we can write this equation as $\frac{1}{4}t = 7$, in which form you can see clearly that the fraction $\frac{1}{4}$ is the coefficient of *t*.

What is the multiplicative inverse of $\frac{1}{4}$? Do you see that, if we multiply $\frac{1}{4}t$ by 4, the result is *t*? We can use this idea to transform an equation in which the coefficient of the variable is a fraction. We call it the **multiplication property of equality**, and state it as follows:

For each *r*, each *s*, and each *t* that is not zero,

if *r* = *s*, then *tr* = *ts*.

Study the solutions in the following examples.

EXAMPLE 1.

$\frac{m}{3} = 7$ Given equation   Check: $\frac{m}{3} = 7$

$3 \cdot \frac{m}{3} = 3 \cdot 7$ Multiply each term by 3 $\frac{21}{3}$ | 7

$m = 21$ Substitution principle   7 | 7

Therefore, the solution set is {21}.

EXAMPLE 2.

$\frac{3k}{5} = 12$ Given equation   Check: $\frac{3k}{5} = 12$

$5 \cdot \frac{3k}{5} = 12 \cdot 5$ Multiply each term by 5 $\frac{3 \cdot 20}{5}$ | 12

$3k = 60$ Substitution principle $\frac{60}{5}$ | 12

$\frac{3k}{3} = \frac{60}{3}$ Divide each term by 3   12 | 12

$k = 20$ Substitution principle

Therefore, the solution set is {20}.

In the case of Example 2, could we have transformed the equation into $k = 20$ by *first* dividing by **3**, and *then* multiplying by **5**? If you try it for yourself, you will find that it can be done in that order, but that the work is more complicated.

ORAL EXERCISES

Name the coefficient of the variable in each equation. Then tell by what number the coefficient must be multiplied to give a product of 1. Do not solve the equation.

SAMPLE: $\dfrac{n}{3} = 9$ *What you say:* The coefficient of n is $\tfrac{1}{3}$; $\tfrac{1}{3}$ must be multiplied by **3** to give a product of **1**.

1. $\dfrac{b}{4} = 3$
2. $\dfrac{n}{7} = 10$
3. $\dfrac{k}{5} = 2$
4. $\dfrac{x}{10} = 5$
5. $\dfrac{a}{3} = 3$
6. $\dfrac{d}{4} = \dfrac{1}{2}$
7. $\dfrac{h}{35} = 1$
8. $13 = \dfrac{y}{5}$
9. $10 = \dfrac{c}{8}$

Name the solution set for each equation. Begin by telling the number by which each member is to be multiplied.

SAMPLE: $\dfrac{x}{7} = 3$ *What you say:* Multiply each member by **7**. The solution set is {**21**}.

10. $\dfrac{y}{2} = 2$
11. $\dfrac{m}{4} = 1$
12. $\dfrac{t}{2} = 5$
13. $\dfrac{a}{3} = 2$
14. $\dfrac{s}{10} = 3$
15. $\dfrac{c}{5} = 6$
16. $8 = \dfrac{b}{5}$
17. $2\tfrac{1}{2} = \dfrac{x}{4}$
18. $\tfrac{1}{3}x = 6$
19. $\tfrac{1}{5}w = 10$
20. $\tfrac{1}{2}z = \tfrac{1}{4}$
21. $\tfrac{1}{4}y = \tfrac{1}{5}$
22. $b \cdot \tfrac{1}{3} = 7$
23. $m \cdot \tfrac{1}{2} = 4$
24. $g \cdot \tfrac{1}{5} = 4$

WRITTEN EXERCISES

Solve each equation. Check each answer.

 1. $\dfrac{m}{10} = 18$ 2. $m \cdot \tfrac{1}{6} = 0.15$ 3. $\dfrac{2k}{3} = 4$

4. $\dfrac{b}{15} = 35$

5. $\dfrac{w}{17} = 15$

6. $25 = \dfrac{c}{3}$

7. $4.2 = \dfrac{a}{9}$

8. $\tfrac{1}{3}(f) = 11$

9. $\tfrac{1}{4}(k) = 8.1$

10. $(r)\tfrac{1}{8} = 75$

11. $h \cdot \tfrac{1}{3} = 1.5$

12. $\dfrac{k}{3} = (10 + 4)$

13. $\dfrac{b}{7} = (3 + 9)$

14. $\dfrac{a}{5} = \dfrac{2}{5}$

15. $\dfrac{s}{10} = \dfrac{3}{5}$

16. $m \cdot \tfrac{2}{3} = 10$

17. $\dfrac{2x}{3} = 12$

18. $\tfrac{3}{5}t = 9$

19. $\tfrac{5}{8}y = 15$

20. $\dfrac{3w}{2} = 9$

21. $\dfrac{5a}{2} = 10$

B 22. $56 = \tfrac{7}{8}(m)$

23. $40 = \tfrac{4}{5}(b)$

24. $\dfrac{m}{3} = 0$

25. $\dfrac{k}{14} = 0$

26. $3a \div 2 = 24$

27. $2d \div 5 = 10$

28. $\tfrac{1}{2}v = 4\tfrac{1}{2}$

29. $\dfrac{k}{0.5} = 14$

30. $\dfrac{c}{0.3} = 80$

C 31. $\dfrac{12y}{5} = 24$

32. $\tfrac{1}{3}m = 1.25$

33. $\tfrac{3}{4}t = 75$

34. $\dfrac{15a}{4} = 0$

35. $\dfrac{9n}{13} = 0$

36. $\tfrac{1}{2}x = \tfrac{5}{19}$

37. $\tfrac{1}{5}y = \tfrac{1}{12}$

38. $n \div 10 = 3.8$

39. $40 = w \div 2\tfrac{1}{2}$

Working with Equations

7–5 More about Solving Equations

Until now our work with equations has emphasized a variety of uncomplicated concepts and skills. We are now ready to apply these concepts and skills to solve more complicated equations. It will be important that you watch for the various strategies that will be used.

Consider the equation $5n + 3n + 2 = 20 + 6$. Do you have any ideas as to how we might proceed to solve for n? **First**, we will look for ways to simplify the equation by **combining similar terms**. **Second**, we will **transform** the equation by using the properties of equalities in such a way that the variable and its coefficient stand alone as one member. **Finally**, we will use the properties of equalities to transform the equation into one in which the coefficient of the variable is **1**. Follow each step of this solution carefully.

Equations and Problem Solving

$5n + 3n + 2 = 20 + 6$ Given equation
$8n + 2 = 26$ Combine similar terms
$8n + 2 - 2 = 26 - 2$ Subtract 2 from each member
$8n = 24$ Substitution principle
$\dfrac{8n}{8} = \dfrac{24}{8}$ Divide both members by 8
$n = 3$ Substitution principle

Check: $5n + 3n + 2 = 20 + 6$

$5(3) + 3(3) + 2$	$20 + 6$
$15 + 9 + 2$	26
26	26

Therefore, the solution set is $\{3\}$.

Now let us use a similar approach to solve this equation:

$$x - \tfrac{2}{3}x + 3 = 12 - 7.$$

$x - \tfrac{2}{3}x + 3 = 12 - 7$ Given equation
$\tfrac{1}{3}x + 3 = 5$ Combine similar terms
$\tfrac{1}{3}x + 3 - 3 = 5 - 3$ Subtract 3 from each member
$\tfrac{1}{3}x = 2$ Substitution principle
$3 \cdot \tfrac{1}{3}x = 3 \cdot 2$ Multiply each member by 3
$x = 6$ Substitution principle

Check: $x - \tfrac{2}{3}x + 3 = 12 - 7$

$6 - \tfrac{2}{3}(6) + 3$	$12 - 7$
$6 - 4 + 3$	5
5	5

Therefore, the solution set is $\{6\}$.

ORAL EXERCISES

Name the fraction that is the coefficient of the variable in each expression.

SAMPLE: $\dfrac{3t}{7}$ What you say: $\dfrac{3}{7}$

1. $\dfrac{m}{2}$ 2. $\tfrac{5}{3}x$ 3. $\dfrac{10w}{3}$

4. $\dfrac{2r}{3}$ 6. $\dfrac{2n}{5}$ 8. $\dfrac{4k}{7}$

5. $\tfrac{1}{5}k$ 7. $\dfrac{5y}{3}$ 9. $\dfrac{b}{10}$

Simplify each equation by combining similar terms. Do not solve the equation.

SAMPLE: $8n + n + 5 = 10 + 3$ *What you say:* $9n + 5 = 13$

10. $3x + 10x = 8$
11. $y + 5y = 25$
12. $4a + a = 6 + 7$
13. $8m + 2m = 12 - 3$
14. $k + 3k + 4k = 9 + 2$
15. $8w + w + 5w = 18 - 3$
16. $12d - 7d = 35$
17. $19d - 10d = 3 + 7$

18. $\tfrac{1}{2}x + 3x = 10$
19. $t + \tfrac{1}{3}t = 24$
20. $\tfrac{2}{3}m + m = 5 + 2$
21. $\tfrac{3}{2}x + 2x = 9 - 4$
22. $2w - \tfrac{1}{2}w = 3\tfrac{1}{2} + 4$
23. $z - \tfrac{2}{3}z = (4)(5)$
24. $\tfrac{1}{2}w + \tfrac{1}{2}w + \tfrac{1}{2}w = 42$
25. $\dfrac{2w}{3} + \dfrac{w}{3} = 5$

WRITTEN EXERCISES

Solve each equation.

SAMPLE: $3t + t - 5 = 12 + 7$

Solution: $3t + t - 5 = 12 + 7$
$4t - 5 = 19$
$4t - 5 + 5 = 19 + 5$
$4t = 24$
$\dfrac{4t}{4} = \dfrac{24}{4}$
$t = 6$

Check: $\begin{array}{r|l} 3t + t - 5 = 12 + 7 \\ 3(6) + 6 - 5 & 19 \\ 18 + 6 - 5 & 19 \\ 19 & 19 \end{array}$

The solution set is $\{6\}$.

1. $5m + 2m = 21$
2. $4k + k = 35$
3. $3a + 6a = 90$
4. $5y + 3y = 18 + 24$
5. $14x - 6x = 40 + 8$
6. $22w - 9w = 30 + 9$
7. $3b + b + 2 = 18$
8. $5k + 4k - 3 = 33$

9. $30 + 5 = 3n + 10n + n$
10. $40 - 3 = x + 5x + 2$
11. $15b - 10 = 0$
12. $7t - 4 = 0$
13. $4s + 4 = 4$
14. $1.8r + 1.2r = 16 - 4$
15. $0.75y + 1.25y = 4 + 3 + 3$
16. $2.8v - 1.3v = 3 + 0$

17. $10c + c - 5 = 19 + 31$
18. $16m - 4m + 2 = 12 + 50$
19. $0.3x + 0.5x + x = 3.6$
20. $0.9z + 2.6z = 9 + 1 + 0.5$

B 21. $\frac{1}{2}b + \frac{1}{2}b = 9\frac{1}{2} - 2\frac{1}{2}$
22. $\frac{2}{3}m + \frac{1}{3}m = \frac{1}{2} + \frac{1}{4}$
23. $\frac{3t}{5} + \frac{2t}{5} = 6 + \frac{1}{3}$
24. $\frac{1}{4}b + \frac{1}{2} = 3\frac{1}{2}$
25. $\frac{1}{2}a - \frac{1}{3} = 1 + \frac{2}{3}$
26. $\frac{3a}{2} + 5 - \frac{1}{2}a = 14 - 3\frac{1}{2}$
27. $4k + 9\frac{1}{2} + 3k = 10\frac{1}{2}$
28. $1\frac{2}{5}n + 1\frac{3}{5}n - 1 = 3^2$

29. $10 - 2\frac{1}{2} = \frac{x}{3} + 7 + \frac{2x}{3}$
30. $2(m + 3) + 5m = 40 + 8$
31. $3(2t + 1) - 4t = 3 \cdot 11$
32. $67 - 48 = 7(k - 3) - 6k$
33. $\frac{1}{3}c + \frac{1}{3}c + \frac{1}{3}c = 15 + 5$
34. $\frac{5t}{8} - 7 - \frac{1}{8} = 0$
35. $0.8t + 0.1t - 4 = 5$
36. $0.5b - 3 + 0.3b = 4.2$

C 37. $11.4 + 3n + 3n = 12$
38. $4.4 + 10x - 4x = 10.4$
39. $3d - 7.4 - 0.5d = 32.6$
40. $2r + 4 + r + 4r + 5 = 72$

41. $5z + 7 - z + 1 + 3z = 28$
42. $69 - 2 = 11 + 4x - 2x$
43. $6t + 25 + 7 + t = 40$
44. $\frac{s}{2} + \frac{s}{3} + \frac{s}{3} = 4\frac{1}{3} - 2\frac{1}{2}$

PROBLEMS

For each problem stated below, write an equation using a variable to represent the unknown number. Then solve the equation and answer the question asked in the problem.

SAMPLE: The sum of three times some number and 15 is 90. Find the number.

Solution: Let n stand for the number.
$$3n + 15 = 90$$
$$3n + 15 - 15 = 90 - 15$$
$$3n = 75$$
$$\frac{3n}{3} = \frac{75}{3}$$
$$n = 25 \qquad \text{The number is } \mathbf{25}.$$

1. The sum of twice a number and 5 is 12. What is the number?
2. If five times a certain number is increased by 6, the result is equal to 51. Find the number.

3. The sum of twice a certain number and 10 is 46. Find the number.

4. If twice some number is decreased by 3, the result is equal to 19. Find the number.

5. The length of a rectangular plot of ground is four times its width. The perimeter of the plot is 340 feet. How long and how wide is the plot?

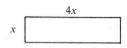

6. The sum of one-half of a certain number and one-third of the same number is equal to 20. What is the number?

7. The sum of some whole number and the next larger whole number is 31. Find the numbers.

8. William has a certain number of pennies in his pocket. Ken has three times as many pennies in his pocket. Altogether they have 28 pennies. How many does each boy have?

9. The sum of an odd number and the next larger odd number is 32. What are the numbers?

10. The relative lengths of the sides of a triangle are shown in the illustration. The perimeter of the triangle is 48 inches. Find the length of each side.

11. A large metal drum contains a certain number of gallons of oil. Each of two smaller drums contains half as much as the large drum. Altogether they contain 120 gallons of oil. How much oil is in each drum?

12. Jack walked from his house to Tom's house. When Jack left Tom's house he walked to school, a distance one-third of the distance from his house to Tom's. Had Jack walked directly to school he would have walked 4 blocks less than he actually did.

How many blocks is it from Jack's house to Tom's? From Tom's to school?

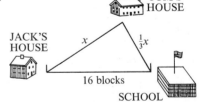

13. One mathematics class has a certain number of students, a second class has 5 more than the first, and a third has 3 more than the first. Altogether there are 71 students in the three classes. How many students are in each class?

7–6 Equations with the Variable in Both Members

The equation $5k = 10 + 3k$ is somewhat different from those we have been working with, since the variable appears in both members. The same properties of equality we have been using to transform equations will help us in solving this kind of equation. Study the solution shown here.

$5k = 10 + 3k$ Given equation
$5k - 3k = 10 + 3k - 3k$ Subtract $3k$ from each member
$2k = 10$ Substitution principle
$\dfrac{2k}{2} = \dfrac{10}{2}$ Divide both members by 2
$k = 5$ Substitution principle

Check: $5k = 10 + 3k$

$5 \cdot 5$	$10 + 3 \cdot 5$
25	$10 + 15$
25	25

The solution set is $\{5\}$.

The solution of a slightly more complicated equation is completed as follows:

$3m + m - 2 = 4 + 3 + m$ Given equation
$4m - 2 = 7 + m$ Combine similar terms
$4m - 2 - m = 7 + m - m$ Subtract m from each member
$3m - 2 = 7$ Substitution principle
$3m - 2 + 2 = 7 + 2$ Add 2 to each member
$3m = 9$ Substitution Property
$\dfrac{3m}{3} = \dfrac{9}{3}$ Divide both members by 3
$m = 3$ Substitution principle

Check: $3m + m - 2 = 4 + 3 + m$

$3(3) + 3 - 2$	$4 + 3 + 3$
$12 - 2$	10
10	10

The solution set is $\{3\}$.

WRITTEN EXERCISES

Solve each equation. Check your answer.

SAMPLE: $4t = 15 + t$

Solution:
$$4t = 15 + t$$
$$4t - t = 15 + t - t$$
$$3t = 15$$
$$t = 5$$

Check:

$4t = 15 + t$	
$4 \cdot 5$	$15 + 5$
20	20

The solution set is $\{5\}$.

A
1. $3x = 8 + x$
2. $2r = 3 + r$
3. $9k = 39 + 6k$
4. $6t = 27 - 3t$
5. $2m + 5 = 7m$
6. $1 + 2s = 3s$
7. $6w - 9 = 5w$
8. $4y - 15 = 3y$
9. $7s - 3 = 3s$
10. $15a - 6 = 6a$
11. $5 + 3n = 7n$
12. $12a = 9a + 18$
13. $4z + 20 = 9z$
14. $7y - 9 = 2y$
15. $t + 27 = 6t$
16. $5c - 5 = 2c$
17. $12 + 3k = 9k$
18. $3b = 5b - 6$
19. $12 - x = 5x$
20. $9 + 3x = x + 9$
21. $10x + 3 = 3x + 17$
22. $12m + 8 = 4m + 16$
23. $5b + 2 = b + 6$
24. $10t - 3 = 17 - 2t$
25. $3r + 2 = 27 - 2r$
26. $8k - 1 = 35 + 2k$

B
27. $12r - 3 = 4 - 2r$
28. $4d + 2 = 2d + 8$
29. $5m + 2 = m + 7$
30. $7s - 7 = 15 - s$
31. $12b - 3 = 4 - b$
32. $16 + 4k = 10k - 20$

C
33. $4b + b - 5 = 7 + 2b$
34. $3a + 2a + 4 = a + 20$
35. $2(b + 3) = 15 - b$
36. $5(x - 2) = 3x + 7$
37. $(2n + 1)3 = 4n + 19$
38. $\frac{1}{2}(a + 6) = a + 3$
39. $\frac{4}{3}c - \frac{1}{2} = \frac{1}{3}(c + 12)$
40. $2d + 12 + 4d = 10 + 18d - 2$
41. $9w + 3 - 2w = 12 - 6w + 5$
42. $4(x + 1) + 9 = 2(3x - 4)$

PROBLEMS

1. Three times a certain number is the same as the number increased by 10. What is the number?

2. If 18 is added to three times some number, the result is 9 times the number. Find the number.

3. If four times some unknown number is increased by 20, the result is equal to nine times the unknown number. Find the unknown number.

4. The length in feet of a side of a square is some number. The length in feet of a rectangle is that same number increased by 2, and the width of the rectangle is 10 feet. The perimeters of the two figures are equal. Find their dimensions.

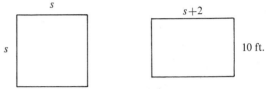

5. If twenty-seven is decreased by three times a certain number, the result is the same as six times the number. What is the number?

6. Five times a certain number is equal to 10 more than three times the number. Find the number.

7. Two less than four times a certain number is equal to seven more than the number. What is the number?

8. The length of one rectangle is 3 yards more than the length of another rectangle. The width of the first rectangle is 5 yards and the width of the second is 8 yards. Find the length of each rectangle if their combined area is 197 square yards.

9. The width of a given rectangle is some number of feet and the length is five feet more than twice the width. Find the length and width of the rectangle if the perimeter is 70 feet.

CHAPTER SUMMARY

Inventory of Structure and Concepts

1. Terms that have the same variable factor are called **like terms** or **similar terms**. Application of the distributive property permits the combining of similar terms.

2. The properties of equality are applicable in solving equations. They include the following:

Addition property of equality: if $r = s$, then $r + t = s + t$.

Subtraction property of equality: if $r = s$, and $r > t$, then $r - t = s - t$.

Division property of equality: if $r = s$, and $t \neq 0$, then $\frac{r}{t} = \frac{s}{t}$.

Multiplication property of equality: if $r = s$, then $rt = st$.

Vocabulary and Spelling

similar terms (*p. 191*)
like terms (*p. 191*)
unlike terms (*p. 192*)
properties of equality (*p. 193*)
addition property of equality (*p. 193*)
subtraction property of equality (*p. 194*)
transformation (*p. 194*)
division property of equality (*p. 198*)
multiplication property of equality (*p. 200*)

Chapter Test

Solve each equation. Check your answer.

1. $x - 48 = 71$
2. $5x - 15 = 225$
3. $\frac{1}{2}y + 16 = 41$
4. $2n + 3n = 75$
5. $3p - 5 = p + 17$
6. $3(3x - 1) = 159$
7. $15 - \frac{3}{2}x = \frac{1}{2}x$
8. $x - \frac{x}{4} = 18$
9. $135 = 3x$
10. $\frac{t}{7} = 51$
11. $\frac{2}{3}k = 36$
12. $7x - x = 78$
13. $\frac{x + 5}{2} = 30$
14. $14 - x = 3x + 2$
15. $2(k - 3) = k + 1$
16. $1.2h + 5 = 0.2h + 6$

Write an equation for each problem stated below. Explain what number the variable represents. Do not solve the equations.

17. If a number is increased by 15, the result is equal to four times the number. What is the number?
18. The sum of two consecutive even numbers is 14. What is the smaller number?
19. If fifteen is increased by one-half a given number, the resulting sum is 19. What is the number?
20. The length of a rectangle is four times its width, and the perimeter of the rectangle is 200 feet. Find its width.

Chapter Review

7–1 Combining Similar Terms

Combine similar terms to simplify each expression where possible.

1. $12 \cdot x - 5 \cdot x$
2. $3k + k$
3. $7x + 2y$
4. $7(k + 3) + 2(2 - k)$
5. $5y - 1y$
6. $3g - 2g + 7g$
7. $5(x + 2) + 8(x + 3)$
8. $3k + 2y - k + 5y$

7–2 Addition and Subtraction Properties of Equality

Solve each equation.

9. $x - 5 = 15$
10. $3.8 + x = 11.5$
11. $c + \frac{1}{4} = 1\frac{1}{4}$
12. $\frac{2}{3} = x - \frac{5}{3}$

7–3 The Division Property of Equality

Solve each equation.

13. $16k = 80$
14. $7n + 5 = 40$
15. $3y - 2 = 5$
16. $0.8x = 56$
17. $4.5x = 54$
18. $4t = \frac{3}{2}$

7-4 The Multiplication Property of Equality

Solve each equation.

19. $\dfrac{x}{6} = 17$

20. $\dfrac{3t}{4} = \dfrac{4}{3}$

21. $24 = \dfrac{8}{3}z$

22. $\dfrac{r}{8} = 1.25$

23. $\dfrac{n}{9} - 2 = 11$

24. $40 = x \div 3\dfrac{1}{3}$

7-5 More about Solving Equations

Solve each equation.

25. $13s - 5s = 72$

26. $\dfrac{3}{4}x - \dfrac{1}{4}x = 54$

27. $40 = 3x + x + 8$

28. $5n - 7 + n - 2n = 29$

Write an equation for each problem. Explain what number the variable represents. Then solve the equation and answer the question.

29. John's father is twice as old as John. If the sum of their ages is 126 years, how old is John?

30. The sum of three consecutive odd numbers is 111. What are the numbers?

31. If five times a certain number is increased by 2, the result is 67. What is the number?

7-6 Equations with the Variable in Both Members

Solve each equation.

32. $3y = 15 - 2y$

33. $8t + 16 = 3t + 17$

34. $x = 5x - 3$

35. $12 + x = 5x - 8$

36. $2(3x - 5) = 5x + 1$

37. $\dfrac{3}{4}x + 2 = \dfrac{1}{4}x + 5$

Write an equation for each problem. Explain what number the variable represents. Then solve the equation and answer the question.

38. If a number is increased by 10, the result is twice the original number. What is the number?

39. If 35 is decreased by four times a certain number, the result is equal to the certain number. What is the number?

40. Twice the difference of two numbers is equal to the sum of the numbers. If the smaller number is 10, what is the larger number?

Review of Skills

Tell what letter on the number line below names the point which might represent each number described.

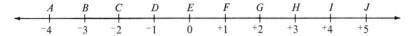

1. Three degrees above zero.
2. Two degrees below zero.
3. Fifth floor above ground level.
4. One floor below ground level.
5. Four miles above sea level.
6. Four miles below sea level.

Answer the following questions, assuming that 3 miles to the east is written $^+3$ and 5 miles to the west is written as $^-5$.

7. If one began at home and walked 12 miles east and then 4 miles west where would he be in reference to his home? (Answer with a number preceded by a + or − symbol.)
8. If one began at home and walked 4 miles east and 7 miles west, where would he be in reference to his home?
9. Starting at Indianapolis, Indiana, and driving 200 miles west and then 340 miles east, where would you be in reference to Indianapolis?

At a weather station there is a thermometer which measures the temperature above and below 0°F. A temperature of 5 degrees above zero is recorded as $^+5°$ and a temperature of 5 degrees below zero is recorded as $^-5°$.

10. The temperature at 5 a.m. on a winter day was $^+4°$. If it rose 11° in the next seven hours, what was the temperature at noon?
11. A temperature of $^-3°$ was recorded the next morning. If the weather forecast predicted a rise of 10° by noon, what temperature reading was expected for noon?
12. On a certain day, the lowest recorded temperature was $^-6°$ at 4 a.m., and the highest temperature for the day was $^+8°$. How many degrees did the temperature rise during the day?
13. One day, temperatures of $^-4°$ and $^-7°$ were recorded at 2 a.m. and at 5 a.m., respectively. Did the temperature rise or fall during the three hours? How many degrees did it change?
14. On a certain day, temperature was recorded at 3 p.m., and again at 9 p.m. If the 9 p.m. reading was $^-3°$ and this represented a drop of 7 degrees since the 3 p.m. reading, what must the temperature have been at 3 p.m.?

CHECK POINT FOR EXPERTS

Polyhedrons

The illustration below shows a figure called a **tetrahedron**. A tetrahedron is one kind of **polyhedron**, or three-dimensional figure whose surface is made up of a number of polygonal regions. The prefix **tetra** means **four** and tells us that a tetrahedron has four **faces**. Each face of a tetrahedron is triangular. Pairs of faces intersect in line segments which are called **edges**. Do you agree that a tetrahedron has six edges? Each point where edges meet is called a **vertex**. Do you count four vertices in the illustration?

Each of these figures represents a polyhedron. We classify polyhedrons according to the number of their faces.

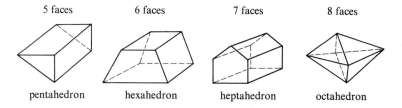

The five figures below are the **regular polyhedrons**. In the case of a regular polyhedron all the faces are **congruent regular** polygons. That is, they are all the same size and shape.

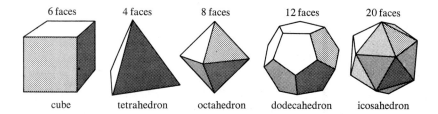

The 18th century mathematician Euler (1707–1783) developed a formula describing the relationship between the number of faces, edges, and vertices of any polyhedron. It is as follows:

$$\text{Number of Faces} + \text{Number of Vertices} = \text{Number of Edges} + 2$$

Questions

1. What is the smallest number of faces a polyhedron may have? the smallest number of edges? the smallest number of vertices?
2. The area of one face of a cube is 14 square centimeters. What is the total surface area?
3. The length of one edge of a regular tetrahedron is 23 inches. What is the total length of all its edges?
4. A regular polyhedron has 6 vertices and 12 edges. What is the name of the regular polyhedron?
5. The area of one face of a regular icosahedron is 3.2 square inches. What is the total area?
6. A dodecahedron has 30 edges. How many vertices does it have?
7. An icosahedron has 12 vertices. How many edges does it have?
8. What is the number of faces, vertices, and edges for each figure shown below? Use Euler's formula to check your answers.

a.

c.

b.

d.

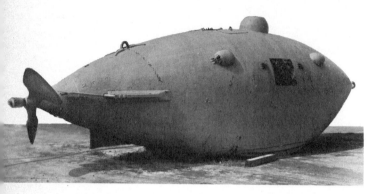

The "Intelligent Whale" . . .

A Gulf Stream explorer . . .

Negative Numbers

The first submarine officially sponsored by the United States Navy was the "Intelligent Whale," built in 1866. This undersea craft failed her only diving test, but such early efforts led to the development of submarines with their present capabilities. In 1969 the submarine "Ben Franklin" made important contributions to oceanography with its Gulf Stream Drift Mission. The 145-ton submersible drifted for 31 days in the Gulf Stream, covering a distance of 1444 nautical miles, from Florida to a point 300 miles south of Halifax, Nova Scotia. In the course of the journey, the crew of six gained surprising information about the speed of the current and its changes of direction, as well as other oceanographic data.

Directed Numbers

8–1 Directed Numbers and the Number Line

The following illustration represents the portion of the number line that we have used in our study of the numbers of arithmetic.

Now we propose to introduce another set of numbers, of which the numbers of arithmetic form a subset. You will recall that the point on the number line whose coordinate is zero is called the **origin**. You have probably seen the number line pictured with numbers assigned to points to the **left** of zero as well as to the **right** of zero. To accomplish this, think of the number line as extending to both the left and the right of the origin. Numbers which name points to the right of zero we choose to call **positive numbers**; those naming points to the left of zero we choose to call **negative numbers**.

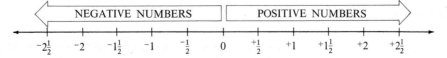

218 *Chapter 8*

To indicate whether the point corresponding to a number is to the left or to the right of the origin we shall use a small, raised $^+$ or $^-$ sign. The small raised $^-$ sign to the left of a numeral tells us that it names a point that lies to the **left** of the origin, or zero point; similarly a small raised $^+$ sign tells us that the point represented is to the **right** of the origin. It is important that you do not confuse the small, raised $^+$ and $^-$ signs with the larger symbols $+$ and $-$ used to indicate the operations of addition and subtraction.

The symbol $^+5$ is read "positive five" and names the point that is 5 units to the **right** of the origin; $^-5$ is read "negative five" and names the point that is five units to the **left** of the origin. The distance between a point and the origin is the **magnitude** of the number naming the point. Both $^+5$ and $^-5$ name points that lie at the same distance (5 units) from the zero point, so each has a magnitude of **5**. Since the idea of **direction** is important in understanding these numbers we shall refer to them as **directed numbers**. Although zero is neither negative nor positive, it is considered to be a directed number.

Thus each directed number is the **coordinate** of a point on the number line, and the associated point is called the **graph** of the number. However, when the meaning is clear, we shall merely refer to a point by naming its coordinate. For example:

If we say:	*We mean:*
The point $^-7$ is 7 units units to the left of **0**.	$^-7$ is the coordinate of the point 7 units to the left of the origin, or the point 7 units to the left of the origin is the graph of $^-7$.
The point $^+2\frac{1}{2}$ is $2\frac{1}{2}$ units to the right of **0**.	The graph of $^+2\frac{1}{2}$ is $2\frac{1}{2}$ units to the right of the graph of **0**.

Figure 8–1

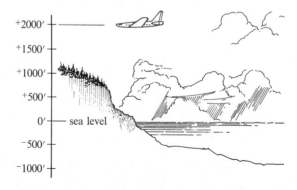

Figure 8–2

Negative Numbers   219

The idea of positive and negative numbers is often applied in ways familiar to you. For example, the temperature reading for the thermometer in Figure 8–1 is 10 degrees below zero and can be indicated by the directed number ⁻10. The position of the airplane in Figure 8–2 can be described as 2000 feet above sea level. Do you see that this is indicated by the directed number ⁺2000?

ORAL EXERCISES

Tell in words the meaning of each of the following.

SAMPLE 1: ⁺7   *What you say:* positive seven
SAMPLE 2: ⁻2¼   *What you say:* negative two and one-fourth

1. ⁺10
2. ⁺390
3. ⁻16
4. ⁻3.8
5. ⁺2$\frac{1}{10}$
6. ⁻3$\frac{5}{7}$
7. ⁺34.6
8. ⁺11
9. ⁻π
10. ⁻$\frac{7}{8}$
11. ⁺$\frac{3}{4}$
12. ⁺1.027

Name the directed number whose graph is located as described in each of the following.

SAMPLE: 5 units to the left of zero   *What you say:* negative five

13. 6 units to the right of 0
14. 3 units to the left of 0
15. 2⅛ units to the left of 0
16. 4⅔ units to the right of 0
17. 9½ units to the right of 0

Name the directed number described by each of the following.

18. The positive number of magnitude 7
19. The negative number of magnitude 1½
20. The directed number that is neither positive nor negative

Express each thermometer reading by using a directed number.

21.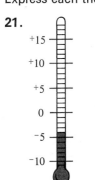
22.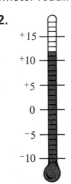
23.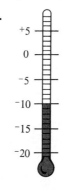
24.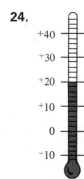

Chapter 8

For each number line, name the directed number coordinates of the points that are labeled with letters.

25.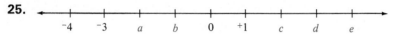

26.

27.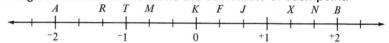

WRITTEN EXERCISES

Use the given number line to name the coordinate of each point.

SAMPLE: *R*   **Solution:** $-1\frac{1}{3}$

A
1. *T*
2. *F*
3. *J*
4. *A*
5. *N*
6. *X*
7. *B*
8. *M*
9. *K*

Use a directed number to express each of the following.

SAMPLE: 5 degrees above zero   **Solution:** $+5$

10. 72 degrees above zero
11. 8 degrees below zero
12. 95 feet above sea level
13. 300 feet below sea level
14. 3.9 degrees above zero
15. 18.2 degrees below zero
16. $12\frac{1}{4}$ degrees below zero
17. $\frac{3}{4}$ degree above zero
18. zero degrees
19. $4\frac{1}{3}$ degrees below zero
20. 4 degrees Fahrenheit above freezing (Freezing is 32°F.)
21. 12 degrees Fahrenheit below freezing
22. 4 degrees Centigrade below freezing (Freezing is 0°C.)

Name each directed number that is described.

SAMPLE: A positive number 6 units from 0   **Solution:** $+6$

23. A positive number 15 units from 0
24. A negative number 3 units from 0

25. A negative number 10.1 units from 0
26. A positive number 2.7 units from 0

[B] 27. A number that lies the same distance from 0 as ⁺9
28. A number that lies the same distance from 0 as ⁻34
29. A number that lies the same distance from 0 as ⁻4⅝
30. A number that lies the same distance from 0 as ⁺2.75
31. A number that lies the same distance from 0 as ⁻0.025
32. Two directed numbers such that each is 6 units from 0
33. Two directed numbers such that each is 17 units from 0
34. Two directed numbers such that each is 3⅛ units from 0
35. Two directed numbers such that each is 7.2 units from 0

[C] 36. Two directed numbers such that each is 3 units from ⁺18
37. Two directed numbers such that each is 10 units from ⁺14
38. Two directed numbers such that each is 5 units from ⁺2
39. Two directed numbers such that each is 5 units from ⁻2
40. Two directed numbers such that each is 3½ units from ⁺3

8–2 Moves on the Number Line Using Directed Numbers

The idea of "moves" on the number line can be illustrated by the yards gained or yards lost by a football team. As an example suppose the team that has the ball **gains 3** yards on the first play and **loses 5** yards on the next play. This information is shown on the number line, for which we assume that the first play originated at **0**.

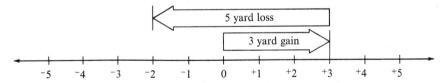

Do you agree that the result of the two plays can be represented by the directed number ⁻2?

Since positive numbers lie to the right of zero on our number line, we shall consider that a move towards the **right** is a move in the **positive** direction; similarly, we shall call a move to the **left** a move in the **negative** direction.

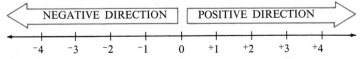

Do you see how this agrees with the football team example, in which the arrow representing the **gain** points to the **right**? In which direction does the arrow point that represents the **loss**?

The number line illustration in Figure 8–3 shows that if, beginning at **0**, a move of **2** units in the **positive** direction is followed by a move of **3** units in the **positive** direction, we finish at ⁺5.

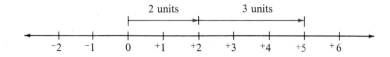

Figure 8–3

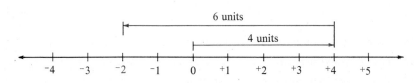

Figure 8–4

Figure 8–4 also shows two successive moves, beginning at **0**. A move of **4** units in the **positive** direction is followed by a move of **6** units in the **negative** direction. Do you see why we finish at ⁻2?

ORAL EXERCISES

Tell the location of each directed number with respect to 0.

SAMPLE: ⁺20   *What you say:* twenty units in the positive direction

1. ⁺3
2. ⁺5¼
3. ⁻10
4. ⁻8
5. ⁻2⅕
6. ⁺2⅞
7. ⁺3.04
8. ⁺14
9. ⁻0.17

Use the number line to complete each statement.

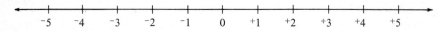

SAMPLE: A move from 0 to ⁺3 is __?__ units to the __?__.

What you say: A move from 0 to ⁺3 is 3 units to the *right*.

10. A move from 0 to ⁺5 is __?__ units to the __?__.
11. A move from 0 to ⁻2 is __?__ units to the __?__.

12. A move from 0 to ⁻4½ is __?__ units to the __?__.
13. A move from ⁺2 to ⁺5½ is __?__ units to the __?__.
14. A move from ⁻1 to ⁺3 is __?__ units to the __?__.
15. A move from ⁺2 to ⁻2 is __?__ units to the __?__.

WRITTEN EXERCISES

Make a number line sketch to show the moves given in each exercise. Tell where you finish.

SAMPLE: Start at 0; move 3 units in the positive direction; then move 4 units in the negative direction.

Solution:

Finish at ⁻1.

A 1. Start at 0; move 3 units in the positive direction; then move 2 units in the negative direction.

2. Start at 0; move 2 units in the negative direction; then move 3 units in the positive direction.

3. Start at 0; move 1 unit in the positive direction; then move 3 units in the negative direction.

4. Start at 0; move 3 units in the negative direction; then move 1 unit in the negative direction.

5. Start at 0; move 2½ units in the positive direction; then move 2½ units in the negative direction.

6. Start at 0; move 3½ units in the negative direction; then move 3½ units in the positive direction.

Name the directed number suggested by each arrow drawn in the figure.

SAMPLE:

Solution: ⁻3

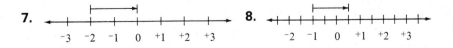

9.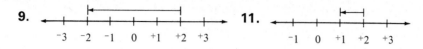

11.

10.

12.

Each statement refers to moves on the number line. Make a number line sketch and tell where you finish.

SAMPLE: Start at ⁻3 and move 5 units in the positive direction.

Solution: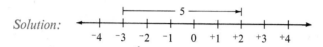

Finish at ⁺2.

13. Start at ⁺2 and move 5 units in the negative direction.
14. Start at ⁺3 and move 2 units in the negative direction.
15. Start at ⁻2 and move 7 units in the positive direction.
16. Start at ⁻4 and move 3 units in the positive direction.
17. Start at ⁻1 and move 3 units in the negative direction.
18. Start at ⁺3½ and move 5 units in the negative direction.

B 19. Start at ⁻2, move 5 units in the positive direction, and then move 2 units in the positive direction.
20. Start at ⁺4, move 3 units in the negative direction, and then move 3 units in the positive direction.

Complete each statement with the word *positive* or *negative*.

21. A move from ⁺5 to ⁺10 is in the __?__ direction.
22. A move from ⁺8 to ⁺3 is in the __?__ direction.
23. A move from ⁻6 to ⁻1 is in the __?__ direction.
24. A move from ⁻3 to 0 is in the __?__ direction.
25. A move from ⁻2 to ⁻7½ is in the __?__ direction.
26. A move from ⁺4½ to ⁻4½ is in the __?__ direction.

Name the coordinate of each point described.

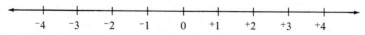

C 27. The point half the distance from ⁻3 to ⁺1.
28. The point half the distance from ⁻4 to ⁻2.
29. The point half the distance from ⁻3 to ⁺3.

30. The point one-third the distance from ⁻1 to ⁺2.
31. The point one-fourth the distance from ⁺½ to ⁻³⁄₂.
32. The point one-third the distance from ⁻⁵⁄₂ to ⁺⁷⁄₂.

8–3 Comparing Directed Numbers

You will probably recall how we interpreted the ideas of "is less than" and "is greater than" during our work with the numbers of arithmetic and the number line.

EXAMPLE 1.　We said:　Since $12\frac{1}{2}$ is to the **right** of 5, we know that $12\frac{1}{2}$ is greater than 5.
　　　　　　　We wrote:　$12\frac{1}{2} > 5$

EXAMPLE 2.　We said:　Since $4\frac{1}{8}$ is to the **left** of 9, we know that $4\frac{1}{8}$ is less than 9.
　　　　　　　We wrote:　$4\frac{1}{8} < 9$

This same idea applies to directed numbers. The arrows in the following diagram point to the directed numbers ⁻3 and ⁺2 on the number line.

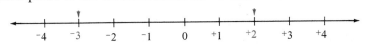

Since ⁻3 is to the **left** of ⁺2 we can say that **⁻3 is less than ⁺2**. Also, since ⁺2 is to the **right** of ⁻3 we can say that **⁺2 is greater than ⁻3**. This same line of reasoning leads us to conclude that **⁻4 is less than ⁻2**, since ⁻4 is to the **left** of ⁻2.

Study these number sentences to see if you agree that each is a true statement.

$$⁻3 > ⁻10 \qquad 0 > ⁻2 \qquad ⁺3 \geq ⁺1$$
$$⁺7 < ⁺8 \qquad ⁻5 < 0 \qquad ⁺1\frac{1}{2} \geq ⁻6$$
$$⁻6\frac{1}{2} < ⁻2 \qquad ⁻4 \leq ⁺2 \qquad ⁻\pi \leq ⁻3$$

ORAL EXERCISES

Tell whether the word *right* or *left* is required to complete each statement correctly.

SAMPLE:　⁻6 is to the __?__ of ⁺3.

What you say: ⁻6 is to the **left** of ⁺3.

1. ⁻2⅔ is to the __?__ of ⁻1.　　2. ⁺3.2 is to the __?__ of 0.

226 *Chapter 8*

3. $^+1\frac{1}{2}$ is to the _?_ of $^-1\frac{1}{2}$.
4. $^+18$ is to the _?_ of $^-35$.
5. $^-2.6$ is to the _?_ of $^-1.5$.
6. $^-10$ is to the _?_ of 0.
7. $^-19$ is to the _?_ of 0.
8. 0 is to the _?_ of $^-14$.
9. $^-49.5$ is to the _?_ of $^+0.005$.
10. $^-4$ is to the _?_ of $^-\pi$.

11. Any negative number is to the _?_ of any positive number.
12. 0 is to the _?_ of every positive number.

WRITTEN EXERCISES

Make each statement true by filling the first blank with either "right" or "left" and the second blank with either ">" or "<".

SAMPLE: $^+3$ is to the _?_ of $^-6$. So, $^+3$ _?_ $^-6$.

Solution: $^+3$ is to the **right** of $^-6$. So, $^+3 > {}^-6$.

1. $^-4$ is to the _?_ of $^-2$. So, $^-4$ _?_ $^-2$.
2. $^-5$ is to the _?_ of $^-10$. So, $^-5$ _?_ $^-10$.
3. $^+3\frac{1}{3}$ is to the _?_ of $^-4$. So, $^+3\frac{1}{3}$ _?_ $^-4$.
4. $^-8$ is to the _?_ of $^+1$. So, $^-8$ _?_ $^+1$.
5. $^+6$ is to the _?_ of $^-\frac{1}{2}$. So, $^+6$ _?_ $^-\frac{1}{2}$.
6. $^+\frac{3}{5}$ is to the _?_ of $^-\frac{3}{5}$. So, $^+\frac{3}{5}$ _?_ $^-\frac{3}{5}$.

Use > or < to complete each sentence so it will be a true statement.

7. $^+5$ _?_ $^-5$
8. $^-3$ _?_ $^+9$
9. $^+4$ _?_ $^+5$
10. 0 _?_ $^+4\frac{1}{3}$
11. 0 _?_ $^-\frac{2}{3}$
12. $^-7$ _?_ 0
13. $^+1$ _?_ 0
14. $^-9$ _?_ $^+9$
15. $^-\pi$ _?_ $^+\pi$
16. $^-35$ _?_ $^+210$
17. $^-62.8$ _?_ $^+0.55$
18. $^+3.27$ _?_ $^+0.89$
19. $^+0.05$ _?_ $^-0.003$
20. $^-3.95$ _?_ $^+1.25$
21. $^+0.001$ _?_ $^-34.8$

Tell whether each sentence is true or false.

B
22. $^-6 < {}^-3$
23. $^+6 > {}^-10$
24. $0 \geq {}^-\frac{1}{3}$
25. $^+5 \geq {}^-5$
26. $^-3 \leq {}^-3$
27. $^-7 \leq 0$
28. $^-10 \geq {}^-2$
29. $^-5 \neq {}^+5$
30. $0 \neq {}^-1$
31. $^-3 \leq {}^-7$
32. $^+0.03 < {}^+0.3$
33. $^-1.04 < {}^-1.03$
34. $^-0.007 > {}^-0.001$
35. $0 < {}^+0.0045$
36. $^-\pi \neq {}^+\pi$

Refer to the accompanying number line. Tell which statements are true and which are false, if *W, S, A, T, R,* and *M* are directed numbers, and if positive numbers, as is customary, are to the right of 0.

[C] 37. *A* is a negative number.
38. *T* is a positive number.
39. *W* is greater than zero.
40. *S* is a negative number.
41. *T* is greater than zero.
42. *W* is not a positive number.

43. $T > A$
44. $R \leq M$
45. $W \geq O$
46. $S \geq W$
47. $M = W$
48. $W \geq A$

8–4 Representations of Directed Numbers

You may have already discovered that a directed number may be represented in many different ways. For example, all these numerals represent the same directed number: $-\frac{1}{2} = -\frac{3}{6} = -0.5 = -\frac{5}{10}$. The number line below illustrates the idea that a coordinate for a point may be represented by many different numerals.

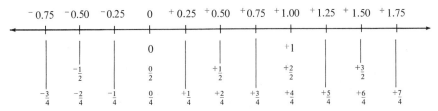

Numbers assigned to unit marks on the number line, such as -6, -4, -1, 0, $+5$, $+19$, and so on, are called **integers**. All other directed numbers, like $-\frac{3}{2}$, -1.05, π, $+3.7$, and $+\frac{7}{5}$, are not integers.

The small raised $-$ sign before the parentheses in Example 1 below indicates that the expression represents a negative number, -18. The expression in Example 2 represents a positive number, $+2$.

EXAMPLE 1. $-(13 + 5)$ is the same as -18.

EXAMPLE 2. $+(4 \cdot \frac{1}{2})$ is the same as $+2$.

In these examples, notice that we simplified each expression by first carrying out the indicated operation and then assigning the positive or negative sign to the result.

ORAL EXERCISES

Name the directed number represented by each of the following.

SAMPLE: $^-(\frac{5}{9} + \frac{2}{9})$ *What you say:* $^-\frac{7}{9}$

1. $^+(3 + 9)$
2. $^+(16 - 9)$
3. $^-(10 + 8)$
4. $^-(\frac{1}{2} + 3)$
5. $^+(3 + 8 + 6)$
6. $^-(23 - 9)$
7. $^+(5^2)$
8. $^-(2 \cdot 2 \cdot 2)$
9. $^-(3^3)$
10. $^-(10^2)$
11. $^+(1.3 + 1.5)$
12. $^+(\frac{3}{5} - \frac{2}{5})$
13. $^+(3 \cdot 15)$
14. $^+(10^3)$
15. $^-(100 + 40 + 5)$
16. $^+(57 \div 3)$
17. $^+(\frac{36}{3})$
18. $^-(18 + 7 + 0)$

WRITTEN EXERCISES

Tell which statements are true and which are false.

SAMPLE 1: $^+(19 - 7) = {^+}12$ *Solution:* $^+12 = {^+}12$; true

SAMPLE 2: $^+(19 - 5) \leq {^-}(6 + 8)$ *Solution:* $^+14 \leq {^-}14$; false

A
1. $^-(4 + 9) = {^+}(4 + 9)$
2. $^+(13 + 14) = {^+}(8 + 9)$
3. $^-(15 + 7) \geq {^-}(2 \cdot 11)$
4. $^-(18 - 6) = {^-}(6 + 18)$
5. $^+(100 + 40) = {^+}(90 + 50)$
6. $^+(3 \cdot 2 \cdot 1) = {^+}(1 \cdot 6)$
7. $^+(\frac{1}{2} + 4) > {^+}(1 + 2\frac{1}{2})$
8. $^-(\frac{4}{5} - \frac{1}{5}) = {^-}(\frac{2}{5} + \frac{1}{5})$

9. $^-(1.07 + 0.25) = {^-}(1.95 - 0.63)$
10. $^+(4 \cdot 4 \cdot 4) < {^+}(4 \cdot 3)$
11. $^-(10.3 + 7.1 + 4.4) = {^-}(4.5 + 16.3)$
12. $^-(10^2 + 40) = {^-}(12^2 - 4)$
13. $^+(300 + 40 + 3) \neq {^+}(300 + 30 + 4)$
14. $^-(10^2 + 10^3) < {^+}(10^2 + 10^3)$
15. $0 > {^-}(3 + 5 + 14)$
16. $^+(10 + \pi) \geq {^-}(10 + \pi)$
17. $^+(\frac{1}{8} + \frac{2}{8} + \frac{5}{8}) = {^+}(\frac{1}{8} + \frac{1}{4} + \frac{5}{8})$
18. $^-(\frac{1}{2} + \frac{3}{4} + \frac{1}{4}) = {^-}(0.5 + 0.75 + 0.25)$
19. $^-(10 \cdot 10 \cdot 10) = {^-}(10^3 \cdot 1)$
20. $^-\left(\dfrac{1 + 2 + 5}{8}\right) = {^-}\left(\dfrac{1 + 7}{8}\right)$

Replace each question mark by a number that makes the statement true.

SAMPLE: $^-(12 - 5) = {^-}(6 + ?)$ Solution: $^-(12 - 5) = {^-}(6 + 1)$

21. $^+(3 + 14) = {^+}(8 + ?)$
22. $^+(16 + 17) = {^+}(30 + ?)$
23. $^-(12 + 9) = {^-}(17 + ?)$
24. $^-(? + 12) = {^-}(9 + 5)$
25. $^+(17 - ?) = {^+}(4 \cdot 3)$
26. $^-(1.38 + 2.44) = {^-}(1 + ?)$
27. $^-(\frac{1}{5} + \frac{3}{5}) = {^-}(? + \frac{2}{5})$
28. $^+(70 + 5) = {^+}(? - 15)$
29. $^+(4.26 - 1.05) = {^+}(0.21 + ?)$
30. $^-(3.80 - ?) = {^-}(2.15 + 1)$

Show that each statement is true.

SAMPLE: $^+(10 + \frac{6}{3}) = {^+}(2^2 \cdot 3)$ Solution:
$$\frac{^+(10 + \frac{6}{3}) = {^+}(2^2 \cdot 3)}{\begin{array}{c|c} ^+(10 + 2) & ^+(4 \cdot 3) \\ ^+12 & ^+12 \end{array}}$$

B
31. $^+(100 + 30 + 5) = {^+}(90 + 45)$
32. $^+(5 \cdot 5 \cdot 5) = {^+}(100 + 20 + 5)$
33. $^-(4 + 4 + 4) = {^-}(2 \cdot 2 \cdot 3)$
34. $^+(10 \cdot 10) = {^+}(9^2 + 19)$
35. $^+(\frac{1}{10} + \frac{4}{10}) = {^+}(4 - 3\frac{1}{2})$
36. $^-(1 + 0.75 + 0.25) = {^-}(10 \div 5)$
37. $^-(13 + \frac{9}{3}) = {^-}(4 \cdot 4)$
38. $^+(2 \cdot 2 \cdot 2 \cdot 2) = {^+}(8 + 0 + 8)$
39. $^+(100 - 36) = {^+}(4^3)$
40. $^-(10^2 + 69) = {^-}(13 \cdot 13)$

C
41. $^-[5 + 3(7 + 2)] = {^-}[(3 + 2)6 + 2]$
42. $^+[3(4 - 2)] = {^+}[(2 + 1)2]$

Name the following directed numbers, where "largest" means "lying farthest to the right on the number line." Assume that each numeral is in standard form.

43. The largest positive integer that can be represented by a two-digit numeral.
44. The smallest positive integer that can be represented by a two-digit numeral.
45. The largest negative integer that can be represented by a two-digit numeral.
46. The largest negative integer that can be represented by a three-digit numeral.
47. The smallest negative integer that can be represented by a three-digit numeral.

Inequalities and Directed Numbers

8-5 Inequalities with Directed Number Solutions

In our earlier work with the numbers of arithmetic we found that an open inequality such as $n > 7$ is neither true nor false until the variable is replaced by a number taken from a designated replacement set. This applies when the replacement set consists of directed numbers, just as it did for numbers of arithmetic.

Let us consider the open sentence $x < {}^+2$ where the designated replacement set is $\{{}^-2, {}^-1, 0, {}^+1, {}^+2\}$. *Each directed number in the replacement set that can be used to replace the variable x and result in a true statement, is a* **solution**; *all* the directed numbers in the replacement set that result in true statements make up the **solution set** or **truth set**. A **truth table** is shown below.

x-replacement	$x < {}^+2$	True/False
${}^-2$	${}^-2 < {}^+2$	True
${}^-1$	${}^-1 < {}^+2$	True
0	$0 < {}^+2$	True
${}^+1$	${}^+1 < {}^+2$	True
${}^+2$	${}^+2 < {}^+2$	False

Therefore ${}^-2, {}^-1, 0,$ and ${}^+1$ are solutions. The solution set is written $\{{}^-2, {}^-1, 0, {}^+1\}$. The graph of this set is as follows.

ORAL EXERCISES

Name the set of directed numbers graphed in each number line picture.

SAMPLE:

What you say: $\{{}^-3, {}^-2, {}^-1\}$

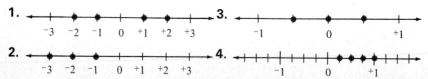

5. 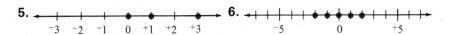 6.

Tell whether each sentence is true, is false, or is neither true nor false.

7. $^+3 < {}^-5$ 10. $-\frac{1}{3} \neq 0$ 13. $t > {}^+47$
8. $^-2 < n$ 11. $^+6 < x$ 14. $y \geq {}^-16$
9. $k > {}^+2\frac{1}{2}$ 12. $^-10 > {}^+3$ 15. $^+35 \geq {}^-35$

WRITTEN EXERCISES

Show whether a true or a false statement results when the variable in the sentence is replaced by each member of the replacement set. Then state the solution set.

SAMPLE: $n \geq {}^-2$; $\{{}^-6, {}^-4, {}^+2, 0\}$

Solution: $^-6 \geq {}^-2$, false; $^-4 \geq {}^-2$, false; $^+2 \geq {}^-2$, true;
$\quad\quad\quad\;\; 0 \geq {}^-2$, true
$\quad\quad\quad$ Solution set: $\{{}^+2, 0\}$

A

1. $w > {}^-3$; $\{{}^-10, {}^-5, {}^+5\}$ 7. $k \leq 0$; $\{{}^-1, -\frac{1}{2}, 0, {}^+\frac{1}{2}\}$
2. $t < {}^+2$; $\{{}^-2, {}^-1, 0, {}^+1\}$ 8. $^+5 > x$; $\{{}^-6, {}^-3, {}^+6\}$
3. $^-3 < n$; $\{{}^-8, {}^-6, {}^-4\}$ 9. $^+20 > t$; $\{{}^+30, {}^+25, {}^+15\}$
4. $^+17 > a$; $\{{}^-10, 0, {}^+10, {}^+20\}$ 10. $r \geq {}^-3$; $\{{}^-3, {}^-2, {}^-1, 0\}$
5. $^-31 < k$; $\{{}^+10, {}^+30, {}^-10, {}^-30\}$ 11. $^-35 < h$; $\{{}^-40, {}^-41, {}^-42\}$
6. $b \geq {}^-2$; $\{{}^-3, {}^-2, {}^+3, {}^+2\}$ 12. $^+\frac{3}{5} \geq m$; $\{0, {}^+1, {}^+2\}$

Write the solution set for each open sentence if the replacement set is $\{{}^-3, {}^-2, {}^-1, 0, {}^+1, {}^+2\}$. Graph the solution set.

SAMPLE: $x \geq {}^-2.5$ Solution: $\{{}^-2, {}^-1, 0, {}^+1, {}^+2\}$

13. $w < {}^+1$ 15. $x \leq {}^+2$ 17. $^+1.5 > r$
14. $m \leq {}^-2$ 16. $y \geq 0$ 18. $-\frac{1}{2} < a$

B

19. $^-1 \geq s$ 21. $^-2.1 < n$ 23. $0 \geq b$
20. $^-3 \leq d$ 22. $k \geq {}^+0.5$ 24. $0 \leq v$

For each open sentence, make a truth table using the designated replacement set. Then write the solution set.

25. $^+2.4 > t$; $\{{}^+3, {}^+2.5, {}^+1.5, {}^+1.0\}$
26. $k > {}^-0.003$; $\{{}^+0.002, {}^+0.003, {}^-0.001\}$

27. $^+1.7 > a$; $\{^-0.5, 0, ^+1.5, ^+2.3\}$
28. $^-(\frac{3}{4}) \leq m$; $\{^+(\frac{1}{8}), ^+(\frac{1}{2}), ^-(\frac{3}{8}), ^-(\frac{3}{5})\}$

[C] 29. $z \geq {}^-2.3$; $\{^-3.1, ^-2.4, ^-1.9, ^+2.2\}$
30. $^+3.7 < t$; $\{^-4.5, ^-2.5, 0, ^+3.5, ^+4\}$
31. $^-\frac{4}{5} < m$; $\{^-1, ^-\frac{3}{5}, ^+\frac{1}{2}, ^+\frac{2}{3}, 0\}$
32. $^-\frac{2}{3} \leq y$; $\{^-1\frac{1}{2}, ^-1, ^-\frac{1}{3}, 0, ^+\frac{1}{3}, ^+\frac{2}{3}\}$

8–6 More about Inequalities and Directed Numbers

Suppose you are to name the solution set of the inequality $x > {}^-2$ when the replacement set is {the directed numbers}. Do you agree that the solution set is {the directed numbers greater than $^-2$}? The graph of this solution set will include all of the numbers to the **right** of $^-2$, and is as shown here.

The open dot in the graph indicates that the directed number $^-2$ is *not* a member of the solution set.

Now let us consider the inequality $t \leq {}^+\frac{1}{2}$ when the replacement set is {the directed numbers}. As you have probably guessed, the solution set includes $^+\frac{1}{2}$ and all the numbers to the **left** of $^+\frac{1}{2}$. The graph looks like this:

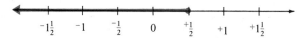

The solid dot at $^+\frac{1}{2}$ tells us that $^+\frac{1}{2}$ is a member of the solution set.

Previously we studied the idea of **between** in connection with the numbers of arithmetic. The illustration in Figure 8–5 below shows the graph of all of the numbers of arithmetic between $\frac{1}{2}$ and 3. Figure 8–6 shows all of the directed numbers between $^-1$ and $^+1$.

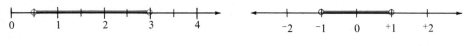

Figure 8–5   Figure 8–6

To solve the inequality $^-2 < y < {}^+1$, you should recall that it means "What are the numbers between $^-2$ and $^+1$?" If the replacement set is {the directed numbers} we can graph the solution in this way:

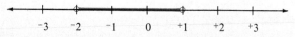

Negative Numbers 233

ORAL EXERCISES

State a question suggested by each inequality. The replacement set is {the directed numbers}.

SAMPLE 1: $k \geq {}^-3$ *What you say:* What directed numbers are equal to or greater than $^-3$?

SAMPLE 2: $^-10 \leq y < {}^-5$ *What you say:* What directed numbers are between $^-10$ and $^-5$?

1. $m < {}^-2$
2. $t > {}^+1$
3. $r \geq {}^-\frac{1}{3}$
4. $x \leq {}^+\frac{1}{4}$
5. $^-10 > w$
6. $^+3 < s$
7. $^+\frac{5}{8} \leq n$
8. $a \geq {}^+4$
9. $^-1 < b < {}^+3$
10. $^+2 < s < {}^+10$
11. $^-7 < w < {}^-4$
12. $^-\frac{3}{5} < a \leq 0$

Match each inequality in Column 1 with the graph of its solution set in Column 2. The replacement set is {the directed numbers}.

COLUMN 1   COLUMN 2

13. $n > {}^-\frac{1}{3}$
14. $t \leq {}^+1$
15. $a < {}^+\frac{1}{2}$
16. $^-1.5 < m < {}^+2$

A.

B.

C.

D.

E.

F.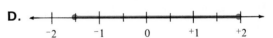

WRITTEN EXERCISES

Graph each set of directed numbers.

SAMPLE: {the directed numbers less than $^+\frac{1}{3}$}

Solution:

A   1. {the directed numbers greater than $^-4$}

2. {the directed numbers less than 0}
3. {the directed numbers greater than $^+\frac{1}{3}$}
4. {the directed numbers greater than or equal to $^-2$}
5. {the directed numbers less than or equal to $^+1\frac{1}{2}$}
6. {the directed numbers between $^-2.5$ and $^+1$}

Name the set of directed numbers described by each number line graph.

SAMPLE 1: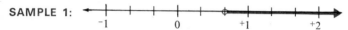

Solution: {the directed numbers greater than $^+\frac{2}{3}$}

SAMPLE 2:

Solution: {$^-2$, $^+1$, and the directed numbers between $^-2$ and $^+1$}

7.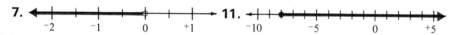

8.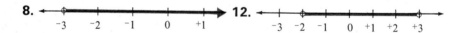

9.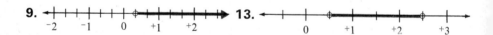

10.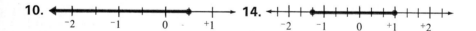

B 15.

16.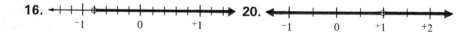

17.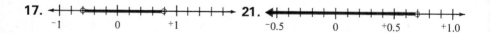

18.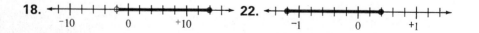

Name the solution set for each inequality if the replacement set is the set of directed numbers.

SAMPLE: $^-3 \leq k < {}^+4$

Solution: {$^-3$ and the directed numbers between $^-3$ and $^+4$}

[C]
23. $^+2 \geq r$
24. $^-7 < t \leq {}^+1$
25. $^-2\frac{1}{2} < w < {}^+5$
26. $q \neq {}^-3$
27. $m > {}^-(3 + 9)$
28. $k \leq {}^+(\frac{1}{3} + 5)$
29. $0 < m < {}^+3\frac{1}{3}$
30. $^-4 \leq x \leq {}^+3.4$
31. $^-3.75 \leq k < {}^+1.04$
32. $^-0.001 \leq y < 0$
33. $^-6(4 + 2) \geq t$
34. $w \neq 3 + 2(4 + \frac{1}{2})$

CHAPTER SUMMARY

Inventory of Structure and Concepts

1. Numbers can be assigned to points to the left as well as to the right of zero on the number line. The **magnitude** of a number is determined by its distance from the zero point, which is called the **origin**.

2. Zero and the set of numbers that correspond to the points of the number line on each side of the origin are called the **directed numbers**.

3. The small raised $^+$ sign assigned to a numeral for a directed number indicates that the number represented is **positive** and names a point to the **right** of the origin on the number line. Similarly, the small raised $^-$ sign indicates that the number represented is **negative** and names a point to the **left** of zero on the number line.

4. A given directed number is **larger** than a second directed number if it lies to the **right** of the second number on the number line.

   A given directed number is **smaller** than a second directed number if it lies to the **left** of the second number on the number line.

5. A directed number that corresponds to a point on the number line is the **coordinate** of that point; the point is called the **graph** of the number.

6. A move to the **right** on the number line is a move in the **positive** direction; a move to the **left** is a move in the **negative** direction.

7. Any directed number may be represented by many different numerals.

8. The directed numbers $\ldots, {}^-3, {}^-2, {}^-1, 0, {}^+1, {}^+2, {}^+3, \ldots$ correspond to unit markings on the number line and are called **integers**.

Vocabulary and Spelling

origin (*p. 217*)
positive number (*p. 217*)
negative number (*p. 217*)
magnitude (*p. 218*)
direction (*p. 218*)
directed number (*p. 218*)
negative direction (*p. 221*)

positive direction (*p. 221*)
integer (*p. 227*)
solution (*p. 230*)
solution set (*p. 230*)
truth set (*p. 230*)
between (*p. 232*)

Chapter Test

1. Name the directed number 3 units to the left of $^+1$.
2. Name the directed numbers that are each 12 units from $^-4$.
3. Name the directed number at which you arrive if you start at $^+7$ and move six units in the negative direction.
4. Name the directed number at which you arrive if you start at $^-3$ and move 8 units in the negative direction and then move 3 units in the positive direction.
5. Name the coordinate of the point three-fourths of the distance from $^-1$ to $^+3$.

Use one of the symbols $<$ or $>$ to make each sentence in Questions 6–8 true.

6. $^-7 \underline{?} ^-2$
7. $^-2.5 \underline{?} ^-2.6$
8. $^+1.1 \underline{?} ^-3.1$

9. The largest negative number whose numeral has two digits is $\underline{?}$.

Write the solution set for each open sentence if the replacement set is $\{^-9, ^-8, ^-7, \ldots, ^-1, 0, ^+1, \ldots, ^+9\}$.

10. $x < ^-5$
11. $^-3 \leq t < ^+5$
12. $^-1.5 > k$
13. $^-12 < p < ^-7$

Graph each set of directed numbers.

14. {the directed numbers less than or equal to $^-1$}
15. $\{^+1.5 < x \leq ^+5\}$, where the replacement set for x is {the directed numbers}.

Negative Numbers

Chapter Review

8–1 Directed Numbers and the Number Line

Name the directed number indicated by each letter on this number line.

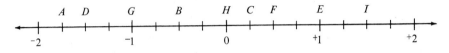

1. A
2. B
3. C
4. D
5. E
6. F
7. G
8. H
9. I

Name each directed number that is described below.

10. A number 15 units to the left of 0.
11. A number in the opposite direction from $^+3\frac{3}{4}$ but with the same magnitude.
12. Two numbers, each 16 units from $^+12$.
13. Two numbers, each 10 units from $^-6$.

8–2 Moves on the Number Line Using Directed Numbers

Questions 14–16 refer to moves on the number line. Name the directed number at which you finish.

14. Start at 0 and move 8 units in the positive direction; then move 5 units in the negative direction.
15. Start at $^+4$ and move 17 units in the negative direction.
16. Start at $^-7$ and move 13 units in the negative direction, and then move 6 units in the positive direction.
17. A move from $^-9$ to a $^+3$ is a move in the __?__ direction.
18. Name the coordinate of the point one-fourth the distance from $^-1$ to $^-3$.

8–3 Comparing Directed Numbers

Use < or > to complete each sentence and make it a true statement.

19. $^+4$ __?__ $^-2$
20. $^-7$ __?__ $^-3$
21. $^-1$ __?__ $^-\frac{1}{2}$

238 *Chapter 8*

Tell whether each sentence is true or false.

22. $^-1.04 < {}^-1.05$ **23.** $^-1.25 \leq {}^+1.25$ **24.** $^-0.5 < {}^-0.55$

8–4 Representations of Directed Numbers

In Questions 25–28, tell which statements are true and which are false.

25. $^+(9 - 3) = {}^+6$
26. $^-(3 \cdot 3 \cdot 3) = {}^-(3 + 3 + 3)$
27. $^-(15 + 9) < {}^-(14 + 8)$
28. $^+(15 + 3) > {}^-(5^2)$

29. What is the smallest negative number that can be named with a standard two-digit numeral?

8–5 Inequalities with Directed Number Solutions

Write the solution set for each open sentence if the replacement set is $\{^-4, {}^-3, {}^-2, {}^-1, 0, {}^+1, {}^+2, {}^+3\}$.

30. $w \leq {}^-2$
31. $^+2 > x$
32. $k > {}^-1$
33. $^-3 \geq y$

8–6 More about Inequalities and Directed Numbers

Graph each of the following.

34. {the directed numbers greater than or equal to $^-4$}
35. {the directed numbers between $^-3\frac{1}{4}$ and $^+1\frac{1}{2}$}
36. $\{^+2 \leq x\}$ where the replacement set is {the directed numbers}.
37. $\{^-3 < x \leq {}^+2\}$ where the replacement set is {the directed numbers}.

Name the set of directed numbers represented by each number line graph.

38.

39.

40.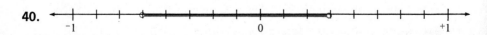

Review of Skills

Using a directed number line as a guide, answer the following with corresponding directed numbers.

1. A gain of 6 yards on the first down is followed by a loss of 9 yards on the second down. Where is the ball then, in relation to its location at the beginning of the first down?
2. Walking 20 miles east (positive direction) is followed by walking 13 miles to the west (negative direction). Where are you then, in relation to where you started?
3. You get on an elevator on the ninth floor above ground ($^+9$) and then ride the elevator ten floors in the downward direction. Where are you then, in relation to the ground level?

In Questions 4–9, name the opposite of each number or expression.

4. Left hand
5. $^+12$
6. Up
7. $^-5$
8. East
9. 0

10. If a football team gained 3 yards on one play and then lost 3 yards on the following play, where is the ball in relation to where it started?
11. What would you follow a move of $^+4$ by, to have as an end result no change (0)?
12. What would you follow a move of $^-7$ by, to have as an end result no change?

Simplify each expression.

13. $0 + 5$
14. $x + 0$
15. $(12 + 8) + 15$
16. $12 + (8 + 15)$
17. $x + (2 + 3)$
18. $(x + 2) + 3$
19. $15 + 4 - 4$
20. $x + 2 - 2$
21. $13 + 15 - 5$
22. $n + 13 - 4$
23. $15 + 3 + 7 - 4$
24. $x + 5 + 7 - 5$

For each of the following, try to determine a value of n that will make the statement true.

25. $n = 5 + 7$
26. $54 - n = 6$
27. $5 + n = 12$
28. $38 + n = 77$
29. $12 - n = 7$
30. $4 + n = 3$

Complete each set of number pairs according to the given function rule.

31. $x + 5 = f(x)$: $\{(2, 7), (5, ?), (10, ?), (?, 12), (?, 20)\}$
32. $x - 7 = f(x)$: $\{(9, 2), (15, 8), (22, ?), (45, ?), (?, 3), (?, 17)\}$

240   *Chapter 8*

■ ■

CHECK POINT
FOR EXPERTS

Symmetry

You have probably folded a piece of paper and then cut a figure from it. The illustration below shows how a tree shape can be cut from a piece of paper.

 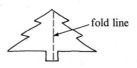

The fold line (broken line in the right-hand illustration) of the piece of paper is called a **line of symmetry** of our tree-shaped figure. If we fold the tree shape along this line, the two parts will fit over each other exactly. We can say that the figure is **symmetric** about the fold line.

Study the figure of a moth shown below. Do you agree that the broken line is a line of symmetry? We can test this conclusion by folding the figure along this line to see if the parts fit over each other exactly. Another way to test for a line of symmetry is to place a mirror along the broken line, as

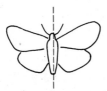 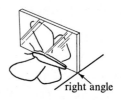

shown. If the mirror image, the part of the moth that you see in the mirror, is exactly like the part that is hidden behind the mirror, the figure is symmetric about the line determined by the edge of the mirror.

While one figure may not have *any* lines of symmetry, another may have *one*, and another, *more than one* line of symmetry.

Study these illustrations.

        No line of symmetry   One line of symmetry   More than one
                                                                  line of symmetry

A figure consisting of more than a single shape may also have one or more lines of symmetry. In Figure 1 the broken line is a line of symmetry.

Do you see that the broken line in Figure 2 is also a line of symmetry?

Questions

Which of the following figures have at least one line of symmetry?

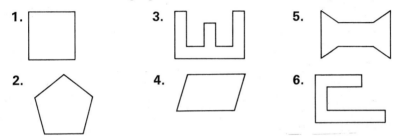

How many lines of symmetry can you find in each figure? Make a copy of each to illustrate your answer.

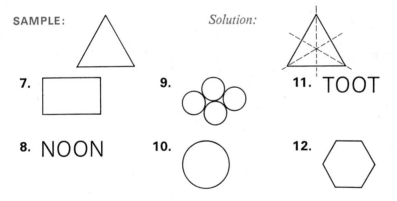

Copy each figure and sketch its mirror image, using the broken line as a line of symmetry for the completed figure.

One-tube radio, early 1920's . . .

World-wide Apollo 11 telecast . . .

Addition and Subtraction of Directed Numbers

At the start of the 1920's, the person who owned a one-tube, battery-operated radio set was the envy of his friends. Since amplifiers and loudspeakers were not available, the listener had to use earphones to enjoy this mode of communication. Through continual development and research, radios were improved, to be followed by the miracle of television. By 1969 this had reached the stage where a live telecast from the surface of the moon was possible. In a park in Seoul, Korea, a crowd of people are seen as they watch the start of the Apollo 11 mission. It was estimated that this momentous voyage and man's first moonwalk were seen and heard round the world by an audience of over 500 million people.

Addition

9–1 Number Line Addition and the Commutative Property

You will recall that we have used arrows in connection with the number line to represent directed numbers. By building on this idea we can interpret the addition of directed numbers through the use of the number line.

As you have probably discovered, zero and the positive directed numbers are the same as the numbers of arithmetic that you have already used. Therefore, the easiest place at which to begin the discussion of addition of directed numbers is the addition of two positive numbers. To find the sum of $^+3$ and $^+5$ we can think of the equation $^+3 + {^+5} = ?$. (Notice that the addition operation is indicated by the large $+$ sign. Do you remember that the small raised $^+$ sign tells us that a number is **positive**?)

To solve the equation $^+3 + {}^+5 = ?$, we begin at **0** and draw an arrow **3** units long in the positive direction. The second arrow begins where the first one finished and is drawn **5** units long, also in the positive direction.

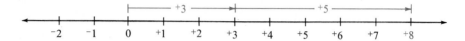

The combined lengths (displacements) of the arrows show that we finish at $^+8$. So, $^+3 + {}^+5 = {}^+8$. From our work with the numbers of arithmetic, we also know that $^+5 + {}^+3 = {}^+8$.

Now let us consider the sum of the directed numbers $^+6$ and $^-2$. In equation form we can write $^+6 + {}^-2 = ?$. To find a number line solution, we begin at **0** and draw an arrow **6** units long in the **positive** direction. Then, a second arrow begins where the first one finished and is drawn **2** units long in the **negative** direction.

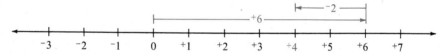

We finish at $^+4$. So, $^+6 + {}^-2 = {}^+4$.

The following number line illustration shows that in a similar manner $^-2 + {}^+6 = {}^+4$.

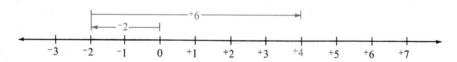

Can you visualize how to add two negative numbers on the number line? To add $^-4$ and $^-3$ we can think of the equation $^-4 + {}^-3 = ?$. Study this number line solution.

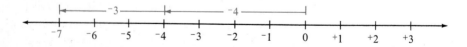

Thus, we see that $^-4 + {}^-3 = {}^-7$. Use the number line to check that $^-3 + {}^-4 = {}^-7$, also. The order in which two directed numbers are added does not affect the sum. In general, we can say that **addition of directed numbers is commutative.**

ORAL EXERCISES

Name the equation suggested by each number line illustration.

SAMPLE: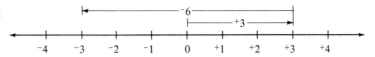

What you say: $^{+}3 + {^{-}6} = {^{-}3}$

1. 4.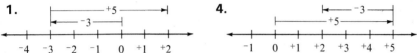

2. 5.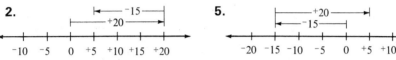

3. 6.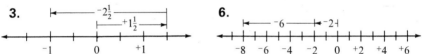

WRITTEN EXERCISES

Sketch a number line solution for each pair of equations. Complete each equation to make a true statement.

SAMPLE: $^{-}3 + {^{+}4} = ?$
$^{+}4 + {^{-}3} = ?$

Solution: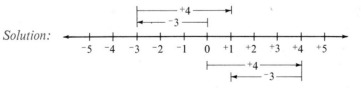

$$^{-}3 + {^{+}4} = {^{+}1} \quad \text{and} \quad {^{+}4} + {^{-}3} = {^{+}1}$$

A 1. $^{-}5 + {^{-}1} = ?$
$^{-}1 + {^{-}5} = ?$

2. $^{+}5 + {^{+}6} = ?$
$^{+}6 + {^{+}5} = ?$

3. $? = {^{+}8} + {^{-}4}$
$? = {^{-}4} + {^{+}8}$

4. $? = {^{-}3\frac{1}{2}} + {^{-}2\frac{1}{2}}$
$? = {^{-}2\frac{1}{2}} + {^{-}3\frac{1}{2}}$

5. $^+4 + {}^-5 = ?$
   $^-5 + {}^+4 = ?$
6. $^-8 + {}^+8 = ?$
   $^+8 + {}^-8 = ?$
7. $? = {}^-3 + {}^-6$
   $? = {}^-6 + {}^-3$
8. $? = {}^-7 + {}^+2\frac{1}{2}$
   $? = {}^+2\frac{1}{2} + {}^-7$
9. $^+6\frac{1}{2} + {}^-1 = ?$
   $^-1 + {}^+6\frac{1}{2} = ?$
10. $^+25 + {}^-15 = ?$
    $^-15 + {}^+25 = ?$

Tell whether each expression names a positive or a negative number. Then simplify the expression. Use the number line for help if necessary.

SAMPLE: $^+8 + {}^-3$ Solution: Positive; $^+8 + {}^-3 = {}^+5$

11. $^+7 + {}^+8$
12. $^+15 + {}^+9$
13. $^+25 + {}^-3$
14. $^+68 + {}^-10$
15. $^-10 + {}^+68$
16. $^+7 + {}^-9$
17. $^-17 + {}^+5$
18. $^+8 + {}^-19$
19. $^+14 + {}^-6$
20. $^-10 + {}^+11$
21. $^+7.5 + {}^+2.1$
22. $^-6.3 + {}^-1.5$
23. $^+2\frac{1}{2} + {}^+3$
24. $^-1\frac{1}{3} + {}^-2\frac{1}{3}$
25. $^-2\frac{1}{2} + {}^+1\frac{1}{2}$
26. $^+4\frac{1}{3} + {}^-2$
27. $^-3 + {}^+6\frac{1}{4}$
28. $^+3\frac{1}{4} + {}^-3\frac{1}{4}$

Copy and complete each addition table.

29.
+	$^+1$	$^-2$	$^+3$	$^-4$
$^+1$				
$^-2$				
$^+3$				
$^-4$				

30.
+	$^+5$	$^-5$	$^+10$	$^-10$
$^+5$				
$^-5$				
$^+10$				
$^-10$				

Find each sum.

B 31. $^+135$
       $^+26$

32. $^-167$
    $^+32$

33. $^+129$
    $^-116$

34. $^-15$
    $^+27$

35. $^-140$
    $^-356$

36. $^+234$
    $^-571$

37. $^-34$
    $^-47$

38. $^+150$
    $^-150$

Name the directed number that can replace each question mark and make the resulting statement true.

SAMPLE: $^+10 = {}^+15 + ?$ Solution: $^+10 = {}^+15 + {}^-5$

C 39. $^+23 = {}^+9 + ?$
40. $^-17 = {}^-5 + ?$
41. $^+15 = {}^-3 + ?$
42. $^+10 = {}^-9 + ?$
43. $^-16 = {}^+5 + ?$
44. $^-12 + ? = {}^-5$
45. $^+18 + ? = {}^+12$
46. $^-15 + ? = {}^-34$
47. $^+27 + ? = {}^+41$
48. $? + {}^-30 = {}^+60$
49. $? + {}^+30 = {}^-75$
50. $? + {}^-18 = 0$

PROBLEMS

Express each problem as the indicated sum of two directed numbers. Find the sum and answer the question.

SAMPLE: Mr. Thomas opens a bank account. He deposits $75.00. The next day he withdraws $47.00. How much money remains in the bank?

Solution: $^+75 + {}^-47 = ?$
Since $^+75 + {}^-47 = {}^+28$, $28.00 remains in Mr. Thomas' account.

1. William opens a checking account. He deposits $35.00. He writes a check for $16.00. How much remains in the account?
2. A football player makes a 12-yard gain on the first play. He loses 7 yards on the next play. How many yards had he gained or lost after the second play?
3. An airplane takes off and climbs to an altitude of 4500 feet. Next, the pilot descends 1800 feet. What is his new altitude?
4. On a cold winter night the temperature dropped 5 degrees before midnight. By morning it had dropped another 6 degrees. What was the total temperature change?
5. Ken opens a checking account with a $25 deposit. Later he writes a check for $30. What is the status of his account after the check is written?
6. A scuba diver descended 40 feet into the water. He then rose 18 feet toward the surface. At what depth was he then?
7. A messenger in a skyscraper office building got into an elevator at the third floor. He went down two floors. He then went up nine floors. At what floor was he then?
8. During a heavy spring rain a river rose $8\frac{1}{2}$ feet above flood level. The water then receeded $3\frac{1}{4}$ feet. At what level was the water then?
9. A stockmarket investor makes a gain of $225 on Monday. However, he has a loss of $220 on Tuesday. What is his profit or loss?
10. Mr. Kendall is in debt $1575. He pays off $490 of his debt. What is his financial position now?

9–2 Opposites of Directed Numbers

If you have been a sharp observer during our work with directed numbers you will have noticed that each positive number has a corresponding negative number. We say that corresponding pairs of numbers like ⁺6 and ⁻6, or ⁺3½ and ⁻3½, are **opposites** of each other. The number line figure below shows that numbers that are **opposites** of each other are equally distant from zero but lie on opposite sides of zero. Zero is its own opposite.

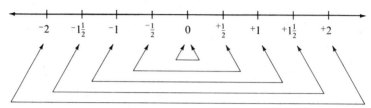

Read each of these statements carefully.

⁻5 is the opposite of ⁺5.   ⁺9 is the opposite of ⁻9.
0 is the opposite of **0**.   ⁻5½ is the opposite of ⁺5½.

The symbol −(⁻6) means "the opposite of ⁻6," or ⁺6; the symbol −(⁺3) means "the opposite of ⁺3," or ⁻3. Note that the symbol for "the opposite of" is the minus sign in its lowered position. For the present, we shall continue to use the small raised ⁻ sign to indicate a negative number.

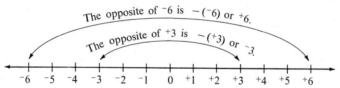

Check that each of the following statements is true:

−(⁻10) is the opposite of ⁻10. So, −(⁻10) = ⁺10.
−(⁺7) is the opposite of ⁺7. So, −(⁺7) = ⁻7.
−(⁺2½) is the opposite of ⁺2½. So, −(⁺2½) = ⁻2½.
−(⁻5.3) is the opposite of ⁻5.3. So, −(⁻5.3) = ⁺5.3.

ORAL EXERCISES

Name the opposite of each directed number.

SAMPLE: Negative ten   *What you say:* Positive ten

1. Positive seven   3. ⁺25   5. ⁻1.5
2. Positive twenty   4. ⁻48   6. ⁺⅝

Addition and Subtraction of Directed Numbers **249**

7. Negative fifteen
8. Zero
9. Negative one-half
10. ⁻3.01
11. ⁺π
12. ⁺0.75
13. ⁻340
14. ⁻0.35
15. ⁻π

WRITTEN EXERCISES

Express each of the following in symbols in two ways.

SAMPLE: The opposite of positive 5   Solution: −(⁺5); ⁻5

A
1. The opposite of positive 3
2. The opposite of positive $1\frac{1}{2}$
3. The opposite of negative 12
4. The opposite of negative 6
5. The opposite of negative 25
6. The opposite of ⁺13
7. The opposite of ⁻$2\frac{1}{8}$
8. The opposite of ⁺8.03
9. The opposite of 0
10. The opposite of −(⁻1)
11. The opposite of the number 8 units to the left of zero
12. The opposite of the number $5\frac{1}{3}$ units to the right of zero

Name each pair of directed numbers with reference to the number line.

SAMPLE: The numbers 4 units from zero   Solution: ⁺4, ⁻4

13. The numbers 17 units from zero
14. The numbers $\frac{1}{3}$ unit from zero
15. The numbers 12.5 units from zero
16. The numbers 0.025 units from zero
17. The numbers $6\frac{1}{8}$ units from zero
18. The numbers π units from zero

Give the meaning of each of the following; then write it in simplest form.

SAMPLE 1: −(⁺10)   Solution: The opposite of ⁺10; ⁻10
SAMPLE 2: −(⁻15)   Solution: The opposite of ⁻15; ⁺15

B
19. −(⁻21)
20. −(⁻75)
21. −(⁺4)
22. −(⁺9)
23. −(⁻$3\frac{1}{5}$)
24. −(⁺$8\frac{1}{4}$)
25. −(0)
26. −(⁺0.87)
27. −(⁺9.2)

Simplify each of the following.

SAMPLE: −(−(⁻3))   Solution: ⁻3

C
28. −(⁻5)
29. −(−(⁻5))
30. −(⁺2)
31. −(−(⁺2))
32. −(−(⁻6))
33. −(−(⁺14))

9–3 Addition: Additive Inverses

As we have pointed out, the set of directed numbers consists of the familiar numbers of arithmetic and their opposites. The result of adding a number to its opposite is of special importance in the study of algebra. Consider the number line illustrations in Figures 9–1 and 9–2.

Figure 9–1   Figure 9–2

In Figure 9–1, we see that $^+3 + {}^-3 = 0$, and in Figure 9–2 that $^-4 + {}^+4 = 0$. If we examined other cases on the number line, it would soon become clear that the **sum** of any **directed number** and its **opposite** is **zero**. Verify the truth of each of the following statements:

$^+2 + {}^-2 = 0$ $^-15 + {}^+15 = 0$ $0 = {}^-5\tfrac{1}{7} + {}^+5\tfrac{1}{7}$

$^+4\tfrac{1}{3} + {}^-4\tfrac{1}{3} = 0$ $0 + 0 = 0$ $0 = {}^-0.03 + {}^+0.03$

On the basis of the illustrations we have seen, we now state the **additive property of opposites**:

For every directed number r,

$$r + (-r) = -r + r = 0.$$

Because of this property, a directed number that is the opposite of another directed number is called its **additive inverse**.

$^-11$ is the **additive inverse** of $^+11$.

$^+11$ is the **additive inverse** of $^-11$.

So, $^+11$ and $^-11$ are additive inverses of each other.

$^-3\tfrac{1}{5}$ is the **additive inverse** of $^+3\tfrac{1}{5}$.

$^+3\tfrac{1}{5}$ is the **additive inverse** of $^-3\tfrac{1}{5}$.

So, $^+3\tfrac{1}{5}$ and $^-3\tfrac{1}{5}$ are additive inverses of each other.

Note that if the variable t represents a directed number, then "the opposite of t" is written $-t$, and

(1) when t is **positive**, $-t$ is **negative**,

(2) when t is **negative**, $-t$ is **positive**,

(3) when t is 0, $-t$ is also 0.

Addition and Subtraction of Directed Numbers

Think of selecting a directed number on the number line and then moving *no* units. It is obvious that you would finish at the starting point; that is, at the number first selected. Thus you can see that **0** is the **identity element** for **addition** of directed numbers, just as it was for numbers of arithmetic. The sentences $^+3 + 0 = {}^+3$ and $^-5 + 0 = {}^-5$ are true statements. In general, we now state the **additive property of zero** as follows:

For any directed number *r*,

$$r + 0 = 0 + r = r.$$

ORAL EXERCISES

Tell how to complete each of the following to make a true statement.

SAMPLE: $^+6$ is the additive inverse of __?__.

What you say: $^+6$ is the additive inverse of $^-6$.

1. $^-5\frac{1}{2}$ is the additive inverse of __?__.
2. $^+5\frac{1}{2}$ is the additive inverse of __?__.
3. $^+35$ is the additive inverse of __?__.
4. $^-1.6$ is the opposite of __?__.
5. __?__ is the additive inverse of $^-2\frac{1}{5}$.
6. __?__ is the additive inverse of $^+0.14$.

Give a simpler name for each expression.

SAMPLE: $^-23 + {}^+23$ *What you say:* Zero

7. $^-17 + {}^+17$
8. $^-26 + {}^+26$
9. $^+14 + {}^-14$
10. $^-62 + 0$
11. $^+\frac{3}{8} + {}^-\frac{3}{8}$
12. $^+0.76 + {}^-0.76$
13. $^+\pi + {}^-\pi$
14. $^+(3^2) + {}^-(3^2)$
15. $0 + {}^+10$

WRITTEN EXERCISES

Write the number statement suggested by each number line illustration.

SAMPLE:

Solution: $^+2 + {}^-2 = 0$

1. 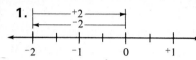   5.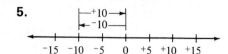

2.   6.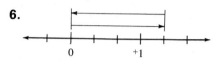

3.   7.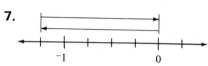

4.   8.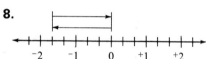

Tell which pairs of numbers are additive inverses of each other and which are not.

9. ⁻5, ⁺5
10. ⁺17, ⁻17
11. 0, ⁻3
12. 0, 0
13. −(⁻2), 2
14. ⁻75, ⁺75
15. ⁻6, −(⁺6)
16. ⁻5, −(⁻5)
17. −b, b

Find the solution set of each equation. The replacement set is the set of directed numbers.

SAMPLE 1: $m = {}^+8 + {}^-8$ Solution: $m = {}^+8 + {}^-8$
$m = 0$; solution set: $\{0\}$

SAMPLE 2: $a = {}^-12 + 0$ Solution: $a = {}^-12 + 0$
$a = {}^-12$; solution set: $\{{}^-12\}$

18. $k = {}^-40 + {}^+40$
19. $r = {}^+19 + 0$
20. $t = {}^-7\frac{1}{2} + {}^+7\frac{1}{2}$
21. $^+21 + {}^-21 = x$
22. $^-13 + w = {}^-13$
23. $^+3 + s = 0$

B 24. $^-15 + b = {}^-15$
25. $^-1\frac{1}{8} + w = 0$
26. $m + {}^+45 = 0$
27. $y + {}^+\frac{1}{9} = 0$
28. $0 = a + {}^-33$
29. $d = 0 + {}^+27$
30. $0 = {}^-44 + r$
31. $h = {}^-2.3 + {}^-2.3$
32. $^-3.9 + n = 0$

Show which sentences are true and which are false for the following replacements for the variables: $a = {}^+6; b = {}^-6; c = {}^+5; d = {}^-5$.

SAMPLE: $c + (-c) = 0$ Solution:

$c + -c = 0$	
$^+5 + (-({}^+5))$	0
$^+5 + {}^-5$	0
0	0

The sentence is true.

Addition and Subtraction of Directed Numbers 253

C 33. $b + a = 0$ 36. $d + c = 0$ 39. $-c + (-d) = 0$
   34. $a + b = 0$ 37. $-b + a = 0$ 40. $c + d = a + b$
   35. $d + (-d) = 0$ 38. $-a + b = 0$ 41. $c + (-c) = 0$

9–4 Using Additive Inverses to Simplify Expressions

Since the set of directed numbers consists of the numbers of arithmetic and their opposites, it seems logical to write positive directed numbers in the same manner as numbers of arithmetic. That is, $^+5$ can be written simply as 5; $^+7\frac{1}{4}$ can be written simply as $7\frac{1}{4}$. Let us agree that from now on, as a matter of convenience, we will write positive directed numbers without the small raised $^+$ sign.

Consider the expression $-(7 + 5)$, which means "the opposite of $(7 + 5)$."

$$\text{Since} \qquad 7 + 5 = 12,$$
$$\text{we know that} \qquad -(7 + 5) = -12.$$
$$\text{It is also true that} \quad -7 + (-5) = {^-7} + {^-5}$$
$$= {^-12}$$
$$= -12.$$
$$\text{Thus} \qquad -(7 + 5) = -7 + (-5).$$

Study the following examples:

EXAMPLE 1. $-(^-2 + 7) = -5$
              $-(^-2) + (-7) = 2 + {^-7}$
              $\qquad\qquad\quad = {^-5} = -5$
              So, $-(^-2 + 7) = -(^-2) + (-7)$

EXAMPLE 2. $-4 + (-10) = {^-4} + {^-10}$
            $\qquad\qquad = {^-14} = -14$
            $-(4 + 10) = -14$
            So, $-4 + (-10) = -(4 + 10)$

In general, it appears to be true that:

For any directed numbers r and s,

$$-(r + s) = -r + (-s).$$

ORAL EXERCISES

Give two meanings for each expression.

SAMPLE 1: $-(2 + 5)$ *What you say:* The opposite of 7, or -7; the opposite of 2 plus the opposite of 5, or $-2 + (-5)$

SAMPLE 2: $-(2 + {}^-5)$ *What you say:* The opposite of $^-3$, or $-({}^-3)$; the opposite of 2 plus the opposite of $^-5$, or $-2 + (-({}^-5))$

1. $-(6 + 1)$
2. $-(3 + 8)$
3. $-(4 + \frac{1}{2})$
4. $-(3 + 7 + 2)$
5. $-(5 + {}^-2)$
6. $-(10 + 6)$
7. $-(8 + 4 + 5)$
8. $-({}^-3 + 5)$
9. $-(6 + {}^-9)$
10. $-(4 + {}^-10)$
11. $-(3 + {}^-2)$
12. $-(12 + {}^-3)$

Name each sum.

13. $^-3 + 5$
14. $^-10 + 6$
15. $7 + {}^-3$
16. $10 + {}^-4$
17. $^-5 + 3$
18. $^-12 + 15$
19. $^-7 + {}^-6$
20. $^-3 + 11$
21. $^-18 + 6$
22. $^-20 + 11$
23. $^-15 + {}^-12$
24. $^-19 + {}^-35$

WRITTEN EXERCISES

Give a simpler name for each expression.

SAMPLE 1: $-(5 + 4)$ *Solution:* $-(5 + 4) = -(9) = {}^-9$

SAMPLE 2: $^-5 + {}^-3$ *Solution:* $^-8$

A
1. $-(6 + 5)$
2. $-(10 + 7)$
3. $-(13 + 19)$
4. $-(4\frac{1}{8} + 2\frac{1}{4})$
5. $-(\frac{5}{9} + \frac{2}{9})$
6. $^-3 + {}^-10$
7. $^-8 + {}^-20$
8. $^-17 + {}^-9$
9. $^-42 + {}^-21$
10. $^-0.35 + {}^-0.27$
11. $^-1.2 + {}^-3.4$
12. $^-\frac{1}{2} + {}^-3\frac{1}{4}$
13. $-({}^-3 + 6)$
14. $-(4 + {}^-2)$
15. $-({}^-2 + {}^-7)$

Show that each of the following statements is true.

SAMPLE 1: $-(9 + 5) = {}^-9 + {}^-5$ Solution: $-(9 + 5) = {}^-9 + {}^-5$
$$\begin{array}{c|c} -(14) & {}^-14 \\ {}^-14 & {}^-14 \end{array}$$

SAMPLE 2: $-({}^-5 + 4) = 5 + {}^-4$ Solution: $-({}^-5 + 4) = 5 + {}^-4$
$$\begin{array}{c|c} -({}^-1) & 1 \\ 1 & 1 \end{array}$$

16. $-(15 + 7) = {}^-15 + {}^-7$
17. $-(9 + 14) = {}^-9 + {}^-14$
18. $-24 + {}^-17 = {}^-(24 + 17)$
19. ${}^-35 + {}^-26 = -(35 + 26)$
20. $-({}^-7 + 6) = -({}^-7) + {}^-6$
21. $-(10 + {}^-3) = {}^-10 + 3$

B 22. $-(\frac{2}{3} + \frac{1}{2}) = {}^-\frac{2}{3} + {}^-\frac{1}{2}$
23. $-(1\frac{1}{4} + 1\frac{1}{2}) = {}^-1\frac{1}{4} + {}^-1\frac{1}{2}$
24. ${}^-3\frac{1}{5} + 2\frac{1}{10} = -(3\frac{1}{5} + {}^-2\frac{1}{10})$
25. ${}^-5 + 12 = -(5 + {}^-12)$
26. $-(7 + {}^-9) = {}^-7 + 9$
27. $-({}^-0.25 + 0.13) = 0.25 + {}^-0.13$

Find the solution set for each equation by making a number line sketch. The replacement set is the set of directed numbers.

SAMPLE: ${}^-2 + k = 3$

Solution:

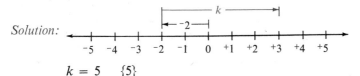

$k = 5 \quad \{5\}$

28. $3 + t = 7$
29. $5 + n = 2$
30. ${}^-3 + b = 4$
31. ${}^-7 + b = {}^-2$
32. $6 + m = 3$
33. ${}^-3 = 4 + x$
34. $5 = {}^-2 + y$
35. $b + 6 = 4$
36. $1 = w + 3$
37. $8 = {}^-6 + d$
38. ${}^-7 = {}^-3 + h$
39. ${}^-6 = {}^-1 + b$
40. $k = {}^-7 + {}^-4$
41. $r = {}^-9 + {}^-2$
42. $5 = t + 3\frac{1}{2}$

Find the solution set for each equation. The replacement set is the set of directed numbers.

SAMPLE: $-(9 + 7) = a$ Solution: $-(9 + 7) = a$
$ {}^-16 = a \quad \{{}^-16\}$

43. $-(3 + 5) = r$
44. $-(8 + 2) = n$
45. $-(4 + {}^-6) = b$
46. $-(5 + {}^-7) = d$
47. $-({}^-2 + 4) = x$
48. $-({}^-3 + 10) = s$
49. $t = -(8 + {}^-9)$
50. $a = -({}^-7 + {}^-12)$
51. $w = -(3 + 7 + 14)$
52. $y = -({}^-17 + 15)$

9–5 Number Line Addition and the Associative Property

The number line illustrations in Examples 1 and 2 show the sum $^-2 + {}^-3 + 6$ found in two different ways.

EXAMPLE 1. $(^-2 + {}^-3) + 6$ **EXAMPLE 2.** $^-2 + (^-3 + 6)$

Do you see that in each illustration the final arrow ends at 1? Thus

$$(^-2 + {}^-3) + 6 = 1 \quad \text{and} \quad ^-2 + (^-3 + 6) = 1.$$

Then the transitive property of equality permits us to conclude that $(^-2 + {}^-3) + 6 = {}^-2 + (^-3 + 6)$. The **addition of directed numbers is associative**.

The combined use of the associative property and the additive property of opposites helps us to add directed numbers easily without the help of a number line. Suppose that we are to do the addition $^-9 + 5 = ?$. Follow carefully the steps shown below. The tricky part is in the first step, in which the number having the greater magnitude is renamed to take advantage of the additive property of opposites.

$^-9 + 5 = (^-4 + {}^-5) + 5$ Substitution principle ($^-9$ is written as $^-4 + {}^-5$)
$ = {}^-4 + (^-5 + 5)$ Associative property
$ = {}^-4 + 0$ Additive property of opposites and substitution
$ = {}^-4$ Additive property of zero

We have now had a good deal of experience adding directed numbers. Although we have not discussed the properties of existence and uniqueness in connection with the set of directed numbers, you have probably assumed that when any two directed numbers are added the result is a unique directed number. Be certain you understand the following list of the **properties** of **addition** as they apply to directed numbers.

When *r*, *s*, and *t* are directed numbers:

$r + s$ is a unique directed number	**Closure property**
$r + s = s + r$	**Commutative property**
$(r + s) + t = r + (s + t)$	**Associative property**
$r + 0 = 0 + r = r$	**Additive property of zero**
$-r + r = r + (-r) = 0$	**Additive property of opposites**
$-(r + s) = -r + (-s)$	**Property of the opposite of a sum**

ORAL EXERCISES

Tell which of the following sentences are true and which are false.

1. $5 + {}^-5 = 0$
2. $7 + {}^-7 = 0$
3. ${}^-8 + 8 = 0$
4. ${}^-3 + {}^-3 = 0$
5. $0 = {}^-3\frac{1}{3} + 3\frac{1}{3}$
6. $0 = 0.15 + {}^-15$
7. ${}^-12 + 12 \neq 0$
8. $0 = 36 + {}^-36$
9. $(3 + {}^-3) + (6 + {}^-6) = 0$
10. ${}^-4 + 4 + 9 + {}^-7 \neq 0$
11. $8 + {}^-8 + {}^-3 + {}^-3 = 0$
12. $1 + {}^-1 + {}^-5 + 5 = 0$

For each sum tell which number has the greater magnitude. Then tell how that number should be renamed to make use of the property of opposites to simplify the expression.

SAMPLE: ${}^-8 + 2$ *What you say:* ${}^-8$ has the greater magnitude; rename ${}^-8$ as ${}^-6 + {}^-2$.

13. ${}^-12 + 3$
14. ${}^-4 + 2$
15. $5 + {}^-2$
16. $10 + {}^-4$
17. $5 + {}^-7$
18. $12 + {}^-4$
19. ${}^-3 + 10$
20. ${}^-5 + 17$
21. $4 + {}^-19$
22. $8 + {}^-15$
23. $10 + {}^-19$
24. $4\frac{5}{8} + {}^-4$
25. ${}^-2\frac{2}{3} + 2$
26. $10.85 + {}^-3.85$
27. ${}^-9.3 + 2.3$

WRITTEN EXERCISES

For each statement, name the property that it illustrates.

SAMPLE 1: $3 + {}^-10 = {}^-10 + 3$ *Solution:* Commutative property

SAMPLE 2: $3 + ({}^-5 + 5) = (3 + {}^-5) + 5$

Solution: Associative property

Chapter 9

SAMPLE 3: $^-9 + 9 = 0$ *Solution:* Additive property of opposites

A
1. $^-6 + 5 = 5 + {}^-6$
2. $3 + {}^-18 = {}^-18 + 3$
3. $^-6 + 6 = 0$
4. $8 + ({}^-2 + 2) = (8 + {}^-2) + 2$
5. $0 = {}^-45 + 45$
6. $0 + {}^-7 = {}^-7$
7. $(7 + {}^-3) + 6 = 7 + ({}^-3 + 6)$
8. $({}^-6 + 6) + 10 = {}^-6 + (6 + 10)$
9. $^-25 + 0 = {}^-25$
10. $4 + ({}^-9 + 8) = ({}^-9 + 8) + 4$

Copy and complete each of the following to make a true statement.

SAMPLE 1: $^-18 + ? = 0$ *Solution:* $^-18 + 18 = 0$
SAMPLE 2: $6 + ({}^-3 + ?) = 6$ *Solution:* $6 + ({}^-3 + 3) = 6$

11. $? + {}^-35 = 0$
12. $? + {}^-15 = {}^-15 + 7$
13. $19 + ? = 0$
14. $0 + {}^-11 = ?$
15. $45 + ? = 45$
16. $17 + {}^-6 = ? + 17$
17. $({}^-4 + 4) + (5 + {}^-5) = ?$
18. $5 + ({}^-8 + ?) = 5$
19. $({}^-7 + 7) + {}^-16 = ?$
20. $({}^-2 + 2) + ? = 23$
21. $^-18 = ({}^-1 + 1) + ?$
22. $({}^-8 + 9) + 3 = 3 + (9 + ?)$
23. $? + ({}^-3 + 3) = 8$
24. $^-10 + {}^-6 + 10 = ?$

Add.

25. 18
    2
    $+\ {}^-2$

26. ${}^-12$
    12
    $+\ 19$

27. ${}^-14$
    ${}^-5$
    $+\ 5$

28. ${}^-39$
    16
    $+\ {}^-16$

29. ${}^-6$
    6
    $+\ 17$

30. 12
    ${}^-12$
    $+\ 23$

31. ${}^-21$
    14
    $+\ 21$

32. 15
    ${}^-3$
    $+\ {}^-15$

Simplify each expression by using the additive property of opposites.

33. $7 + {}^-7 + 18$
34. $^-3 + 3 + 25$
35. $^-14 + 2 + {}^-2$
36. $31 + {}^-19 + 19$
37. $3\tfrac{1}{2} + {}^-3\tfrac{1}{2} + 7\tfrac{1}{8}$
38. $^-17 + 24 + {}^-24$
39. $6 + {}^-6 + 9 + {}^-9 + 37$
40. $^-2\tfrac{1}{5} + 2\tfrac{1}{5} + 3 + 10$

Copy and complete each of the following.

41. $20 + {}^-12 = (8 + 12) + {}^-12$
    $\phantom{20 + {}^-12} = 8 + (? + ?)$
    $\phantom{20 + {}^-12} = 8 + ?$
    $\phantom{20 + {}^-12} = ?$

Addition and Subtraction of Directed Numbers

42. $^-15 + 3 = (^-12 + {}^-3) + 3$
    $= {}^-12 + (? + ?)$
    $= {}^-12 + ?$
    $= ?$

43. $^-5.2 + 8 = {}^-5.2 + (5.2 + 2.8)$
    $= (? + ?) + 2.8$
    $= ? + 2.8$
    $= ?$

44. $17 + {}^-8 = (9 + 8) + {}^-8$
    $= 9 + (? + ?)$
    $= 9 + ?$
    $= ?$

Show that each statement is true.

SAMPLE: $^-3 + 16 = 23 + {}^-10$

Solution: $\quad ^-3 + 16 = 25 + {}^-12$

$^-3 + (3 + 13)$	$(13 + 12) + {}^-12$
$(^-3 + 3) + 13$	$13 + (12 + {}^-12)$
$0 + 13$	$13 + 0$
13	13

B
45. $^-3 + 8 = {}^-7 + 12$
46. $6 + {}^-10 = 5 + {}^-9$
47. $^-6\frac{1}{3} + 3 = 2\frac{1}{3} + {}^-5\frac{2}{3}$

48. $^-12 + 8 = 19 + {}^-23$
49. $^-6 + 11 + 6 = 18 + {}^-7$
50. $6 + 2 + {}^-3 = 8\frac{2}{3} + {}^-3\frac{2}{3}$

Simplify each expression.

SAMPLE: $^-6 + 3 + 2 + {}^-8$

Solution: $^-6 + 3 + 2 + {}^-8 = (^-6 + {}^-8) + (3 + 2)$
$= {}^-14 + 5$
$= {}^-9$

51. $^-8 + 3 + 6 + 2$
52. $^-5 + {}^-2 + 3 + 10$

53. $6 + {}^-3 + {}^-12$
54. $10 + {}^-9 + 2 + {}^-15$

C
55. $^-15 + 35 + {}^-12 + 6$
56. $10 + {}^-7 + {}^-5 + 3 + 2$
57. $^-21 + {}^-16 + 8 + 9 + {}^-3$

58. $^-2.4 + {}^-3.9 + 1.4 + {}^-4.6$
59. $3.2 + 8.6 + {}^-6.9 + 10.2$
60. $25 + {}^-23 + 10 + 18 + {}^-13 + {}^-1$

Add.

61. 35
    $^-7$
    16
    10

62. $^-12$
    10
    $^-15$
    19

63. $^-16$
    12
    44
    $^-17$
    30

64. 9.4
    $^-2.8$
    3.6
    $^-1.7$
    $^-2.5$

Subtraction

9–6 Subtraction of Positive Directed Numbers

We have seen that addition and subtraction are inverse operations, with the addition statement

addend + addend = sum

equivalent to the subtraction statement

sum − addend = addend.

Thus for the subtraction problem **12 − 7 = ?** we can write the equivalent addition problem **7 + ? = 12**. A number line diagram for this would be

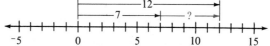

from which we see that each statement is true when the **?** is replaced by **5**.

Compare this with the addition diagram

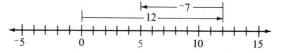

which represents the statement **12 + ⁻7 = 5**. It appears that **12 − 7 = ?** is equivalent to **12 + ⁻7 = ?**, since each is true for the same replacement of the **?**, namely, **5**.

Let us look at a similar situation for the subtraction problem **5 − 9 = ?**. We write the equivalent addition problem **9 + ? = 5**, and from the following diagram we note that the statement is true

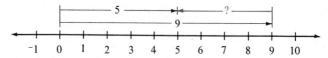

when **?** is replaced by **⁻4**. Thus we know that **5 − 9 = ⁻4**.

Again, comparing the above diagram with this addition diagram:

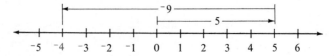

which represents the statement **5 + ⁻9 = ⁻4**, we see that the problems **5 − 9 = ?** and **5 + ⁻9 = ?** are equivalent.

Addition and Subtraction of Directed Numbers

The general principle emerging is that **subtracting** a directed number is equivalent to **adding the opposite** of the number.

Study the following examples. If you question any of them, check it on the number line.

EXAMPLE 1. $8 - 3 = 5$ and $8 + {}^-3 = 5$ are equivalent.

EXAMPLE 2. $12 - 4 = 8$ and $12 + {}^-4 = 8$ are equivalent.

EXAMPLE 3. $5 - 7 = {}^-2$ and $5 + {}^-7 = {}^-2$ are equivalent.

EXAMPLE 4. $3 - 9 = {}^-6$ and $3 + {}^-9 = {}^-6$ are equivalent.

In general, we state that for all positive numbers a and b, $a - b = a + (-b)$.

ORAL EXERCISES

For each number line illustration, state the subtraction sentence and the corresponding addition sentence.

SAMPLE: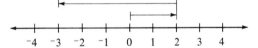

What you say: $2 - 5 = {}^-3$
$2 + {}^-5 = {}^-3$

1.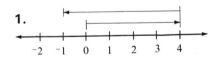

3.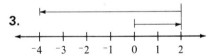

2.

4.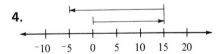

5.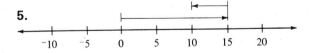

6.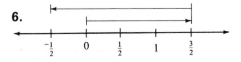

WRITTEN EXERCISES

Write an equivalent addition sentence for each of the following. Then solve both sentences and write their solution sets. Each replacement set is {the directed numbers}.

SAMPLE 1: $12 - 8 = t$ Solution: $8 + t = 12$ $12 - 8 = t$
$t = 4$ $4 = t$
{4}   {4}

SAMPLE 2: $4 - 10 = k$ Solution: $10 + k = 4$ $4 - 10 = k$
$k = {}^-6$ ${}^-6 = k$
{${}^-6$}   {${}^-6$}

A
1. $15 - 6 = a$
2. $4 - 7 = x$
3. $20 - 8 = y$
4. $18 - 7 = d$
5. $3 - 10 = r$
6. $7 - 12 = w$
7. $14 - 9 = b$
8. $24 - 6 = v$
9. $30 - 20 = n$
10. $4 - 14 = s$
11. $c = 14 - 10$
12. $h = 19 - 6$
13. $k = 3 - 12$
14. $m = 2 - 9$
15. $z = 15 - 18$

Simplify each expression. Make a number line sketch of the solution.

SAMPLE: $3 - 5$
Solution: $3 - 5 = {}^-2$

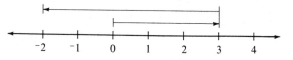

16. $10 - 6$
17. $4 - 9$
18. $8 - 3$
19. $7 - 10$
20. $6 - 6$
21. $8\frac{1}{2} - 5$
22. $2\frac{1}{2} - 5\frac{1}{2}$
23. $4\frac{1}{3} - 3$
24. $3\frac{3}{4} - 5\frac{1}{4}$
25. $3\frac{2}{3} - 3\frac{2}{3}$
26. $3.5 - 7.5$
27. $6.4 - 1.3$

Write each subtraction equation as an addition equation, as shown in the sample. Then find the solution set for the equation.

SAMPLE: $19 - 3 = t$ Solution: $19 + {}^-3 = t$
$t = 16$
{16}

B
28. $12 - 7 = n$
29. $24 - 10 = b$
30. $4 - 9 = m$
31. $12 - 17 = x$
32. $r = 5 - 15$
33. $w = 9 - 9$
34. $8\frac{4}{5} - 6\frac{1}{5} = k$
35. $3\frac{1}{3} - 7\frac{2}{3} = s$
36. $28 - 39 = y$
37. $4.8 - 2.6 = c$
38. $135 - 62 = w$
39. $18.2 - 3.7 = s$

Use $=$ or $\neq$ to replace each question mark to make a true statement.

C
40. $19 - 12 \ ? \ 19 + {}^-12$
41. $8 - 6 \ ? \ 8 + {}^-6$
42. $3 + {}^-10 \ ? \ 3 - 10$
43. $8 + {}^-12 \ ? \ 12 - 8$

Addition and Subtraction of Directed Numbers

44. $15 + {}^-7 \; ? \; 15 - 7$ **46.** $7 + {}^-2 \; ? \; 7 - 2$
45. $10 - 6 \; ? \; 10 + 6$ **47.** $5 + {}^-15 \; ? \; 5 - 15$

Subtract. Check your answer by adding your result to the number subtracted.

SAMPLE 1: $\begin{array}{r}16\\-10\\\hline\end{array}$ *Solution:* $\begin{array}{r}16\\-10\\\hline 6\end{array}$ *Check:* $10 + 6 = 16$

SAMPLE 2: $\begin{array}{r}3\\-8\\\hline\end{array}$ *Solution:* $\begin{array}{r}3\\-8\\\hline {}^-5\end{array}$ *Check:* $8 + ({}^-5) = 3$

48. $\begin{array}{r}18\\-13\\\hline\end{array}$ **49.** $\begin{array}{r}5\\-7\\\hline\end{array}$ **50.** $\begin{array}{r}10\\-19\\\hline\end{array}$ **51.** $\begin{array}{r}14\\-6\\\hline\end{array}$

9–7 Subtraction of Negative Directed Numbers

The basic ideas we have used for subtracting positive directed numbers also apply to the subtraction of negative directed numbers. Let us consider the subtraction problem $15 - {}^-8 = ?$ and the equivalent addition problem.

$\underbrace{15}_{\text{sum}} - \underbrace{{}^-8}_{\text{addend}} = \underbrace{?}_{\text{addend}} \qquad \underbrace{?}_{\text{addend}} + \underbrace{{}^-8}_{\text{addend}} = \underbrace{15}_{\text{sum}}$

Do you agree that the question mark in the second equation must be replaced by **23** to make a true statement? Then, since the two equations are equivalent, the correct replacement for the question mark in the first must also be **23**.

Thus we see that

$$15 - {}^-8 = 15 + (-({}^-8)) = 15 + 8 = 23$$

and it appears that, for negative numbers as well as for positive numbers, **subtracting** a directed number is equivalent to **adding the opposite** of the number.

Study these examples to be certain that you understand this important idea.

EXAMPLE 1. $6 - {}^-2 = 8$ and $6 + 2 = 8$ are equivalent.

EXAMPLE 2. $7 - {}^-3 = 10$ and $7 + 3 = 10$ are equivalent.

EXAMPLE 3. $3 - {}^-8 = 11$ and $3 + 8 = 11$ are equivalent.

EXAMPLE 4. $6 - {}^-10 = 16$ and $6 + 10 = 16$ are equivalent.

We are now ready to extend our previous statement about subtraction to include all directed numbers:

For any directed numbers *a* and *b*,

$$a - b = a + (-b).$$

ORAL EXERCISES

Name an equivalent addition statement for each of the following.

SAMPLE 1: $5 - {}^-3 = 8$ *What you say:* ${}^-3 + 8 = 5$
SAMPLE 2: ${}^-9 - {}^-4 = {}^-5$ *What you say:* ${}^-4 + {}^-5 = {}^-9$

1. $7 - {}^-2 = 9$
2. $10 - {}^-4 = 14$
3. $3 - {}^-9 = 12$
4. ${}^-2 - {}^-4 = 2$
5. $15 - {}^-6 = 21$
6. $24 - {}^-5 = 29$
7. $30 = 25 - {}^-5$
8. $16 = 4 - {}^-12$
9. ${}^-5 - 3 = {}^-8$
10. ${}^-6 - {}^-4 = {}^-2$
11. ${}^-1 - {}^-5 = 4$
12. ${}^-3 - {}^-15 = 12$

Tell what directed number can replace each question mark to make a true statement.

SAMPLE: $5 - {}^-3 = 5 + ?$ *What you say:* $5 - {}^-3 = 5 + 3$

13. $10 - {}^-8 = 10 + ?$
14. $15 - {}^-3 = 15 + ?$
15. $3 - {}^-4 = 3 + ?$
16. ${}^-10 - 2 = {}^-10 + ?$
17. ${}^-6 - {}^-9 = {}^-6 + ?$
18. $4 - ? = 4 + 8$
19. $9 - ? = 9 + 5$
20. ${}^-10 - ? = {}^-10 + 8$

WRITTEN EXERCISES

Write an equivalent addition sentence for each of the following. Then solve both sentences and write the solution sets. Each replacement set is {the directed numbers}.

SAMPLE 1: $4 - {}^-5 = x$ *Solution:* ${}^-5 + x = 4$ $4 - {}^-5 = x$
                                           $x = 9$ $9 = x$
                                           $\{9\}$ $\{9\}$

Addition and Subtraction of Directed Numbers 265

SAMPLE 2: $^-8 - {}^-6 = y$ Solution: $^-6 + y = {}^-8$ $^-8 - {}^-6 = y$
$y = {}^-2$ $^-2 = y$
$\{^-2\}$ $\{^-2\}$

A
1. $3 - {}^-18 = m$
2. $10 - {}^-4 = b$
3. $15 - {}^-6 = c$
4. $^-7 - {}^-19 = n$
5. $5 - {}^-11 = a$
6. $r = 8 - {}^-1$
7. $k = {}^-2 - {}^-5$
8. $^-16 - {}^-12 = y$
9. $^-11 - 3 = w$
10. $^-2 - 17 = r$
11. $^-3\frac{1}{2} - 1 = t$
12. $^-4\frac{2}{3} - {}^-1\frac{1}{3} = s$

Subtract. Check your answer by adding your result to the number subtracted.

SAMPLE: 12   Solution: 12   Check: $^-8 + 20 = 12$
        $- {}^-8$ $- {}^-8$
                              20

13. 10   16. 3   19. 35   22. $^-24$
    $- {}^-2$ $- {}^-7$ $- {}^-12$ $- {}^-10$

14. 19   17. 5   20. $^-10$ 23. $^-6$
    $- {}^-8$ $- {}^-8$ $- {}^-5$ $- {}^-13$

15. $^-36$ 18. $^-15$ 21. $^-19$ 24. $^-3$
    $- {}^-22$ $- 5$ $- {}^-12$ $- 18$

Simplify each expression.

SAMPLE 1: $4 - {}^-9$ Solution: $4 - {}^-9 = 4 + 9$
$= 13$

SAMPLE 2: $^-10 - {}^-15$ Solution: $^-10 - {}^-15 = {}^-10 + 15$
$= 5$

B
25. $8 - 15$
26. $13 - {}^-6$
27. $10 - {}^-9$
28. $2 - {}^-16$
29. $3 - {}^-5$
30. $^-2 - 9$
31. $^-10 - 24$
32. $-6 - {}^-21$
33. $^-20 - {}^-30$
34. $^-0.8 - {}^-0.2$
35. $^-0.30 - 0.10$
36. $^-10 - {}^-10$

SAMPLE 3: $10 - {}^-(2 + 4)$ Solution: $10 - {}^-(2 + 4) = 10 - {}^-6$
$= 10 + 6$
$= 16$

C
37. $3 - {}^-(10 + 4)$
38. $15 - (2 + 5)$
39. $^-10 - (1 + 2)$
40. $^-2 - (6 + 7)$
41. $(3 + 5) - {}^-7$
42. $^-(10 + 6) - 4$
43. $(4 + 8) - 18$
44. $^-(3 + 9) - {}^-4$
45. $^-(10 - 4) - {}^-4$
46. $^-12 - {}^-(7 - 4)$
47. $^-(3 + 4) - {}^-(6 + 5)$
48. $^-10 - {}^-(7 + 10)$

Functions and Nomographs

9-8 Functions and Directed Numbers

It should be interesting to see what happens with our familiar function machine when it is used for directed numbers. The function rule for the machine in Figure 9-3 is "**Add 3.**" Recall that the *x*-values, shown in the first column of the table, are the **input** numbers. The $f(x)$ values in the second column are the **output** numbers from the machine and in this case each is equal to $x + 3$. The function equation is $f(x) = x + 3$.

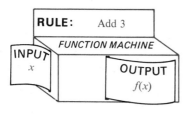

x	$f(x)$	$(x, f(x))$
⁻2	1	(⁻2, 1)
⁻1	2	(⁻1, 2)
0	3	(0, 3)
1	4	(1, 4)
2	5	(2, 5)

Figure 9-3

The set of number pairs consisting of the entries in the last column of the table is the function

$$\{(-2, 1), (-1, 2), (0, 3), (1, 4), (2, 5)\}.$$

That is, no two different number pairs have the same first number. Notice that, if we used {the directed numbers} as replacement set, the machine would go on indefinitely giving out values for $f(x)$. The resulting function would be an infinite set of number pairs, of which our function is a subset.

Now suppose the function rule for the machine is changed to **Add ⁻6**. The function equation for this rule is $f(x) = x + ^-6$. Study the table that accompanies the machine in Figure 9-4 to be sure that you see where the number pairs, $(x, f(x))$, come from.

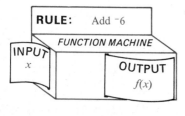

x	$f(x)$	$(x, f(x))$
0	⁻6	(0, ⁻6)
1	⁻5	(1, ⁻5)
2	⁻4	(2, ⁻4)
⁻1	⁻7	(⁻1, ⁻7)
⁻2	⁻8	(⁻2, ⁻8)

Figure 9-4

Do you believe that the set of number pairs consisting of the entries in the last column is a function? Check to verify that each number pair has a different first number.

$$\{(1, {}^-6), (1, {}^-5), (2, {}^-4), ({}^-1, {}^-7), ({}^-2, {}^-8)\}$$

The machine in Figure 9–5 uses the rule "**Subtract 3.**" Check the $f(x)$ values and the $(x, f(x))$ pairs recorded in the table.

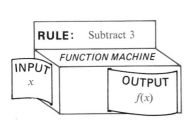

Figure 9–5

x	$f(x)$	$(x, f(x))$
0	$^-3$	$(0, {}^-3)$
1	$^-2$	$(1, {}^-2)$
2	$^-1$	$(2, {}^-1)$
3	0	$(3, 0)$
4	1	$(4, 1)$
5	2	$(5, 2)$
⋮	⋮	⋮

ORAL EXERCISES

Tell how to complete the table that accompanies the function machine shown below.

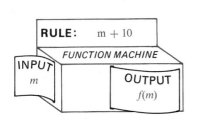

	m	$f(m)$	$(m, f(m))$
SAMPLE:	$^-1$	9	$(^-1, 9)$
1.	$^-2$	8	?
2.	$^-3$	?	?
3.	$^-4$	?	?
4.	$^-5$	?	?
5.	$^-6$	?	?

Use the given function rule to tell how to complete each set of number pairs.

6. Subtract 8: $\{(5, {}^-3), (4, {}^-4), (3, {}^-5), (2, ?), (1, ?), (0, ?)\}$
7. Add $^-2$: $\{(6, 4), (4, 2), (2, ?), (0, ?), ({}^-2, ?), ({}^-4, ?)\}$
8. Subtract $^-3$: $\{(3, 6), (2, ?), (1, ?), (0, ?), ({}^-1, ?), ({}^-2, ?)\}$

WRITTEN EXERCISES

Copy and complete the table according to the pictured function machine.

A

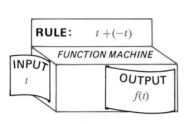

	t	f(t)	(t, f(t))
SAMPLE:	4	0	(4, 0)
1.	3	?	?
2.	2	?	?
3.	1	?	?
4.	0	?	?
5.	⁻1	?	?
6.	⁻2	?	?

Complete each set of number pairs according to the given function rule.

7. $s + 2 = f(s)$: {(⁻6, ⁻4), (⁻4, ?), (⁻2, ?), (0, ?), (2, ?), (4, ?)}
8. $x + {}^-4 = f(x)$: {(⁻4, ⁻8), (⁻3, ⁻7), (⁻2, ?), (⁻1, ?), (0, ?)}
9. $m - 5 = f(m)$: {(0, ⁻5), (2, ⁻3), (4, ?), (6, ?), (8, ?)}
10. $w - {}^-3 = f(w)$: {(15, 18), (13, ?), (11, ?), (9, ?), (?, 10)}
11. $s + {}^-10 = f(s)$: {(6, ⁻4), (4, ?), (⁻3, ?), (2, ?), (⁻5, ?)}
12. $^-7 + t = f(t)$: {(⁻2, ⁻9), (⁻3, ?), (⁻4, ?), (7, ?), (8, ?), (10, ?)}
13. $n - 0 = f(n)$: {(5, 5), (?, 6), (?, 9), (?, ⁻3), (?, ⁻5)}

Tell which sets of number pairs are functions and which are not.

14. {(18, 3), (15, 3), (12, 3), (9, 3), (6, 3), (3, 3)}
15. {(1, 4), (1, 5), (2, 3), (2, 4), (3, 2), (3, 3)}
16. {(⁻3, 3), (⁻2, 2), (3, ⁻3), (2, ⁻2), (0, 0), (1, ⁻1)}
17. {(30, ⁻3), (40, ⁻4), (50, ⁻5), (10, ⁻1), (⁻10, 1), (⁻20, 2)}
18. {(⁻4, 1), (⁻3, 1), (⁻2, 1), (⁻1, 1), (0, 1), (1, 1), (2, 1), ...}

Match each set of number pairs in Column 1 with its function equation in Column 2.

COLUMN 1

B 19. {(4, 3), (3, 2), (2, 1), (1, 0), (0, ⁻1)}
20. {(⁻2, ⁻1), (⁻1, 0), (0, 1), (1, 2)}
21. {(5, ⁻1), (4, ⁻2), (3, ⁻3), (2, ⁻4)}
22. {(⁻2, 2), (⁻1, 3), (0, 4), (1, 5), (2, 6)}
23. {(⁻8, ⁻8), (15, 15), (⁻2½, ⁻2½), (4, 4)}

COLUMN 2

A. $s - 6 = f(s)$
B. $x + {}^-1 = f(x)$
C. $w + 0 = f(w)$
D. $n - {}^-1 = f(n)$
E. $m - {}^-4 = f(m)$

Use each function equation and the given replacement set to write a set of number pairs that is a function.

SAMPLE: $t + {}^-8 = f(t);$ $\{{}^-4, {}^-2, 0, 2, 4\}$

Solution: $\{({}^-4, {}^-12), ({}^-2, {}^-10), (0, {}^-8), (2, {}^-6), (4, {}^-4)\}$

C 24. $m - 10 = f(m);$ $\{14, 12, 10, 8, 6\}$
25. $r + {}^-9 = f(r);$ $\{3, 6, 9, 12, 15, 18\}$
26. $t - 5 = f(t);$ $\{4, 2, 0, {}^-2, {}^-4, {}^-6\}$
27. $s - {}^-10 = f(s);$ $\{{}^-15, {}^-10, {}^-5, 0, 5, 10, 15\}$
28. $3 - (-x) = f(x);$ $\{2, 1, 0, {}^-1, {}^-2, {}^-3, {}^-4\}$

9–9 Nomographs: Addition and Subtraction of Directed Numbers

The idea of using nomographs for adding and subtracting numbers of arithmetic was explored earlier in this book. Recall that our nomographs consisted of three number lines or scales, arranged as shown in Figure 9–6. Two number lines are marked **addend** and the middle one is marked **sum**. The red lines on the nomograph might be used to solve the following equations:

(1) $4 + 6 = ?$ (2) $16 - 7 = ?$

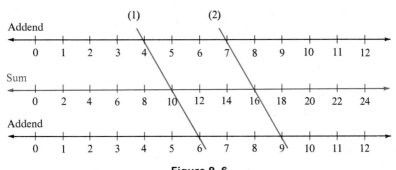

Figure 9–6

A nomograph for adding and subtracting directed numbers is similar to that used with arithmetic numbers, with the number lines extended to include negative numbers.

Using a nomograph for adding directed numbers is based on the statement **addend + addend = sum**. Shown in Figure 9–7 is an illustration of the use of the nomograph for two addition problems:

270 Chapter 9

(1) ⁻3 + ⁻1 = ? and (2) 3 + ⁻4 = ?. The solution of each is found by drawing a line joining the two addends. The sum is the point of intersection of the line drawn and the sum scale.

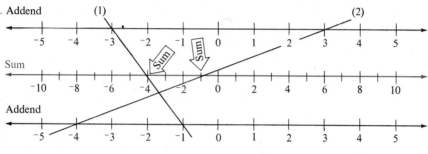

Figure 9–7

Thus according to the nomograph ⁻3 + ⁻1 = ⁻4 and 3 + ⁻4 = ⁻1.

Now let us consider subtracting directed numbers on the nomograph by using the concept that **sum** − **addend** = **addend**. In the first problem, ⁻2 − 3 = ?, the sum is ⁻2, so we find ⁻2 on the sum scale. The addend 3 is located on an addend scale, and the line marked (1) is drawn through these two points. It intersects the other addend scale at ⁻5, so we see that the other addend is ⁻5, and the equation is ⁻2 − 3 = ⁻5.

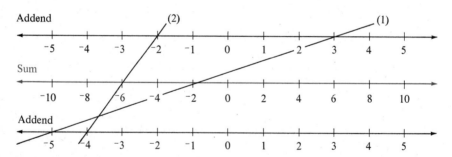

The line marked (2) shows the use of the nomograph for the problem ⁻6 − ⁻2 = ?. Check for yourself that this tells us that ⁻6 − ⁻2 = ⁻4.

ORAL EXERCISES

1. Name the directed numbers needed to replace each letter to complete the nomograph which follows.

SAMPLE: *a*   *What you say:* Negative 20

Addition and Subtraction of Directed Numbers

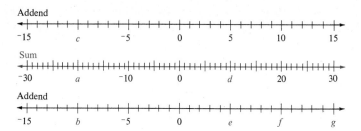

2. Give an addition number statement and two equivalent subtraction statements for each line drawn across the nomograph below.

SAMPLE: Line a   What you say: $^-5 + {}^-3 = {}^-8$; $^-8 - {}^-3 = {}^-5$; $^-8 - {}^-5 = {}^-3$.

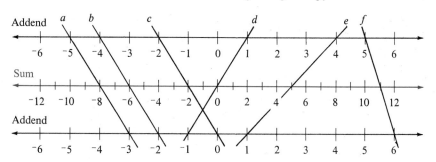

WRITTEN EXERCISES

Use this nomograph to solve each addition equation and state the solution set.

SAMPLE: $^-3 + 5 = m$ Solution: $^-3 + 5 = m$
                                              $2 = m$ $\{2\}$

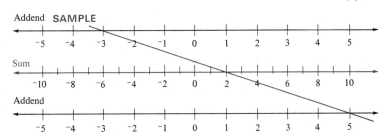

1. $^-4 + 1 = a$
2. $^-3 + {}^-5 = t$
3. $^-2 + 3 = y$
4. $3 + {}^-6 = d$
5. $1 + {}^-4 = m$
6. $0 + 4 = x$
7. $^-3 + 0 = b$
8. $s = 2 + {}^-6$
9. $c = 5 + {}^-5$
10. $0 + {}^-6 = r$
11. $5 + w = 4$
12. $^-3 + k = 2$
13. $4 + n = {}^-1$
14. $^-4 + t = 0$
15. $b + {}^-3 = {}^-2$

272 Chapter 9

Use this nomograph to solve each subtraction equation and state the solution set.

SAMPLE: $^-5 - {^-2} = c$ Solution: $^-5 - {^-2} = c$
$^-3 = c$ $\{^-3\}$

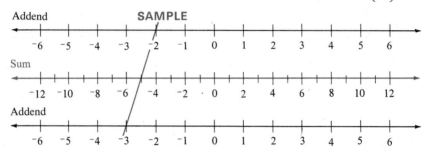

16. $6 - 4 = n$
17. $5 - 5 = b$
18. $3 - 5 = t$
19. $5 - 6 = x$

20. $^-2 - {^-3} = c$
21. $^-3 - {^-5} = s$
22. $2 - {^-3} = v$
23. $4 - {^-1} = a$

24. $^-3 - 2 = y$
25. $^-3 - {^-3} = d$
26. $^-3 - t = {^-1}$
27. $4 - k = {^-1}$

Write an addition statement indicated by each line across the nomograph.

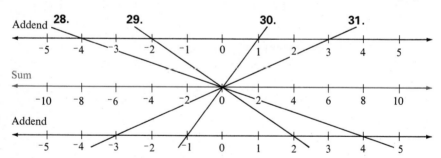

28.–31. (See nomograph above.)

Use the nomograph to solve each equation and state the solution set.

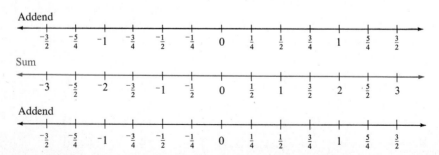

B 32. $\frac{1}{2} + {}^-1 = d$ 37. ${}^-\frac{1}{4} + \frac{3}{4} = r$ 42. $\frac{1}{2} + {}^-\frac{3}{4} = a$
33. ${}^-1 + {}^-1\frac{1}{2} = s$ 38. ${}^-\frac{1}{2} + \frac{1}{4} = n$ 43. ${}^-\frac{3}{4} + {}^-\frac{3}{4} = w$
34. ${}^-\frac{1}{2} + {}^-\frac{1}{2} = c$ 39. $\frac{5}{4} + {}^-\frac{1}{4} = y$ 44. $\frac{5}{4} + {}^-\frac{3}{4} = b$
35. $\frac{3}{4} + 0 = t$ 40. $\frac{3}{4} + {}^-\frac{3}{4} = k$ 45. $\frac{5}{4} + x = 0$
36. $0 + {}^-\frac{5}{4} = x$ 41. $\frac{5}{4} + \frac{1}{4} = z$ 46. $\frac{1}{2} + m = {}^-\frac{1}{2}$

C 47. ${}^-\frac{1}{2} - \frac{1}{4} = t$ 50. ${}^-2 - {}^-1 = w$ 53. ${}^-\frac{1}{2} - {}^-\frac{3}{4} = y$
48. $\frac{1}{2} - \frac{3}{4} = n$ 51. $0 - \frac{1}{4} = k$ 54. ${}^-\frac{1}{2} - {}^-1 = d$
49. $\frac{3}{2} - \frac{1}{4} = c$ 52. $\frac{1}{2} - 0 = r$ 55. $1 - {}^-\frac{1}{4} = x$

CHAPTER SUMMARY

Inventory of Structure and Concepts

1. Zero and the positive directed numbers are essentially the same as the numbers of arithmetic.

2. For every positive directed number there is a corresponding negative directed number that is the same distance from zero. These pairs of numbers are **opposites** of each other. Zero is its own opposite.

3. The addition of directed numbers is **commutative**. That is, for all directed numbers r and s, $r + s = s + r$.

4. Numbers that are opposites of each other are said to be **additive inverses** of each other. The sum of every directed number and its additive inverse (its opposite) is zero. That is, for every directed number r, $r + (-r) = -r + r = 0$.

5. **Zero** is the **additive identity** element for directed numbers. That is, for every directed number r, $r + 0 = 0 + r = r$.

6. The **additive inverse** (opposite) of an indicated sum of two numbers is the **sum** of their **additive inverses** (opposites). That is, for all directed numbers r and s, $-(r + s) = -r + (-s)$.

7. The addition of directed numbers is **associative**. That is, for all directed numbers r, s, and t, $r + (s + t) = (r + s) + t$.

8. The **sum** of any two directed numbers r and s is a **unique directed number**. The set of directed numbers is **closed** under **subtraction**.

9. **Subtracting** a directed number is equivalent to **adding the opposite** of the number. That is, for all directed numbers r and s, $r - s = r + (-s)$.

Vocabulary and Spelling

commutative property of addition (*p. 244*)
positive direction (*p. 244*)
negative direction (*p. 244*)
opposite (*p. 248*)
additive property of opposites (*p. 250*)
additive inverse (*p. 250*)
additive identity element (*p. 251*)
associative property of addition (*p. 256*)
inverse operations (*p. 260*)
equivalent sentence (*p. 261*)
function (*p. 266*)

Chapter Test

Find the solution set for each equation. The replacement set is the set of directed numbers.

1. $^-8 + n = {}^-17$
2. $12 + y = 5$
3. $5 = x - {}^-7$
4. $x - 13 = 7$
5. $^+37 = {}^-57 + x$
6. $^-15 + x = {}^-3$
7. $^-13 = x + 14$
8. $^-18 = k - {}^-7$
9. $x - {}^-14 = {}^-9$
10. $^+73 = x - {}^-38$

Add:

11. ${}^-6$
    $+\ 12$

12. ${}^-17$
    $+\ {}^-8$

Subtract:

13. ${}^-6$
    $-\ 12$

14. ${}^-17$
    $-\ {}^-8$

15. What is the additive inverse of $^-14$?

Indicate whether each of Statements 16–19 is true or false.

16. $^+54 = 54$
17. $(13 + {}^-5) = 13 + {}^-5$
18. $-({}^-2 + 5) = 2 + {}^-5$
19. $^-5 + 3 = 3 + {}^-5$

20. What property of addition in the set of directed numbers is illustrated by the statement $^-5 + ({}^-3 + 2) = ({}^-5 + {}^-3) + 2$?

Addition and Subtraction of Directed Numbers 275

Complete each set of number pairs according to the function rule given.

21. $k + {}^-4 = f(k)$: $\{(1, ?), ({}^-3, ?)\}$
22. $s - 12 = f(s)$: $\{(3, ?), ({}^-5, ?)\}$
23. ${}^-2 - x = f(x)$: $\{(11, ?), ({}^-2, ?)\}$

Simplify each expression.

24. ${}^-5 + {}^-3 - 5$ 25. $8 - {}^-2 + 10$

Chapter Review

9–1 Number Line Addition and the Commutative Property

Sketch a number line solution for each equation, then complete each equation to make a true statement.

1. ${}^+4 + {}^-1 = ?$ 3. ${}^-3 + {}^-2 = ?$
2. ${}^-2 + {}^+5 = ?$ 4. ${}^-6 + {}^+4 = ?$

Find each sum.

5. ${}^+26$ 6. ${}^-34$ 7. ${}^+59$
   ${}^-19$ ${}^-79$ ${}^-118$

Name the directed number that can replace the question mark and make a true statement.

8. ${}^+27 + ? = {}^-5$ 10. ${}^-15 + ? = {}^+7$
9. ${}^-17 + ? = {}^-24$ 11. $? + {}^-47 = {}^-3$

9–2 Opposites of Directed Numbers

Express each of the following in symbols in two ways.

SAMPLE: The opposite of positive 5   Solution: ${}^-({}^+5)$; ${}^-5$

12. The opposite of negative 15   13. The opposite of positive 48

14. The opposite of the number five units to the left of zero

Give the meaning of each of the following in a simpler form of a directed number.

15. $-({}^+13)$ 16. $-({}^-17)$ 17. $-(-({}^-2))$

9–3 Addition: Additive Inverses

Give the additive inverse of each of the following.

18. $^+11$ **19.** $^-14$ **20.** 0   **21.** $-(^+3)$

Find the solution set of each equation. The replacement set is the set of directed numbers.

22. $k = {}^+23 + {}^-23$ **23.** $^+54 + x = 0$ **24.** $x + {}^-32 = 0$

9–4 Using Additive Inverses to Simplify Expressions

Tell which of the following statements are true.

25. $^+15 = 15$
26. $-(7 + 14) = {}^-7 + {}^-14$
27. $(8 + {}^-9) = 8 + {}^-9$
28. $^-31 = 31$
29. $-({}^-3 + 7) = 3 + {}^-7$

Find the solution set for each equation. The replacement set is the set of directed numbers.

30. $^-5 + b = {}^-17$
31. $x + 53 = {}^-1$
32. $-(3 + {}^-5) = x$
33. $14 + x = {}^-38$
34. $35 = k + {}^-3$
35. $y = -({}^-3 + 15)$

9–5 Number Line Addition and the Associative Property

Tell which property of addition justifies each statement.

36. $78 + {}^-78 = 0$
37. $({}^-8 + 10) + {}^-1 = {}^-8 + (10 + {}^-1)$
38. $5 + {}^-8 = {}^-8 + 5$
39. $0 + {}^-5 = {}^-5 + 0 = {}^-5$

Simplify each expression.

40. $7 + {}^-18 + {}^-7$
41. $^-9 + {}^-3 + 7$
42. $^-5.1 + 3.6 + 5.1$
43. $10 + {}^-14 + {}^-27 + 3$

9–6 Subtraction of Positive Directed Numbers

Write an equivalent addition sentence for each of the following. Then solve both sentences and write their solution sets. Each replacement set is {the directed numbers}.

Addition and Subtraction of Directed Numbers   277

44. $7 - 12 = x$ 46. $m = 13 - 15$ 48. $34 - 14 = y$
45. $4.8 - 5.9 = n$ 47. $1\frac{1}{8} - 2\frac{1}{4} = k$

Simplify each expression.

49. $14 - 19$ 50. $28 - 49$ 51. $37 - 54$ 52. $6.4 - 7.1$

9–7 Subtraction of Negative Directed Numbers

Write an equivalent addition sentence for each of the following. Then solve both sentences and write their solution sets. Each replacement set is {the directed numbers}.

53. $3 - {}^-5 = x$ 55. $k = 17 - {}^-25$ 57. ${}^-8 - 9 = y$
54. ${}^-12 - 3 = y$ 56. ${}^-25 - {}^-4 = t$ 58. ${}^-31 - {}^-37 = h$

Simplify each expression.

59. $8 - 13$ 62. $3 - {}^-4$ 65. ${}^-9 - 3$
60. ${}^-11 - 17$ 63. ${}^-29 - {}^-14$ 66. ${}^-32 - {}^-37$
61. $3 - (7 + 4)$ 64. $19 - {}^-(3 + 5)$ 67. ${}^-8 - {}^-(13 + 5)$

9–8 Functions and Directed Numbers

Complete each set of number pairs according to the function rule given.

68. $k + {}^-17 = f(k)$: $\{(1, {}^-16), ({}^-3, {}^-20), ({}^-7, ?), (13, ?)\}$
69. $s - 9 = f(s)$: $\{(1, {}^-8), ({}^-7, {}^-16), (4, ?), (17, ?), ({}^-5, ?)\}$
70. ${}^-12 - x = f(x)$: $\{(1, {}^-13), ({}^-4, {}^-8), (15, ?), (7, ?), ({}^-8, ?)\}$

Use each function equation and the given replacement set to write the set of number pairs that is a function for the given replacement set.

71. $t - 13 = f(t)$; $\{15, 7, {}^-3, {}^-17, 0\}$
72. ${}^-18 - k = f(k)$; $\{3, {}^-5, 24, {}^-31\}$
73. $p + {}^-4 = f(p)$; $\{7, 2, {}^-3, 4, {}^-9\}$

9–9 Nomographs: Addition and Subtraction of Directed Numbers

Use the following nomograph to solve each equation. Write the solution set for each.

Chapter 9

```
Addend
   ──┼────┼────┼────┼────┼────┼────┼────┼────┼────┼──
     ⁻5   ⁻4   ⁻3   ⁻2   ⁻1    0    1    2    3    4    5
Sum
   ──┼────┼────┼────┼────┼────┼────┼────┼────┼────┼──
    ⁻10   ⁻8   ⁻6   ⁻4   ⁻2    0    2    4    6    8   10
Addend
   ──┼────┼────┼────┼────┼────┼────┼────┼────┼────┼──
     ⁻5   ⁻4   ⁻3   ⁻2   ⁻1    0    1    2    3    4    5
```

74. $4 - 3 = s$ **77.** $2 - {}^-1 = x$ **80.** ${}^-6 - {}^-4 = t$
75. ${}^-2 + {}^-5 = x$ **78.** ${}^-4 - x = 1$ **81.** $3 - y = {}^-2$
76. $x - 3 = {}^-1$ **79.** $x + {}^-5 = {}^-3$ **82.** $x - {}^-2 = {}^-5$

Review of Skills

Simplify the following expressions.

1. $0 \cdot 7$ **2.** $15 \cdot 0$ **3.** ${}^-2 \cdot 7$

Complete the following patterns.

4. $4 \cdot 7 = 28$ **5.** $5 \cdot 4 = 20$ **6.** $5 \cdot 1 = 5$
 $3 \cdot 7 = 21$ $5 \cdot 3 = 15$ $4 \cdot 1 = 4$
 $2 \cdot 7 = 14$ $5 \cdot 2 = 10$ $3 \cdot 1 = 3$
 $1 \cdot 7 = 7$ $5 \cdot 1 = \ ?$ $2 \cdot 1 = \ ?$
 $0 \cdot 7 = \ ?$ $5 \cdot 0 = \ ?$ $1 \cdot 1 = \ ?$
 ${}^-1 \cdot 7 = \ ?$ $5 \cdot {}^-1 = \ ?$ ${}^-1 \cdot 1 = \ ?$
 ${}^-2 \cdot 7 = \ ?$ $5 \cdot {}^-2 = \ ?$ ${}^-2 \cdot 1 = \ ?$

7. ${}^-3 + {}^-3 + {}^-3 + {}^-3 + {}^-3 = {}^-15$
 ${}^-3 \cdot 5 = \ ?$

Perform the indicated operations.

8. $1\frac{3}{8} \div 2\frac{3}{4}$ **12.** $13\frac{3}{4} - 5\frac{7}{8}$
9. $2\frac{1}{2} \times 3\frac{3}{4}$ **13.** $12\frac{3}{8} \div 4$
10. $\dfrac{16.2}{0.2}$ **14.** $\dfrac{189}{0.03}$
11. $5 \div 3\frac{1}{2}$ **15.** $(0.33)(15.45)$

Addition and Subtraction of Directed Numbers

Match each expression in Column 1 with an expression from Column 2 that represents the same number.

COLUMN 1	COLUMN 2
16. $3 \cdot 5$	**A.** $(15 \cdot 8) \cdot 21$
17. $15 \cdot (8 \cdot 21)$	**B.** $b \cdot a$
18. $23 \cdot (12 + 15)$	**C.** $a \cdot b + a \cdot c$
19. $a \cdot b$	**D.** $5 \cdot 3$
20. $a \cdot (b \cdot c)$	**E.** $(a \cdot b) \cdot c$
21. $a \cdot (b + c)$	**F.** $23 \cdot 12 + 23 \cdot 15$

Simplify the following expressions.

22. 2^4 **25.** 3^2 **28.** 5^3

23. 4^2 **26.** 2^3 **29.** 3^5

24. 1^4 **27.** 1^{10}

Use the distributive property to give a simpler name for each of the following expressions.

30. $3x + 5x$ **32.** $14k - 8k$ **34.** $15y - 3y + 7y$

31. $\frac{1}{2}n + \frac{3}{4}n$ **33.** $4x + 7y - 3y + 2x$

35. What are the common factors of 15 and 20?

36. What are the common divisors of 15 and 20?

Simplify the following expressions.

37. $15 \div 3$ **39.** $45 \overline{)585}$ **41.** $\frac{72}{6}$

38. $\frac{3}{4} \cdot \frac{2}{3}$ **40.** $\frac{3}{4} \div 2$ **42.** $\frac{2}{3} \div \frac{1}{2}$

For each of the following, find the value of x that will make the statement true.

43. $3x = 24$ **45.** $\dfrac{x}{5} = 18$ **47.** $\frac{3}{4}x = 18$

44. $\dfrac{2}{x} = 10$ **46.** $\dfrac{15}{x} = 3$ **48.** $\frac{4}{3}x = \frac{5}{6}$

Replace each question mark with =, <, or > to make a true statement.

49. 3.55 ? 3.555 **52.** 3^2 ? 2^3

50. 2^4 ? 4^2 **53.** $\dfrac{15}{0.1}$? $\dfrac{15}{0.01}$

51. $3^5 \cdot 2^5$? $(3 \cdot 2)^5$ **54.** 11.01 ? 11.009

CHECK POINT
FOR EXPERTS

Probability

The branch of mathematics that concerns itself with the chance that a given event will occur is called **probability**. The science of probability seems to have originated in the 17th century when gamblers began to analyze games of chance. It makes use of mathematics to measure the *likelihood* of some happening. For example, suppose you were to toss an ordinary quarter. Do you see that there are only two possible outcomes: a **head** or a **tail**? If we assume that the possible outcomes are **equally likely**, then a "head" is just as likely to be the actual outcome as a "tail."

We can say:

(1) The chances are 1 in 2 that a **head** will turn up in any one toss. So, the **probability** of a head turning up is $\frac{1}{2}$. In symbols we write: $P(h) = \frac{1}{2}$.

(2) The chances are 1 in 2 that a **tail** will turn up in any one toss. So, the **probability** of a tail turning up is $\frac{1}{2}$. In symbols we write: $P(t) = \frac{1}{2}$.

Suppose you toss a cube with faces numbered **1** through **6**, as pictured, and suppose you are interested in the numeral that comes up on the top face of the cube. If we again assume that each outcome is equally likely in any one toss, then for one number to turn up is just as likely to be the actual outcome as for some other number to be on top. The chances are **1** in **6** that a specified number will turn up in any one toss. So the probability of a specified number turning up is $\frac{1}{6}$. In symbols we write:

$$P(n) = \frac{1}{6}.$$

Another question we might ask about the cube is: "What is the probability that an **even** number will turn up in any one toss?" To answer this question you must first note that *three* faces, **2**, **4**, and **6**, bear even numbers.

The chances are **3** in **6** that an even number turns up in any one toss. The probability of such an event is $\frac{3}{6}$, or $\frac{1}{2}$. In symbols we write:

$$P(e) = \frac{3}{6} = \frac{1}{2}.$$

Questions

In each of the following exercises, assume that the various outcomes are equally likely.

1. Three cards with the letters A, B, and C are dropped into a box. A person draws one of the cards from the box without looking. What is the probability, $P(A)$, that the card with the letter A will be drawn? The card with the letter B? What is the probability that *either* the card with the letter B *or* the card with the letter C will be drawn?

2. What is the probability that after one toss of a die the face numbered 6 will turn up? The face numbered 3? What is the probability that an odd-numbered face will turn up? What is the probability that either a 4 or a 5 will turn up?

3. Three marbles, two red and one black, are placed in a box. A blindfolded person draws one marble from the box. What is the probability of drawing the black marble? What is the probability of drawing a red marble?

4. Five marbles, three black and two red, are placed in a bowl. What is the probability of drawing a black marble? a red marble?

5. Suppose you draw one card from a standard deck of playing cards. What is the probability that you draw a king? an ace? the queen of hearts? What is the probability that the card is a spade? a heart? a diamond?

6. Consider the spinner shown in the illustration. The areas marked "30" and "40" are equal; those marked "0," "10," and "20" are also equal, and the vertical line halves the circular area. The outcome of a spin is the number indicated by the needle. What do you think is the probability that the needle will point to 40 after one spin? Estimate the probability that the needle will point to some number other than 0 after one spin.

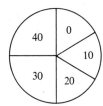

A rotary press of 1845 . . .

A modern rotary press . . .

Multiplication and Division of Directed Numbers

From the time of Gutenberg until the middle of the nineteenth century, printing was done on flat bed presses, a single sheet at a time. Richard Hoe's 1845 rotary press was the first to print multiple sheets successfully. It was sheet-fed by hand operators, and was capable of printing 8000 sheets an hour, in one color, on one side only. By contrast, a modern two-color web offset press, of which a section is shown, can print 8000 sheets an hour, in two colors, on both sides at the same time. This mathematics book was printed on such a press. A continuous roll of paper, called a web, moves through a web offset press at a speed of 800 feet per minute — a speed at which 22,000 copies of a 48-page tabloid newspaper can be printed in one hour.

Multiplication

10–1 Products of Positive and Negative Numbers

The product of two positive numbers is not really a new idea. It is the same as the product of two numbers of arithmetic. That is, the product of two positive numbers is a positive number. For example, $4 \cdot 6$ can be thought of as $6 + 6 + 6 + 6$, or 24. Also, since multiplication of the numbers of arithmetic is commutative, we can write

$$4 \cdot 6 = 6 \cdot 4 = 4 + 4 + 4 + 4 + 4 + 4 = 24$$

What happens in the case of a multiplication problem in which one of the two factors is negative? For the product $3 \cdot {}^-8$, we can write

$$3 \cdot {}^-8 = {}^-8 + {}^-8 + {}^-8 = {}^-24$$

Now consider the product ${}^-6 \cdot 4$. It is difficult to interpret ${}^-6 \cdot 4$ as a repeated addition in its present form. However, if the **commutative**

283

property is to hold true for multiplication of directed numbers also, we can write:

$$^-6 \cdot 4 = 4 \cdot {}^-6$$

Therefore, $\quad {}^-6 \cdot 4 = {}^-6 + {}^-6 + {}^-6 + {}^-6 = {}^-24$

Try some other products, such as $^-5 \cdot 2$ and $7 \cdot {}^-4$. They should help you verify the fact that the product of a positive number and a negative number is always a negative number.

The special properties for **0** and **1**, which we found to hold for the numbers of arithmetic, also hold for directed numbers. Study the following illustrations of these properties.

$$5 \cdot 0 = 0 \cdot 5 = 0 \qquad 1 \cdot 7 = 7 \cdot 1 = 7$$
$$^-3 \cdot 0 = 0 \cdot {}^-3 = 0 \qquad {}^-5 \cdot 1 = 1 \cdot {}^-5 = {}^-5$$
$$0 \cdot {}^-6 = {}^-6 \cdot 0 = 0 \qquad 1 \cdot {}^-10 = {}^-10 \cdot 1 = {}^-10$$

To summarize, we can state that the following properties hold for the products of directed numbers:

(1) The product of **two positive** numbers is a **positive** number.

(2) The product of a **positive** number and a **negative** number is a **negative** number.

(3) The product of **0** and **any directed number** is **0**.

(4) The product of **1** and **any directed number** is the **directed number**.

ORAL EXERCISES

Tell how to complete each of the following to make a true statement.

SAMPLE: $\quad ^-9 + {}^-9 + {}^-9 = ? \qquad$ *What you say:* $\quad ^-9 + {}^-9 + {}^-9 = {}^-27$
$\qquad\qquad\quad 3 \cdot {}^-9 = ? \qquad\qquad\qquad\qquad\qquad\quad 3 \cdot {}^-9 = {}^-27$

1. $7 + 7 + 7 + 7 = ?$
   $4 \cdot 7 = ?$

2. $^-2 + {}^-2 + {}^-2 = ?$
   $3 \cdot {}^-2 = ?$

3. $^-9 + {}^-9 + {}^-9 + {}^-9 = ?$
   $4 \cdot {}^-9 = ?$

4. $^-10 + {}^-10 = ?$
   $2 \cdot {}^-10 = ?$

5. $0 + 0 + 0 = ?$
   $3 \cdot 0 = ?$

6. $^-12r + {}^-12r = ?$
   $2 \cdot {}^-12r = ?$

7. $^-\frac{1}{2} + {}^-\frac{1}{2} + {}^-\frac{1}{2} = ?$
   $3 \cdot {}^-\frac{1}{2} = ?$

8. $^-1 + {}^-1 + {}^-1 = ?$
   $3 \cdot {}^-1 = ?$

Multiplication and Division of Directed Numbers 285

WRITTEN EXERCISES

Simplify each expression.

SAMPLE 1: $^-18 \cdot 4$ Solution: $^-18 \cdot 4 = {}^-72$

SAMPLE 2: $x \cdot \dfrac{^-3}{5}$ Solution: $x \cdot \dfrac{^-3}{5} = \dfrac{^-3}{5} \cdot x = -\dfrac{3x}{5}$

1. $3 \cdot {}^-5$
2. $7 \cdot {}^-9$
3. $12 \cdot 14$
4. $4 \cdot {}^-8k$
5. $6(^-5)$
6. $^-3(12)$
7. $-s \cdot \tfrac{2}{3}$
8. $(7)(^-5)$
9. $(^-6)(k)$
10. $^-9(5)$
11. $t \cdot {}^-\tfrac{1}{2}$
12. $^-3 \cdot \tfrac{1}{2}$
13. $7t(^-8)$
14. $(10)^-1$
15. $^-0.13(1)$
16. $(-\tfrac{1}{2})(\tfrac{2}{3})$
17. $^-4(0.112)$
18. $\tfrac{3}{5}(-\tfrac{1}{7})$
19. $(\tfrac{1}{5})(4)$
20. $^-\tfrac{7}{8} \cdot 0$
21. $0 \cdot {}^-1\tfrac{1}{5}$

Complete each multiplication table for directed numbers.

22.

×	2	4	6	8	10
$^-2$					
$^-4$		$^-24$			
$^-6$					
$^-8$					$^-80$
$^-10$					

23.

×	0	$^-1$	$^-2$	$^-3$
0				
1			$^-2$	
2				$^-6$
3				

Multiply.

24. $^-12$
    8

25. $^-15$
    12

26. 24
    $^-17$

27. 35
    $^-21$

28. $^-125$
    13

Simplify each expression.

SAMPLE 1: $(3 \cdot {}^-5) + (3 \cdot {}^-2)$

Solution: $(3 \cdot {}^-5) + (3 \cdot {}^-2) = {}^-15 + {}^-6 = {}^-21$

SAMPLE 2: $10 + (^-4 \cdot 6)$

Solution: $10 + (^-4 \cdot 6) = 10 + {}^-24 = {}^-14$

29. $(4 \cdot 7) + (4 \cdot {}^-2)$
30. $(^-1 \cdot 8) + 18$

286 Chapter 10

31. $(^-3 \cdot 6) + (3 \cdot {}^-6)$
32. $(2 \cdot {}^-5) + (2 \cdot {}^-10)$
33. $15 + (2 \cdot {}^-10)$
34. $9 + (3 \cdot {}^-5)$
35. $(^-9 \cdot 1) + {}^-10$
36. $(4 \cdot 8) + {}^-12$
37. $(^-2 \cdot 15) + (2 \cdot 15)$
38. $(10 \cdot {}^-4) + (^-10 \cdot 4)$

Solve each equation and write the solution set. The replacement set is {the directed numbers}.

SAMPLE 1: $9 \cdot {}^-6 = t$

Solution: $9 \cdot {}^-6 = t$
${}^-54 = t \quad \{^-54\}$

SAMPLE 2: $3 \cdot x = {}^-24$

Solution: $ 3 \cdot x = {}^-24$
Since $3(^-8) = {}^-24$
$x = {}^-8 \quad \{^-8\}$

B
39. $^-5 \cdot 12 = h$
40. $4 \cdot 17 = b$
41. $15 \cdot {}^-6 = m$
42. $^-18 \cdot 0 = x$
43. $1 \cdot 27 = n$
44. $^-1 \cdot 62 = a$
45. $^-4 \cdot s = {}^-48$

46. $3(k) = {}^-39$
47. $d \cdot {}^-5 = {}^-65$
48. $t \cdot 7 = 42$
49. $r(7) = {}^-42$
50. $^-3\frac{1}{2}y = {}^-3\frac{1}{2}$
51. $7\frac{1}{3}w = {}^-7\frac{1}{3}$
52. $^-6 \cdot \frac{2}{3} = x$

C
53. $m + (3 \cdot {}^-4) = {}^-18$
54. $(^-7 \cdot 5) + r = 25$
55. $^-9 \cdot 12 = 9 \cdot t$

56. $3 + (4 \cdot {}^-8) = k$
57. $^-12 \cdot b = {}^-8 \cdot 12$
58. $a + (^-6 \cdot 5) = 24$

10–2 More about Products of Positive and Negative Numbers

Our work thus far in the multiplication of directed numbers has dealt with products of only two factors. Of course, we might wish to find a product of more than two factors. For example, consider $4 \cdot {}^-3 \cdot 6$. It is important for us to look for an associative property for multiplication of directed numbers. Using parentheses to group the factors, we write $(4 \cdot {}^-3)6$ and $4(^-3 \cdot 6)$. Simplifying these expressions, our results are as follows:

$$(4 \cdot {}^-3)6 = (^-12)6 \qquad 4(^-3 \cdot 6) = 4(^-18)$$
$$\phantom{(4 \cdot {}^-3)6\ } = {}^-72 = {}^-72$$

Thus $(4 \cdot {}^-3)6 = 4(^-3 \cdot 6)$.

This result indicates that multiplication with directed numbers is **associative**, just as we have seen to be the case with numbers of arithmetic.

Another basic property of numbers of arithmetic is the distributive property. Study the following examples, which illustrate the **distributive property** for the **directed numbers**.

EXAMPLE 1.
$$3(4 + {}^-6) = (3 \cdot 4) + (3 \cdot {}^-6)$$

$3 \cdot {}^-2$	$12 + {}^-18$
6	6

Thus $3(4 + {}^-6) = (3 \cdot 4) + (3 \cdot {}^-6)$

EXAMPLE 2.
$$({}^-2 \cdot 4) + (5 \cdot 4) = ({}^-2 + 5)4$$

${}^-8 + 20$	$3 \cdot 4$
12	12

Thus $({}^-2 \cdot 4) + (5 \cdot 4) = ({}^-2 + 5)4$

The following list contains the basic properties of multiplication of directed numbers.

For all directed numbers *r*, *s*, and *t*,

$r \cdot s$ is a unique directed number	Closure property
$r \cdot s = s \cdot r$	Commutative property
$(r \cdot s)t = r(s \cdot t)$	Associative property
$r \cdot 0 = 0 \cdot r = 0$	Multiplicative property of zero
$r \cdot 1 = 1 \cdot r = r$	Multiplicative property of one
$r(s + t) = r \cdot s + r \cdot t$	Distributive property

ORAL EXERCISES

For each statement, name the property of multiplication of directed numbers that is illustrated.

SAMPLE: ${}^-3 \cdot 16 = 16 \cdot {}^-3$ *What you say:* Commutative property

1. $(2)(3 \cdot {}^-10) = (2 \cdot 3)({}^-10)$
2. $10({}^-6 + {}^-8) = (10 \cdot {}^-6) + (10 \cdot {}^-8)$
3. $({}^-35)(19) = (19)({}^-35)$
4. $1 \cdot {}^-4\frac{1}{8} = {}^-4\frac{1}{8}$
5. $(2 \cdot {}^-3)4 = 4(2 \cdot {}^-3)$
6. ${}^-25 \cdot 0 = 0$
7. $14 \cdot {}^-1 = {}^-1 \cdot 14$
8. ${}^-3\frac{3}{4} = ({}^-3\frac{3}{4})(1)$
9. $({}^-3 \cdot 7) + ({}^-2 \cdot 7) = ({}^-3 + {}^-2)7$

288 *Chapter 10*

Tell whether each statement is true or false.

10. $(^-3\frac{1}{2})(7) = (7)(^-3\frac{1}{2})$
11. $(^-3)(2 + 8) = (2 + 8)(^-3)$
12. $(^-8 + {}^-6)5 = (^-8 \cdot 5) + (^-6 \cdot 5)$
13. $^-10(6 + {}^-2) = {}^-10(4)$
14. $3 \cdot {}^-9 \neq {}^-3 \cdot 9$
15. $(^-6)(2) > (^-5)(2)$
16. $(^-2)(7)(5) = (5)(^-2)(7)$
17. $0 \cdot 4 > (^-2)(4)$

WRITTEN EXERCISES

For each of the following, use the distributive property to rewrite the expression as a sum of two products. Then simplify each expression to show that they are equal.

SAMPLE: $10(^-5 + 3)$ *Solution:* $\underline{10(^-5 + 3) = 10(^-5) + 10(3)}$
$\qquad\qquad\qquad\qquad\qquad\quad 10(^-2) \;\vert\; ^-50 + 30$
$\qquad\qquad\qquad\qquad\qquad\quad\; ^-20 \;\;\vert\; ^-20$

1. $6(4 + {}^-2)$
2. $8(^-3 + 7)$
3. $4(^-9 + {}^-6)$
4. $^-7(4 + 6)$
5. $12(^-3 + {}^-7)$
6. $(5 + {}^-6)19$
7. $(^-8 + {}^-7)10$
8. $(^-9 + 5)12$
9. $\frac{1}{2}(^-4 + {}^-6)$
10. $\frac{2}{3}(^-9 + 12)$
11. $^-\frac{1}{4}(8 + 12)$
12. $^-1.5(3 + 7)$

Show that each statement is true.

SAMPLE 1: $3(4 \cdot {}^-8) = (3 \cdot 4)\,8$ *Solution:* $\underline{3(4 \cdot {}^-8) = (3 \cdot 4)^-8}$
$\qquad\qquad\qquad\qquad\qquad\qquad\qquad\qquad\quad 3(^-32) \;\vert\; (12)^-8$
$\qquad\qquad\qquad\qquad\qquad\qquad\qquad\qquad\quad\; ^-96 \;\;\;\vert\; ^-96$

SAMPLE 2: $^-6(2 + 7) = (^-6 \cdot 2) + (^-6 \cdot 7)$
Solution: $\underline{^-6(2 + 7) = (^-6 \cdot 2) + (^-6 \cdot 7)}$
$\qquad\qquad\quad ^-6(9) \;\vert\; ^-12 + {}^-42$
$\qquad\qquad\quad\; ^-54 \;\;\vert\; ^-54$

13. $(5)(3 \cdot {}^-7) = (5 \cdot 3)(^-7)$
14. $(6 \cdot 8)(^-2) = (6)(8 \cdot {}^-2)$
15. $4(9 + {}^-5) = (4 \cdot 9) + (4 \cdot {}^-5)$
16. $^-3(10 + 6) = (^-3 \cdot 10) + (^-3 \cdot 6)$
17. $^-7(8 + 12) = (^-7 \cdot 8) + (^-7 \cdot 12)$
18. $(3 + {}^-6)2 = (3 \cdot 2) + (^-6 \cdot 2)$
19. $(^-2 + 10)5 = (^-2 \cdot 5) + (10 \cdot 5)$
20. $(6 \cdot 10) + (^-9 \cdot 10) = (6 + {}^-9)10$

Multiplication and Division of Directed Numbers   289

21. $^-6(10 - 3) = (^-6 \cdot 10) - (^-6 \cdot 3)$
22. $(3 - 7)5 = (3 \cdot 5) - (7 \cdot 5)$
23. $^-2(3 + 6 + 1) = (^-2 \cdot 3) + (^-2 \cdot 6) + (^-2 \cdot 1)$
24. $^-6(\frac{1}{2} + \frac{2}{3}) = (^-6 \cdot \frac{1}{2}) + (^-6 \cdot \frac{2}{3})$

Show that each of the following is true for the replacements $a = 4$, $b = ^-6$, $c = 9$, and $d = ^-\frac{1}{2}$.

B 25. $a(bc) = (ab)c$ 30. $b \cdot 0 = 0 \cdot b$
   26. $ab = ba$ 31. $d \cdot 0 = b \cdot 0$
   27. $a(b + c) = ab + ac$ 32. $10(a + d) = 10a + 10d$
   28. $(ad)c = c(ad)$ 33. $b(c - a) = bc - ba$
   29. $(a + c)b = ab + bc$ 34. $ac - bc = (a - b)c$

Multiply, using the distributive property as shown in the samples.

SAMPLE 1: $3(^-321)$ Solution: $\begin{array}{r} ^-300 + ^-20 + ^-1 \\ 3 \\ \hline ^-900 + ^-60 + ^-3 = ^-963 \end{array}$

SAMPLE 2: $4(^-68)$ Solution: $\begin{array}{r} ^-70 + 2 \\ 4 \\ \hline ^-280 + 8 = ^-272 \end{array}$

C 35. $2(^-434)$ 38. $^-4(88)$ 41. $9(^-241)$
   36. $^-10(256)$ 39. $3(^-2103)$ 42. $^-7(3004)$
   37. $3(^-49)$ 40. $^-3(229)$ 43. $10(^-5356)$

10–3 Products of Negative Numbers

So far, we have seen that the product of two positive numbers is a positive number, and that the product of one positive and one negative number is negative. Now we are ready to discuss the product of two negative numbers. Consider the following examples, and note how the distributive property is used.

EXAMPLE 1.

$^-3(4 + ^-4) = (^-3 \cdot 4) + (^-3 \cdot ^-4)$ Distributive property
$^-3(0) = (^-3 \cdot 4) + (^-3 \cdot ^-4)$ Additive property of opposites
$0 = (^-3 \cdot 4) + (^-3 \cdot ^-4)$ Multiplicative property of zero
$0 = ^-12 + (^-3 \cdot ^-4)$ Substitution principle

For the statement above to be true, $(^-3 \cdot ^-4)$ must equal 12.

EXAMPLE 2. $^-2(7 + {}^-3) = {}^-2(7) + ({}^-2 \cdot {}^-3)$
$^-2(4) = {}^-2(7) + ({}^-2 \cdot {}^-3)$
$^-8 = {}^-2(7) + ({}^-2 \cdot {}^-3)$
$^-8 = {}^-14 + ({}^-2 \cdot {}^-3)$

So, $({}^-2 \cdot {}^-3)$ must equal **6**.

EXAMPLE 3. $^-5({}^-10 + 12) = {}^-5({}^-10) + ({}^-5 \cdot 12)$
$^-5(2) = {}^-5({}^-10) + ({}^-5 \cdot 12)$
$^-10 = {}^-5({}^-10) + ({}^-5 \cdot 12)$
$^-10 = {}^-5({}^-10) + ({}^-60)$

So, $^-5({}^-10)$ must equal **50**.

The three examples illustrate the idea that the **product** of **two negative numbers** is a **positive number**.

Here are some more applications of this idea about the product of two negative numbers:

EXAMPLE 1. $^-6({}^-4 + 7) = {}^-6 \cdot {}^-4 + {}^-6 \cdot 7$
$^-6(3) = 24 + {}^-42$
$^-18 = {}^-18$

EXAMPLE 2. $^-5({}^-6 + {}^-7) = {}^-5 \cdot {}^-6 + {}^-5 \cdot {}^-7$
$^-5({}^-13) = 30 + 35$
$65 = 65$

EXAMPLE 3. $^-2(3 + {}^-15) = {}^-2 \cdot 3 + {}^-2 \cdot {}^-15$
$^-2({}^-12) = {}^-6 + 30$
$24 = 24$

EXAMPLE 4. $({}^-8 + {}^-3){}^-3 = {}^-8 \cdot {}^-3 + {}^-3 \cdot {}^-3$
$^-11({}^-3) = 24 + 9$
$33 = 33$

Can you think of a general rule that tells whether the product of two directed numbers, neither one of which is 0, is positive or negative? Study the following table carefully, and verify the conclusions that are stated.

Multiplication and Division of Directed Numbers

Factors	Are the signs of the factors alike?	Product	Sign of Product
(3)(7)	yes	21	positive
(4)(⁻6)	no	⁻24	negative
(⁻6)(⁻5)	yes	30	positive
(⁻2)(8)	no	⁻16	negative

Do you agree that in general we can state that the product of two directed numbers is **positive** when their signs are **alike**? Do you also agree that the product of two directed numbers is **negative** when their signs are **not alike**?

Is it possible to tell at a glance whether an indicated product of three or more factors, such as (⁻3)(5)(⁻2)(⁻6), represents a positive or a negative number? Study the following table and see what pattern appears that might help you state a general rule.

Factors	Is the number of negative factors odd or even?	Product	Sign of Product
(2)(5)(⁻3)	odd	⁻30	negative
(⁻3)(⁻1)(⁻6)	odd	⁻18	negative
(⁻4)(2)(⁻3)	even	24	positive
(⁻5)(⁻1)(⁻2)(⁻3)	even	30	positive

What pattern did you note? Did you observe that the product of two or more directed numbers is **positive** if the number of **negative factors** is **even**, and that the product is **negative** if the number of **negative factors** is **odd**?

ORAL EXERCISES

Give a simpler name for each product.

SAMPLE: (⁻9)(⁻5)   *What you say:* 45

1. (⁻4)(⁻7)   2. ⁻10(⁻$\frac{1}{5}$)   3. ⁻(2 + 4)(⁻3 + ⁻5)

4. $(^-10)(^-9)$
5. $(^-2a)(\frac{1}{3})$
6. $(^-\frac{3}{2})(b)$
7. $(-x)(^-8)$
8. $(^-4)(^-12)$
9. $(^-\frac{1}{3})(^-\frac{1}{3})$
10. $^-6(^-4 + ^-3)$
11. $(^-5)^2$
12. $(^-6 + ^-2)(5)$

Tell whether each expression represents a positive number, a negative number, or zero. Do *not* simplify.

13. $(^-3)(^-8)(^-15)$
14. $(^-5)(2)(7)(^-3)$
15. $(10)(^-4)(^-6)(^-10)$
16. $^-3 \cdot 7 \cdot 12 \cdot 18 \cdot 2$
17. $(^-5)(^-8)(7)(0)(^-3)$
18. $(^-3)(^-\frac{1}{2})(\frac{1}{5})(\frac{2}{3})(^-\frac{1}{4})$
19. $(^-2)(^-2)(^-2)(^-2)$
20. $(^-3)(^-5)(9)(^-2)(^-1)$
21. $(\frac{1}{8})(\frac{3}{5})(0)(^-1)(\frac{7}{8})(\frac{1}{2})$
22. $(\frac{2}{3})(^-\frac{3}{5})(\frac{3}{2})(^-\frac{7}{8})$
23. $(^-3)^2$
24. $(^-10)^3$
25. $(^-2)^5$
26. $(^-\frac{1}{2})^4$

WRITTEN EXERCISES

Tell whether the number represented is positive, negative, or zero. Then simplify the expression.

SAMPLE: $6(^-2)(^-3)$ *Solution:* Positive; 36

1. $(5)(^-4)(4)$
2. $(^-8)(^-12)$
3. $(^-5)(^-3)(0)$
4. $^-1(6)(2)(^-2)$
5. $(^-3)(^-3)(^-3)$
6. $^-5(^-2)(2)(^-5)$
7. $(3)(7)(4)$
8. $(\frac{1}{2})(4)(^-2)$
9. $(^-3)(^-2)(^-3)(^-5)$
10. $(^-\frac{1}{3})(6)(^-4)$
11. $(^-2)(^-1)(0)(12)$
12. $\frac{1}{2}(^-\frac{1}{2})(^-\frac{1}{2})$
13. $(^-3)^3$
14. $(^-3)^4$
15. $(^-\frac{1}{5})(0)(\frac{3}{8})(\frac{1}{9})$
16. $(3 \cdot ^-2)(3 \cdot ^-2)$
17. $(^-5 \cdot 4)(^-5 \cdot 3)$
18. $^-1(^-3)(^-4 \cdot 2)$

For each of the following, use the distributive property to rewrite the expression as a sum of two products. Then simplify each expression to show that they are equal.

SAMPLE: $^-3(4 + ^-5)$ *Solution:* $\underline{^-3(4 + ^-5) = (^-3)(4) + (^-3)(^-5)}$
$^-3(^-1) \mid ^-12 + 15$
$3 \mid 3$

19. $^-2(^-3 + 10)$
20. $^-5(^-6 + ^-2)$
21. $(^-8 + 4)(^-10)$
22. $(^-5 + ^-4)(^-1)$
23. $(3 + ^-9)2$
24. $(6 + ^-8)(^-5)$

B
25. $(^-4 + 2)(^-\frac{1}{2})$
26. $3(4 + ^-6 + ^-2)$
27. $^-2(^-3 + 4 + ^-6)$
28. $^-\frac{2}{3}(^-9 + ^-6 + 3)$
29. $^-2(^-3 + 0)$
30. $(^-5 + ^-12)(0)$
31. $^-3(4 - ^-4)$
32. $4(^-5 + ^-3 - ^-4)$
33. $^-1(^-3 - ^-5)$

Evaluate each expression. Let $s = 2$, $t = ^-4$, $c = ^-5$, $d = 10$, and $a = ^-2$.

SAMPLE: $(s)(t)(-a)$ Solution: $(2)(^-4)(2) = {}^-16$

34. $(t)(c)(s)$
35. $3(t)(c)(a)$
36. $10(t)(a)$
37. $t \cdot t \cdot t \cdot t$
38. $st + tc$
39. $^-5(a)(c)(t)$
40. $(-t)(d)$
41. $(-c)(-a)s$
42. $2(a)(s)(c)$
43. $ct + ac$
44. t^3
45. c^2
46. $s^2 + a$
47. $(tc)^2$
48. $(sa)(sa)$

Simplify each of the following.

[C] 49. $^-4 + [^-2(^-3 + 5)]$
50. $8 - [^-4(6 + {}^-3)]$
51. $(^-2)[^-1(8 + {}^-3)]$
52. $^-5[^-3(^-6 - {}^-8)]$

Evaluate each of the following expressions.

53. $(^-1)^2$
54. $(^-1)^3$
55. $(^-1)^4$
56. $(^-2)^2$
57. $(^-2)^3$
58. $(^-2)^4$
59. $(^-10)^2$
60. $(^-10)^3$
61. $(^-10)^4$

Tell whether each expression represents a positive or a negative number.

62. $(^-3)^8$
63. $(^-18)^7$
64. $(^-10)^{15}$
65. $(^-375)^{12}$

Which of these sets of numbers are closed under multiplication? If a set is not closed, give an example to prove that it is not.

66. $\{^-1, 0, 1\}$
67. $\{^-1, 1\}$
68. $\{^-\frac{1}{2}, ^-1, 1, \frac{1}{2}\}$
69. {negative numbers}
70. $\{\ldots, {}^-3, {}^-2, {}^-1, 0, 1, \ldots\}$
71. {positive numbers}

10–4 The Distributive Property and Algebraic Expressions

We have been using the distributive property in simplifying numerical expressions because we have seen that it holds for directed numbers, as well as for numbers of arithmetic. Earlier we found that the distributive property could help us to simplify expressions involving variables. Now we shall continue to use it for that purpose when the expressions to be simplified involve directed numbers. Study the following examples.

EXAMPLE 1. $5x + 2y + 3y + 2x = 5x + 2x + 3y + 2y$
$= (5 + 2)x + (3 + 2)y$
$= 7x + 5y$

EXAMPLE 2. $3n + 5m - 8n + 2m = 3n - 8n + 5m + 2m$
$= (3 - 8)n + (5 + 2)m$
$= {}^-5n + 7m$

ORAL EXERCISES

Give a simpler name for each expression by combining similar terms.

SAMPLE: $3a + 5a + 7a - 5$ *What you say:* $15a - 5$

1. $5x + 6x + 3x$
2. $8 + 3s + 8s$
3. $2m + 5m + 4m$
4. $3a + a + 12a$
5. $4k - 3 + 5k$
6. $5b + 7b + 4b + {}^-10b$
7. $2x + 7x + 4y + y$
8. $10t + {}^-3t + 2s + s$
9. $5n + 6r + 4n + 11r$
10. $\frac{2}{3}p + \frac{3}{5}k + \frac{1}{3}p + \frac{2}{5}k$

WRITTEN EXERCISES

Simplify each expression by combining similar terms.

SAMPLE 1: $3a + b + 4a - a - 5b$

Solution: $3a + b + 4a - a - 5b = (3a + 4a + {}^-1a) + (b + {}^-5b)$
$= 6a + {}^-4b = 6a - 4b$

SAMPLE 2: $5(x^2 + 3x) + 2(x^2 - 7x)$

Solution: $5x^2 + 15x + 2x^2 - 14x = (5x^2 + 2x^2) + (15x + {}^-14x)$
$= 7x^2 + x$

A
1. $6x + 5 + x + 2x + 9$
2. $4m + 7 + 9m + 3 + m$
3. $10k - 3k + 2t + 6t$
4. $4s + 5r - r + s + 2r$
5. $12a - 10b + 3a + 7b$
6. $3ab + 6xy - 5ab + 9xy$
7. $2rs - 3 - 5rs + 6rs - 5$
8. $5mn + 2mn - mn + t$
9. $2xyz + 3k - k + 5xyz$
10. $^-8ms + 3ms + 6b + 5b$
11. $5a^2 - 3b + 12a^2 + b - a^2$
12. $^-3x^2 + x^2 + y^2 + 6y^2$

B
13. $3(a + b) + 2(a + b)$
14. $4(x - y) + 9(x + y)$
15. $^-15y + 3y^2 - y^2 + 9y$
16. $\frac{1}{3}k + \frac{1}{2} + \frac{2}{3}k - \frac{3}{2}$
17. $3 + s^2 + 4s - 2s - s^2$
18. $3(r - 2r) + (7r + 2r)$
19. $^-2(a + b) + 5(a - b)$
20. $3ab + 5(a + b) + 5(a - b)$

Multiplication and Division of Directed Numbers

[C] 21. $2[5(x + 1) + 2] + 6$
22. $7[^-5 + 3(m + 4)] + 40$
23. $^-5y + 3[4(1 + 3y) - 6]$
24. $10b + ^-2[3 + ^-4(2b + ^-2)]$

Find the value of each expression, if $r = 8$, $s = ^-2$, $x = ^-3$, and $y = 6$.

25. $9rs - 5y$
26. $4 + (r + 1)^2$
27. $2xy - 3r$
28. $\frac{1}{2}r + 4y + \frac{1}{3}y$
29. $0.2r + 0.3ry$
30. $0.5y - \frac{1}{2}xs$
31. $\frac{1}{2}(xs)^3$
32. $s(2r + y)$
33. $(y + r)^2(x)$
34. $\frac{1}{2}(ry + xs)$
35. $(r + s)^2 + (x - y)^2$
36. $(r - y)^3(s)^2$

Division

10–5 Division of Directed Numbers

You will recall that multiplication and division are inverse operations. That is, for the multiplication sentence

factor × factor = product

we can write an equivalent division sentence in either of these forms:

product ÷ factor = factor, or $\dfrac{\text{product}}{\text{factor}} = \text{factor.}$

Since we already know how to multiply directed numbers, we can easily decide how to divide by a directed number.

Consider the following statements, which involve only positive numbers:

Since $4 \cdot 7 = 28$, you know that $28 \div 7 = 4$ and that $28 \div 4 = 7$.
Since $9 \cdot 5 = 45$, you know that $45 \div 5 = 9$ and that $45 \div 9 = 5$.

Now look at these statements, which follow the same pattern, but involve negative numbers:

Since $4 \cdot ^-5 = ^-20$, you know that $^-20 \div ^-5 = 4$
and that $^-20 \div 4 = ^-5$.

Since $^-6 \cdot 7 = ^-42$, you know that $\dfrac{^-42}{7} = ^-6$ and that $\dfrac{^-42}{^-6} = 7$.

Since $^-3 \cdot ^-5 = 15$, you know that $15 \div ^-5 = ^-3$
and that $15 \div ^-3 = ^-5$.

Since $^-9 \cdot ^-3 = 27$, you know that $\dfrac{27}{^-3} = ^-9$ and that $\dfrac{27}{^-9} = ^-3$.

From the examples shown, verify the following facts concerning division with directed numbers.

$$\text{positive} \div \text{positive} = \text{positive}, \quad \text{or} \quad \frac{\text{positive}}{\text{positive}} = \text{positive}$$

$$\text{positive} \div \text{negative} = \text{negative}, \quad \text{or} \quad \frac{\text{positive}}{\text{negative}} = \text{negative}$$

$$\text{negative} \div \text{positive} = \text{negative}, \quad \text{or} \quad \frac{\text{negative}}{\text{positive}} = \text{negative}$$

$$\text{negative} \div \text{negative} = \text{positive}, \quad \text{or} \quad \frac{\text{negative}}{\text{negative}} = \text{positive}$$

It is important to notice that, according to the rules just stated, the fractions $\frac{^-10}{5}$ and $\frac{10}{^-5}$ name the same number.

$$\frac{^-10}{5} = {^-2} \qquad \frac{10}{^-5} = {^-2}$$

Do you see that the following statements are also true?

$$\frac{^-4}{2} = \frac{4}{^-2} = {^-2};\qquad \frac{^-3}{5} = \frac{3}{^-5} = \frac{^-3}{5}$$

$$\frac{^-1}{2} = \frac{1}{^-2} = \frac{^-1}{2};\qquad \frac{7}{^-8} = \frac{^-7}{8} = \frac{^-7}{8}$$

As you might well have guessed we can write fractions such as $\frac{^-3}{4}$ and $\frac{2}{^-5}$ in decimal form by completing the indicated division.

The symbol $\frac{^-3}{4}$ means $\qquad$ The symbol $\frac{2}{^-5}$ means

$^-3$ divided by 4. $\qquad\qquad\qquad$ 2 divided by $^-5$.

$$\begin{array}{r}^-0.75\\4\overline{)^-3.00}\end{array} \qquad\qquad \begin{array}{r}^-0.40\\^-5\overline{)2.00}\end{array}$$

Therefore, $\frac{^-3}{4} = {^-0.75}$ $\qquad$ Therefore, $\frac{2}{^-5} = {^-0.40}$

It is important to recall that we agreed that symbols such as $\frac{3}{0}$ and $\frac{6}{0}$ are meaningless for arithmetic numbers. That is, division by 0 is not possible. This is also true for directed numbers. Symbols such as $\frac{-3}{0}$, $\frac{-1}{0}$, and $10 \div 0$ are meaningless.

ORAL EXERCISES

Complete each statement to make it true.

1. Because $^-5 \cdot 7 = ^-35$, we know $^-35 \div ^-5 = \underline{\ ?\ }$.
2. Because $6 \cdot 9 = 54$, we know $54 \div 9 = \underline{\ ?\ }$.
3. Because $3 \cdot ^-10 = ^-30$, we know $^-30 \div 3 = \underline{\ ?\ }$.
4. Because $^-8 \cdot ^-9 = 72$, we know $72 \div ^-8 = \underline{\ ?\ }$.
5. Because $^-6 \cdot \underline{\ ?\ } = ^-48$, we know $^-48 \div ^-6 = 8$.
6. Because $\underline{\ ?\ } \cdot ^-11 = 99$, we know $99 \div ^-9 = ^-11$.

WRITTEN EXERCISES

Complete each indicated division.

1. $^-16 \div 8$
2. $^-30 \div ^-2$
3. $^-10 \div 5$
4. $42 \div ^-3$
5. $\frac{21}{^-3}$
6. $\frac{^-45}{15}$
7. $\frac{^-42}{^-7}$

8. $56 \div 7$
9. $98 \div ^-2$
10. $^-6\overline{)72}$
11. $^-8\overline{)64}$
12. $8\overline{)^-64}$
13. $8\overline{)64}$
14. $\frac{60}{^-4}$

15. $\frac{^-125}{5}$
16. $\frac{^-15}{2+3}$
17. $\frac{^-27}{^-6+^-3}$
18. $5.25 \div ^-5$
19. $^-0.78 \div 3$
20. $\frac{10+2}{^-4}$
21. $\frac{^-10+2}{^-4}$

Write the decimal equivalent for each of the following.

SAMPLE: $\frac{^-1}{5}$ Solution: $5\overline{)^-1.0}\ ^-0.2$

22. $\frac{^-1}{2}$
23. $\frac{^-3}{4}$
24. $\frac{^-4}{^-5}$
25. $\frac{^-5}{8}$
26. $\frac{12}{^-5}$
27. $\frac{^-3}{10}$
28. $\frac{^-3}{8}$
29. $\frac{^-5}{4}$

298 Chapter 10

30. $\dfrac{4}{-5}$ 32. $\dfrac{7}{-8}$ 34. $\dfrac{1}{-8}$ 36. $\dfrac{-4 + -8}{15}$

31. $\dfrac{8}{-10}$ 33. $\dfrac{-1}{4}$ 35. $\dfrac{-4}{-25}$ 37. $\dfrac{-49}{100}$

Tell whether each statement is true or false.

B 38. $\dfrac{-1}{2} = \dfrac{1}{2}$ 41. $\dfrac{-5}{16} = \dfrac{5}{16}$ 44. $\dfrac{-5}{3} = \dfrac{-5}{-3}$

39. $\dfrac{-3}{4} = \dfrac{3}{-4}$ 42. $\dfrac{-3}{-5} = \dfrac{3}{5}$ 45. $\dfrac{-8}{10} = -0.8$

40. $\dfrac{1}{-3} = \dfrac{-3}{1}$ 43. $\dfrac{5}{12} = \dfrac{5}{-12}$ 46. $\dfrac{25}{-100} = -0.25$

Divide.

47. $-15\overline{)375}$ 48. $16\overline{)-76.8}$ 49. $-37\overline{)-229.4}$

10–6 Directed Numbers and Reciprocals

Consider the following products. Check to see that each product is **1**.

$\dfrac{1}{3} \cdot 3;$ $\dfrac{-2}{3} \cdot \dfrac{3}{-2};$ $1 \cdot 1;$ $-1 \cdot -1$

$\dfrac{-2}{5} \cdot \dfrac{-5}{2};$ $\dfrac{-1}{5} \cdot -5;$ $-0.5 \cdot -2;$ $\dfrac{a}{b} \cdot \dfrac{b}{a}$, if $a \neq 0, b \neq 0$

If the product of two numbers is **1**, the numbers are said to be **reciprocals** of each other. Since $\dfrac{-3}{4} \cdot \dfrac{4}{-3} = 1$, we say that $\dfrac{-3}{4}$ is the reciprocal of $\dfrac{4}{-3}$, and that $\dfrac{4}{-3}$ is the reciprocal of $\dfrac{-3}{4}$.

It is apparent that every directed number, except zero, has a reciprocal. We state the **reciprocal property** as follows:

For every directed number *r*, excluding 0, there is a number $\dfrac{1}{r}$, such that $r \cdot \dfrac{1}{r} = \dfrac{1}{r} \cdot r = 1.$

The reciprocal property is helpful in solving equations. These examples show how it can be used.

Multiplication and Division of Directed Numbers

EXAMPLE 1.
$$6n = 42$$
$$6 \cdot n = 42$$
$$\frac{1}{6} \cdot 6 \cdot n = 42 \cdot \frac{1}{6}$$
$$1 \cdot n = 42 \cdot \frac{1}{6}$$
$$n = 7$$

EXAMPLE 2.
$$\frac{2b}{3} = {}^-18$$
$$\frac{2}{3} \cdot b = {}^-18$$
$$\frac{3}{2} \cdot \frac{2}{3} \cdot b = {}^-18 \cdot \frac{3}{2}$$
$$1 \cdot b = {}^-18 \cdot \frac{3}{2}$$
$$b = {}^-27$$

EXAMPLE 3.
$${}^-4t = 20$$
$${}^-4 \cdot t = 20$$
$$\frac{1}{-4} \cdot {}^-4 \cdot t = 20 \cdot \frac{1}{-4}$$
$$1 \cdot t = 20 \cdot \frac{1}{-4}$$
$$t = {}^-5$$

EXAMPLE 4.
$$\frac{{}^-5x}{2} = 25$$
$$\frac{{}^-5}{2} \cdot x = 25$$
$$\frac{2}{-5} \cdot \frac{{}^-5}{2} \cdot x = 25 \cdot \frac{2}{-5}$$
$$1 \cdot x = 25 \cdot \frac{2}{-5}$$
$$x = {}^-10$$

Notice that in the second step of each solution the left member is written as the product of the variable and a numerical coefficient. Then, in the third step both the left member and the right member are multiplied by the **reciprocal** of the numerical coefficient. This strategy results in an equation equivalent to the original equation and one in which the **coefficient of the variable** is **1**. Simplifying the right member of the equation gives the solution, or **root**, of the equation.

The pairs of directed-number statements below illustrate another very important idea about reciprocals. Study the statements and verify the fact that **dividing** by a number is equivalent to **multiplying** by its **reciprocal**.

$${}^-6 \div 2 = {}^-3 \qquad 4 \div \frac{1}{2} = 8 \qquad \frac{1}{2} \div \frac{1}{-4} = {}^-2$$

$${}^-6 \cdot \frac{1}{2} = {}^-3 \qquad 4 \cdot \frac{2}{1} = 8 \qquad \frac{1}{2} \cdot \frac{-4}{1} = {}^-2$$

For each directed number *r*, and each directed number *s* except 0,

$$r \div s = r \cdot \frac{1}{s}.$$

ORAL EXERCISES

Name the reciprocal of each directed number.

SAMPLE 1: $\frac{^-7}{2}$ What you say: $\frac{2}{^-7}$

SAMPLE 2: $^-0.2$ What you say: $^-5$ (Note: $^-0.2 = ^-\frac{1}{5}$)

SAMPLE 3: $1\frac{1}{2}$ What you say: $\frac{2}{3}$ (Note: $1\frac{1}{2} = \frac{3}{2}$)

1. $\frac{3}{10}$
2. $\frac{^-7}{8}$
3. $^-\frac{2}{5}$
4. $^-\frac{6}{7}$
5. $1\frac{1}{4}$
6. $^-1\frac{1}{2}$
7. 6
8. $^-3$
9. $\frac{x}{3}$, for $x \neq 0$
10. 0.50
11. $^-0.25$
12. $\frac{^-10}{3}$
13. $\frac{3}{^-8}$
14. $\frac{^-5}{16}$
15. $^-4$
16. $^-5$
17. 3
18. $\frac{^-8}{3}$

Tell how to complete each of the following to make a true statement.

SAMPLE: $\frac{^-2}{3} \div 2 = \frac{^-2}{3} \cdot ?$ What you say: $\frac{^-2}{3} \div 2 = \frac{^-2}{3} \cdot \frac{1}{2}$

19. $^-\frac{3}{5} \div 4 = ^-\frac{9}{5} \cdot ?$
20. $\frac{^-7}{2} \div \frac{1}{3} = \frac{^-7}{2} \cdot ?$
21. $\frac{^-5}{2} \div ? = \frac{^-5}{2} \cdot \frac{2}{3}$
22. $\frac{1}{2} \div ? = \frac{1}{2} \cdot \frac{1}{6}$
23. $\frac{^-5}{16} \cdot \frac{4}{3} = \frac{^-5}{16} \div ?$
24. $^-\frac{7}{8} \div ^-\frac{5}{8} = ^-\frac{7}{8} \cdot ?$
25. $^-0.50 \cdot 2 = ^-0.50 \div ?$
26. $0.25 \div \frac{1}{4} = 0.25 \cdot ?$

WRITTEN EXERCISES

Simplify each expression by completing the indicated division.

SAMPLE 1: $\frac{^-5}{7} \div \frac{2}{3}$ Solution: $\frac{^-5}{7} \div \frac{2}{3} = \frac{^-5}{7} \cdot \frac{3}{2} = \frac{^-15}{14} = \frac{^-15}{14}$

SAMPLE 2: $\frac{1}{^-16} \div \frac{3}{^-4}$ Solution: $\frac{1}{^-16} \div \frac{3}{^-4} = \frac{1}{^-16} \cdot \frac{^-4}{3} = \frac{^-4}{^-48} = \frac{1}{12}$

 1. $\frac{3}{^-10} \div \frac{1}{3}$ 2. $1 \div ^-\frac{1}{5}$ 3. $^-10 \div \frac{7}{1}$

4. $\frac{-12}{20} \div \frac{-9}{5}$ 8. $\frac{-2}{5} \div 1$ 12. $^-3 \div 10$

5. $\frac{4}{-10} \div \frac{2}{-3}$ 9. $\frac{-4}{1} \div \frac{-1}{2}$ 13. $\frac{4}{5} \div {}^-3$

6. $\frac{-5}{3} \div \frac{-5}{3}$ 10. $6 \div \frac{-5}{8}$ 14. $\frac{-3}{5} \div \frac{-5}{2}$

7. $\frac{7}{-8} \div \frac{-7}{8}$ 11. $\frac{-3}{4} \div 5$ 15. $16 \div \frac{-2}{3}$

Complete each of the following to make a true statement.

SAMPLE: $\frac{-4}{9} \cdot \frac{9}{-4} = ?$ Solution: $\frac{-4}{9} \cdot \frac{9}{-4} = \frac{-36}{-36} = 1$

16. $\frac{12}{5} \cdot \frac{5}{12} = ?$ 19. $\frac{-3}{8} \cdot ? = 1$ 22. $\frac{1}{-8} \cdot ? = 1$

17. $\frac{-7}{16} \cdot \frac{-16}{7} = ?$ 20. $\frac{5}{1} \cdot ? = 1$ 23. $^-1 \cdot ? = 1$

18. $\frac{2}{-9} \cdot \frac{-9}{2} = ?$ 21. $? \cdot 6 = 1$ 24. $\frac{-6}{11} \cdot \frac{11}{-6} = ?$

Solve each equation and write its solution set. Each replacement set is {the directed numbers}.

SAMPLE: $6n = {}^-18$ Solution: $6n = {}^-18$
$\frac{1}{6} \cdot 6 \cdot n = {}^-18 \cdot \frac{1}{6}$
$1 \cdot n = \frac{-18}{6}$
$n = {}^-3$ $\{^-3\}$

25. $4k = {}^-8$ 29. $^-5w = {}^-25$ 33. $^-12 = {}^-4x$
26. $^-7s = 35$ 30. $6a = {}^-42$ 34. $^-6 = {}^-6v$
27. $^-3x = 24$ 31. $12m = 30$ 35. $0 = {}^-8k$
28. $2y = {}^-16$ 32. $3c = {}^-10$ 36. $^-3m = 0$

Solve each equation and write its solution set. Each replacement set is the set of directed numbers, excluding 0.

SAMPLE: $\frac{1}{r} = {}^-3$ Solution: $\frac{1}{r} = {}^-3$
$r \cdot \frac{1}{r} = {}^-3 \cdot r$
$1 = {}^-3 \cdot r$
$\frac{-1}{3} = r$ $\{\frac{-1}{3}\}$

B 37. $\frac{1}{t} = 5$ 38. $\frac{2}{k} = {}^-3$ 39. $\frac{2}{3n} = {}^-6$

40. $\frac{1}{m} = {}^-4$ **41.** $\frac{1}{2b} = {}^-2$ **42.** $\frac{{}^-2}{x} = {}^-3$

Complete each indicated division.

SAMPLE: $5x \div \frac{1}{5}$ Solution: $5x \div \frac{1}{5} = 5x \cdot \frac{5}{1} = 25x$

43. $\frac{c}{4} \div 3$ **46.** $\frac{2x}{3} \div \frac{{}^-1}{2}$ **49.** $({}^-3x + 2x) \div 4$

44. ${}^-2m \div 7$ **47.** $5k \div {}^-8$ **50.** $-(5m + m) \div 6$

45. $10b \div {}^-5$ **48.** ${}^-12t \div {}^-17$ **51.** ${}^-\frac{2}{3}h \div {}^-1$

Tell whether or not each set of numbers is closed under division. If a set is not closed, give an example to prove that it is not.

C **52.** $\{{}^-1, 1\}$ **54.** $\{{}^-\frac{1}{2}, {}^-1, 1, 2\}$ **56.** {positive numbers}

53. $\{1\}$ **55.** $\{1, 0, {}^-1\}$ **57.** $\{{}^-1, {}^-2, {}^-3, {}^-4, \ldots\}$

PROBLEMS

Find the unknown numbers in these problems.

1. The product of $\frac{{}^-3}{10}$ and some other number is 1. What is the other number?

2. The number ${}^-5$, divided by some other number, is equal to 15. What is the other number?

3. When 25 is divided by a certain number the result is that same number. Find the number. *Hint:* There are two possible correct answers.

4. The number ${}^-7$ is the same as 35 divided by some number. What is the number?

5. In a multiplication statement the product is ${}^-6$ and one factor is $\frac{2}{3}$. Find the other factor.

6. The reciprocal of some number is 3. Find the number.

7. The reciprocal of some number is ${}^-\frac{2}{3}$. Find the number.

8. Twice the reciprocal of some number is ${}^-6$. Find the number.

10–7 Functions: Multiplication and Division with Directed Numbers

You will recall our use of the function machine as it applied to the addition and subtraction of directed numbers. Now let us consider multiplication and division of directed numbers, as these operations apply to mathematical functions.

Observe that the function rule "Multiply by ⁻3" appears on the machine. Thus, every input number x is multiplied by ⁻3, and the output is the corresponding $f(x)$ value. The function equation $^-3(x) = f(x)$ summarizes this.

If $\{^-2, ^-1, 0, 1, 2\}$ is the replacement set, the resulting table of values is as shown in Figure 10–1.

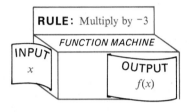

x	$f(x)$	$(x, f(x))$
⁻2	6	(⁻2, 6)
⁻1	3	(⁻1, 3)
0	0	(0, 0)
1	⁻3	(1, ⁻3)
2	⁻6	(2, ⁻6)

Figure 10–1

The set of number pairs $\{(^-2, 6), (^-1, 3), (0, 0), (1, ^-3), (2, ^-6)\}$ is a function, since the first number in each pair is matched with a unique second number.

The rule for the function machine in Figure 10–2 is given by the expression $m \cdot \frac{^-1}{5}$. Can you name the entries missing from the table, for which the function equation is $m \cdot \frac{^-1}{5} = f(m)$?

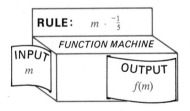

m	$f(m)$	$(m, f(m))$
⁻1	$\frac{1}{5}$	$(^-1, \frac{1}{5})$
$^-\frac{1}{2}$	$\frac{1}{10}$	$(^-\frac{1}{2}, \frac{1}{10})$
0	0	0
$\frac{1}{2}$	?	?
1	?	?

Figure 10–2

Do you agree that the entries opposite $\frac{1}{2}$ should be $\frac{^-1}{10}$ and the number pair $\left(\frac{1}{2}, \frac{^-1}{10}\right)$? Those opposite 1 should be $\frac{^-1}{5}$ and the number pair $\left(1, \frac{^-1}{5}\right)$.

ORAL EXERCISES

To each function rule in Column 1, match the correct function from Column 2. The replacement set for each is $\{-6, -4, -2, -1\}$.

COLUMN 1

1. $x \cdot -\frac{1}{2}$
2. $-3 \div x$
3. $x \div \frac{1}{2}$
4. $x(-1)$
5. $x \cdot \frac{1}{-3}$

COLUMN 2

A. $\{(-6, -12), (-4, -8), (-2, -4), (-1, -2)\}$
B. $\{(-6, -3), (-4, -2), (-2, -1), (-1, -\frac{1}{2})\}$
C. $\{(-6, \frac{1}{2}), (-4, \frac{3}{4}), (-2, \frac{3}{2}), (-1, 3)\}$
D. $\{(-6, 2), (-4, \frac{4}{3}), (-2, \frac{2}{3}), (-1, \frac{1}{3})\}$
E. $\{(-6, 3), (-4, 2), (-2, 1), (-1, \frac{1}{2})\}$
F. $\{(-6, 6), (-4, 4), (-2, 2), (-1, 1)\}$

Tell which sets of ordered number pairs are functions and which are not.

7. $\{(6, -3), (5, -7), (-3, 6), (7, 0)\}$
8. $\{(-10, 3), (-10, 4), (-5, 2), (-5, 1)\}$
9. $\{(-3, 1), (-2, 1), (-4, 1), (-5, 1), (-6, 1)\}$
10. $\{(-\frac{1}{2}, 4), (-\frac{1}{3}, 5), (-\frac{1}{4}, 5)\}$
11. $\{(0, -1), (-1, -2), (-2, -3), (-3, -4)\}$
12. $\{(-1, 3), (3, -1), (-1, 5), (5, -1), \ldots\}$

WRITTEN EXERCISES

Complete the table according to the rule shown on the function machine. The replacement set is $\{-3, -2, -1, 1, 2\}$.

A

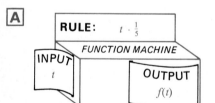

RULE: $t \cdot \frac{1}{5}$
INPUT t
FUNCTION MACHINE
OUTPUT $f(t)$

	t	$f(t)$	$(t, f(t))$
SAMPLE:	3	$-\frac{3}{5}$	$(-3, -\frac{3}{5})$
1.	-2		
2.	-1		
3.	1		
4.	2		

Use the replacement set $\{-3, -\frac{1}{3}, -2, -\frac{1}{2}, 2, \frac{1}{2}\}$, and complete the table that accompanies each function machine. Compare the completed tables, and comment.

Multiplication and Division of Directed Numbers   305

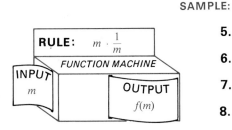

	m	f(m)	(m, f(m))
SAMPLE:	⁻3	1	(⁻3, 1)
5.	$-\frac{1}{3}$		
6.	⁻2		
7.	$-\frac{1}{2}$		
8.	2		
9.	$\frac{1}{2}$		

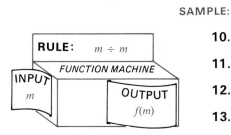

	m	f(m)	(m, f(m))
SAMPLE:	⁻3	1	(⁻3, 1)
10.	$-\frac{1}{3}$		
11.	⁻2		
12.	$-\frac{1}{2}$		
13.	2		
14.	$\frac{1}{2}$		

Complete each set of number pairs by using the given function rule. The replacement set is {⁻5, ⁻3, ⁻1, 0, 1}.

SAMPLE: $2x \cdot -\frac{1}{2} = f(x)$; {(⁻5, 5), (⁻3, 3), (⁻1, ?), (0, ?), (1, ?)}
Solution: {(⁻5, 5), (⁻3, 3), (⁻1, 1), (0, 0), (1, ⁻1)}

15. $-5t \cdot \frac{2}{5} = f(t)$; {(⁻5, 10), (⁻3, 6), (⁻1, ?), (0, ?), (1, ?)}
16. $4n \div 2 = f(n)$; {(⁻5, ⁻10), (⁻3, ?), (⁻1, ?), (0, ?), (1, ?)}
17. $\frac{m}{-15} + 1 = f(m)$; {(⁻5, $\frac{4}{3}$), (⁻3, ?), (⁻1, ?), (0, ?), (1, ?)}
18. $s \div \frac{1}{3} = f(s)$; {(⁻5, ⁻15), (⁻3, ?), (⁻1, ?), (0, ?), (1, ?)}
19. $-t \cdot \frac{1}{2} = f(t)$; {(⁻5, $\frac{5}{2}$), (⁻3, ?), (⁻1, ?), (0, ?), (1, ?)}
20. $-3n \cdot -\frac{1}{3} = f(n)$; {(⁻5, ⁻5), (⁻3, ?), (⁻1, ?), (0, ?), (1, ?)}

Write the number pairs indicated by each given function equation and replacement set.

B 21. $t \cdot t = f(t)$; {⁻6, ⁻4, ⁻2, 0, 2, 4, 6}

22. $(-h)(-h) = f(h)$; $\{^-4, ^-3, ^-2, 2, 3, 4\}$

23. $n(n^2) = f(n)$; $\{^-3, ^-2, ^-1, 0, 1, 2, 3\}$

24. $x \cdot \dfrac{1}{-x} = f(x)$; $\{^-5, ^-3, ^-1, 1, 3, 5, 7\}$

25. $(-s)(s) = f(s)$; $\{0, ^-1, ^-2, ^-3, ^-4, ^-5, ^-6, \ldots\}$

26. $\dfrac{m}{-m} \cdot 1 = f(m)$; $\{^-2, ^-1\tfrac{1}{2}, ^-1, ^-\tfrac{1}{2}, \tfrac{1}{2}, 1, 1\tfrac{1}{2}, 2\}$

27. $^-1.3(b) = f(b)$; $\{^-5, ^-3, 0, 1, 2, 7, 12\}$

28. $\dfrac{3.12}{-t} = f(t)$; $\{^-6, ^-4, ^-2, ^-1, 2, 4, 6\}$

29. $\dfrac{4}{m} = f(m)$; $\{^-\tfrac{1}{5}, ^-\tfrac{1}{4}, ^-\tfrac{1}{3}, ^-\tfrac{1}{2}, \tfrac{1}{2}, \tfrac{1}{3}, \tfrac{1}{4}, \tfrac{1}{5}\}$

30. $\dfrac{-2}{g} = f(g)$; $\{^-1, ^-0.1, ^-0.01, ^-0.001, 0.1, 0.01, 0.001\}$

CHAPTER SUMMARY

Inventory of Structure and Concepts

1. Multiplication involving directed numbers has the following properties:

   For all directed numbers r, s, and t,

   Closure property: $r \cdot s$ is a unique directed number
   Commutative property: $r \cdot s = s \cdot r$
   Associative property: $(r \cdot s)t = r(s \cdot t)$
   Multiplicative property of zero: $r \cdot 0 = 0 \cdot r = 0$
   Multiplicative property of one: $r \cdot 1 = 1 \cdot r = r$
   Distributive property: $r(s + t) = r \cdot s + r \cdot t$
   $r(s - t) = r \cdot s - r \cdot t$

2. For multiplication of directed numbers:

   the product of **two positive** numbers is **positive**;
   the product of a **positive** and a **negative** number is **negative**;
   the product of **two negative** numbers is **positive**.

3. The product of two or more directed numbers is **positive** if it contains an **even** number of **negative** factors, and is **negative** if it contains an **odd** number of **negative** factors.

4. For division with positive and negative numbers:

$$\text{positive} \div \text{positive} = \text{positive}$$
$$\text{positive} \div \text{negative} = \text{negative}$$
$$\text{negative} \div \text{positive} = \text{negative}$$
$$\text{negative} \div \text{negative} = \text{positive}$$

5. **Division** by **zero** is impossible.

6. If the product of two numbers is 1, each number is the **reciprocal** of the other.

7. For every directed number r except 0, there is a number $\frac{1}{r}$ such that $r \cdot \frac{1}{r} = \frac{1}{r} \cdot r = 1$.

8. **Dividing** by a directed number is equivalent to **multiplying** by its **reciprocal**.

9. For every directed number r, and every directed number s except 0, $r \div s = r \cdot \frac{1}{s}$.

Vocabulary and Spelling

commutative property (*p. 283*)
associative property (*p. 286*)
distributive property (*p. 287*)
multiplicative property of one (*p. 287*)
multiplicative property of zero (*p. 287*)

closure property (*p. 287*)
similar terms (*p. 294*)
reciprocal (*p. 298*)
numerical coefficient (*p. 299*)

Chapter Test

Simplify each expression.

1. $^-5 \cdot 7$
2. $^-4 \cdot ^-9$
3. $(^-4)(3)(^-2)$
4. $\frac{81}{^-3}$
5. $\frac{^-51}{^-3}$
6. $12 \cdot ^-5$
7. $(^-2)(^-2)(^-2)(^-2)$
8. $(^-1)^{15}$
9. $\frac{^-42}{7}$
10. $\frac{10 - ^-24}{^-17}$

Solve each equation and write the solution set. The replacement set is {the directed numbers}.

11. $^{-}3k + {^-}4 = 11$

12. $3t - (-t) = {^-}24$

13. $\dfrac{^{-}3}{y} = {^-}15$

14. $^{-}\tfrac{3}{2}x = 2\tfrac{1}{2}$

15. $\dfrac{x}{5} = {^-}2.5$

16. $^{-}3 \cdot x = 12 \cdot {^-}3$

17. What is the reciprocal of $^{-}5\tfrac{1}{4}$?

18. Is the set $\{{^-}1, {^-}2, {^-}3, {^-}4, \ldots\}$ closed under multiplication?

Complete each set of number pairs by using the given function rule.

19. $(^{-}2)^3 + x = f(x)$: $\{(0, ?), (1, ?), (2, ?)\}$

20. $^{-}2t - 5 = f(t)$: $\{(0, ?), ({^-}5, ?), ({^-}1, ?), (1, ?)\}$

Chapter Review

10–1 Products of Positive and Negative Numbers

Simplify each expression.

1. $4 \cdot {^-}7$
2. $^{-}5(2t)$
3. $12 \cdot {^-}5$
4. $^{-}\tfrac{1}{2} \cdot 9$
5. $^{-}8 \cdot 3$
6. $^{-}15 \cdot 0$

7. $({^-}3 \cdot 5) + (4 \cdot {^-}6)$

8. $({^-}1 \cdot 9) + (2 \cdot 4)$

Solve each equation and write the solution set. The replacement set is {the directed numbers}.

9. $2(t) = 14$

10. $x - ({^-}6 \cdot 2) = 3$

11. $^{-}5 \cdot 7 + x = 15$

12. $^{-}7 \cdot k = 5 \cdot {^-}7$

10–2 More about Products of Positive and Negative Numbers

Use the distributive property to rewrite each expression as a sum of two products; then simplify each form of the expression.

13. $3({^-}2 + 5)$

14. $\tfrac{1}{4}(12 + {^-}16 + {^-}4 + 0)$

15. $^{-}8(5 + {^-}6)$

16. $^{-}13({^-}11 + {^-}9)$

10–3 Products of Negative Numbers

Complete the following.

17. $4 \cdot {}^-3 = {}^-12$
    $3 \cdot {}^-3 = ?$
    $2 \cdot {}^-3 = ?$
    $1 \cdot {}^-3 = ?$
    $0 \cdot {}^-3 = ?$
    ${}^-1 \cdot {}^-3 = ?$
    ${}^-2 \cdot {}^-3 = ?$

18. $({}^-1)({}^-1) = 1$
    $({}^-1)({}^-1)({}^-1) = ?$
    $({}^-1)({}^-1)({}^-1)({}^-1) = ?$
    $({}^-1)({}^-1)({}^-1)({}^-1)({}^-1) = ?$

19. $({}^-2)^1 = {}^-2$
    $({}^-2)^2 = ?$
    $({}^-2)^3 = ?$
    $({}^-2)^4 = ?$

Simplify the following expressions.

20. $({}^-3)(5)({}^-2)$
21. ${}^-1(3)({}^-7)$
22. $({}^-2)({}^-4)({}^-5)$
23. $({}^-4 \cdot 2)(3 \cdot {}^-3)$

10–4 The Distributive Property and Algebraic Expressions

Simplify each expression by combining similar terms.

24. $8k - 3k + 7k - k$
25. $4t + {}^-5t$
26. $3(t - 3t) + (4t + t)$
27. $3xy + 5xy - 12xy$
28. $7x + 4x + {}^-1x)$
29. $4(x + y) + 2(x + y)$

10–5 Division of Directed Numbers

Simplify each of the following.

30. $\dfrac{{}^-24}{3}$
31. $\dfrac{({}^-12 + 3)}{{}^-9}$
32. $\dfrac{48}{{}^-8}$
33. $\dfrac{15 - {}^-7}{{}^-4 + 2}$
34. $\dfrac{{}^-35}{{}^-7}$
35. $\dfrac{{}^-12}{16}$

Write a decimal equivalent for each of the following.

36. $\dfrac{{}^-7}{10}$
37. $\dfrac{2}{{}^-5}$
38. $\dfrac{37}{{}^-100}$
39. $\dfrac{{}^-3}{{}^-25}$

10–6 Directed Numbers and Reciprocals

Write the reciprocal of each numeral.

40. $-\tfrac{3}{4}$
41. 0.25
42. ${}^-3$
43. $\dfrac{-a}{b}$, $a \neq 0$, $b \neq 0$
44. $5\tfrac{1}{2}$

Simplify each expression by completing the indicated division.

45. $\dfrac{2}{3} \div \dfrac{^-3}{4}$　　　　**47.** $^-\dfrac{1}{2} \div 4$　　　　**49.** $2\dfrac{1}{2} \div ^-\dfrac{3}{4}$

46. $\dfrac{1\frac{1}{2}}{\frac{3}{4}}$　　　　**48.** $\dfrac{^-\frac{2}{5}}{\frac{2}{3}}$　　　　**50.** $\dfrac{^-\frac{5}{2}}{^-\frac{3}{4}}$

Solve each equation and write the solution set. The replacement set is {the directed numbers}.

51. $^-5k = 20$　　　　**52.** $\dfrac{x}{3} = \dfrac{^-7}{2}$　　　　**53.** $\dfrac{1}{^-4k} = ^-2$

10–7 Functions: Multiplication and Division with Directed Numbers

Complete each set of number pairs by using the given function rule.

54. $^-3k + ^-2 = f(k)$: $\{(^-1, ?), (\frac{2}{3}, ?), (^-4, ?)\}$
55. $t^2 - 5 = f(t)$: $\{(^-1, ?), (^-3, ?), (0, ?)\}$

Review of Skills

Simplify each expression.

1. $^+12 + ^-12$　　　　4. $4 + ^-4$　　　　7. $^-1 + ^+1$
2. $\frac{1}{2} \cdot 2$　　　　5. $\frac{4}{3} \cdot \frac{3}{4}$　　　　8. $1.5 \cdot \frac{2}{3}$
3. $^-5 \cdot ^-2$　　　　6. $^-3 \cdot ^+2$　　　　9. $4 \cdot ^-3$

Name the additive inverse of each of the following directed numbers.

10. $^+7$　　　　**11.** $^-12$　　　　**12.** 4　　　　**13.** $^+\frac{3}{4}$

Name the reciprocal of each of the following.

14. $\frac{7}{8}$　　　　**15.** $\frac{4}{3}$　　　　**16.** 4　　　　**17.** $^-\frac{1}{2}$

Determine the directed number n that makes each statement true.

18. $n + 5 = 7$　　　　**21.** $n + 7 = 5$　　　　**24.** $15 = 3 + n$
19. $18 = 3n$　　　　**22.** $^-2n = 16$　　　　**25.** $3n = ^-24$
20. $\frac{1}{2} \cdot n = 12$　　　　**23.** $\frac{3}{4} \cdot n = 12$　　　　**26.** $5n = 12$

Use the distributive property to simplify each expression.

27. $3n + 2n$ **28.** $12t - 5t$ **29.** $\frac{3}{4}x + \frac{1}{2}x$

Determine the number *n* that makes each statement true.

30. $3n + 2n = 25$ **31.** $18n - 3n = 45$ **32.** $\frac{3}{4}n + \frac{7}{4}n = 10$

Use < or > to complete each sentence so it will be a true statement.

33. 3 ? 4 **36.** ⁻5 ? ⁺3 **39.** ⁻1 ? ⁻2
34. 15 ? 14 **37.** ⁻10 ? ⁻9 **40.** $-\frac{1}{2}$? $-\frac{3}{4}$
35. $2 + 5$? 6 **38.** $3 \cdot {}^-2$? 4 **41.** $3 \cdot 2$? 4

Indicate whether each of the following is true or false.

42. If $5 + 1 = x$ then $x = 5 + 1$.
43. $5 < 7$ and $7 < 5$.
44. If $x = 2 + 10$ and $2 + 10 = 3 \cdot 4$ then $x = 3 \cdot 4$.
45. If $x = 5$ then $x + 3 = 5 + 3$.
46. $2 < 5$ and $2 + 1 < 5 + 1$.
47. If $x = 13$ then $3x = 3 \cdot 13$.
48. $4 > 3$ and $4 \cdot 2 > 3 \cdot 2$.
49. $4 > 3$ and $4 \cdot {}^-2 > 3 \cdot {}^-2$.

■ ■

CHECK POINT
FOR EXPERTS

Rational Numbers

The study of mathematics often deals with special subsets of the set of directed numbers. You are familiar with these subsets:

Whole numbers: $\{0, 1, 2, 3, 4, 5, 6, \ldots\}$
Natural numbers: $\{1, 2, 3, 4, 5, 6, \ldots\}$
Integers: $\{\ldots, {}^-3, {}^-2, {}^-1, 0, 1, 2, 3, \ldots\}$

The set of **rational** numbers is also a subset of the set of directed numbers.

Chapter 10

A **rational number** is a number that can be written in the fractional form $\frac{a}{b}$, where a and b are integers and $b \neq 0$. Do you see that the numbers $^-3$, **15**, and **0.7** are rational numbers, since each of them can be written as a fraction? For example:

$$^-3 = \frac{^-12}{4}; \quad 15 = \frac{30}{2}; \quad 0.7 = \frac{7}{10}.$$

Could you use other fractions than those shown to name the numbers $^-3$, 15, and 0.7?

Sometimes a rational number in the form $\frac{a}{b}$ can be written as a **terminating decimal** by dividing the numerator by the denominator, as shown here, where the division indicated by the fraction $\frac{5}{8}$ is continued until the remainder 0 is reached. Thus the decimal numeral for $\frac{5}{8}$ is **0.625**, which is a **terminating** decimal.

```
    0.625
 8)5.000
   4 8
   ‾‾‾
     20
     16
     ‾‾
      40
      40
      ‾‾
       0
```

$\frac{5}{8} = 0.625$

A rational number that cannot be expressed as a terminating decimal can be written as a **repeating decimal**. The division at the right shows how to find the repeating decimal numeral for $\frac{2}{3}$. Notice that the remainder 2 will keep repeating, hence the digit 6 will repeat over and over, no matter how long the division process is continued. Thus the decimal numeral for $\frac{2}{3}$ is the **repeating** decimal **0.666 . . .** .

```
   0.666 . . .
3)2.000
  1 8
  ‾‾‾
    20
    18
    ‾‾
     20
     18
     ‾‾
      2
```

$\frac{2}{3} = 0.666\ldots$

We often write a repeating decimal by using a bar above the digit, or digits, that repeat. For example:

$$0.666\ldots = 0.\overline{6}; \quad 0.252525\ldots = 0.\overline{25}; \quad 0.6131313\ldots = 0.6\overline{13}.$$

Every repeating decimal can be written in the fractional form $\frac{a}{b}$, where a and b are whole numbers, with b not 0. The computation here shows how we can find the fractional form for the repeating decimal $0.\overline{13}$. We begin by letting $x = 0.\overline{13}$. Then we know that $100x = 0.\overline{13} \cdot 100$, or $13.\overline{13}$. By subtraction, we find that $99x = 13$. Solving this for x, we see that $x = \frac{13}{99}$.

$100x = 13.\overline{13}$
$x = 0.\overline{13}$
$99x = 13$
$x = \frac{13}{99}$

Multiplication and Division of Directed Numbers

Study the following example, which shows how $0.\bar{5}$ can be written as the fraction $\frac{5}{9}$.

Let $x = 0.\bar{5}$. $10x = 5.\bar{5}$
Then $10x = 0.\bar{5} \cdot 10$ $\underline{-\ x = 0.\bar{5}}$
$= 5.\bar{5}$ $9x = 5$
                             $x = \frac{5}{9}$

Questions

What decimal numeral names the same number as each of the following?

1. $\frac{7}{8}$
2. $\frac{2}{5}$
3. $\frac{2}{9}$
4. $\frac{1}{7}$
5. $\frac{-3}{5}$
6. $\frac{5}{6}$
7. $\frac{3}{7}$
8. $\frac{5}{11}$

Express each of the following as a fraction.

9. 0.235
10. 4.5
11. 10.4
12. 38.6
13. ⁻0.25
14. ⁻0.333 . . .
15. 0.0001
16. 0.1
17. 0.01
18. ⁻0.01
19. 0.413
20. ⁻0.0003

Show how to complete each of the following.

21. Write 0.3737 . . . as a fraction.
    If $x = 0.3737\ldots$
    then $100x = 37.37\ldots$
    Subtract: $100x = 37.37\ldots$
    $\underline{-\ x = 0.37\ldots}$
    $99x = ?$
    So $x = ?$ and $0.3737\ldots = ?$

22. Write $1.\bar{4}$ as a fraction.
    If $x = 1.\bar{4}$,
    then $10x = 14.\bar{4}$
    Subtract: $10x = 14.\bar{4}$
    $\underline{-\ x = 1.\bar{4}}$
    $9x = ?$
    So $x = ?$ and $1.\bar{4} = ?$

Find a fraction which names the same number as each repeating decimal.

23. 0.3232 . . .
24. $0.\overline{17}$
25. $2.\overline{75}$
26. 0.3939 . . .
27. $38.\overline{53}$
28. $^-6.\overline{123}$

van Leeuwenhoek's single-lens microscope . . .

Electron microscope with video tube . . .

Solving Equations and Inequalities

Anton van Leeuwenhoek (1632–1723), the "father of microbiology," was the first to leave written records of what he found by using his single-lens microscope. With his hand-ground lenses that had magnifying powers of up to 270 diameters, he was able to study animalcules (tiny animals). The most powerful modern instrument for this purpose is the electron microscope, which uses streams of electrons instead of light rays for magnifying objects. The most powerful lens microscope can magnify an object 2,000 times; an electron microscope can magnify it 200,000 times on photographic film, and as much as 500,000 times on a fluorescent screen. With special equipment, magnification up to about 2 million times is possible.

Types of Equations

11–1 Equations of Type $x + a = b$

If you were asked to solve the equation $x + 4 = 12$, you would probably do so simply by looking at it and saying "x is equal to 8." This particular kind of equation falls into the general classification $x + a = b$, where x is the **variable** and a and b stand for **directed numbers**.

Do you see that each of the following equations is of the $x + a = b$ type?

$$x + 6 = 35 \qquad t - 4 = 9 \qquad 8 + r = 17$$
$$n + {}^-4 = 2 \qquad a + \tfrac{2}{3} = \tfrac{5}{8} \qquad {}^-2 + y = 15$$
$$12 = b + 2\tfrac{1}{2} \qquad {}^-10 = 3 + w \qquad 4 + k = {}^-11$$

At first it may not be clear that the equation $t - 4 = 9$ is of the $x + a = b$ type. However, recall that **subtracting** a directed number

315

is equivalent to **adding** its additive inverse or opposite. Thus we can write $t - 4 = 9$ as the equivalent sentence $t + {}^-4 = 9$, which is clearly of the $x + a = b$ type.

Equations of this type can be solved by applying the **addition property of equality** and the **property of additive inverses** that we have stated previously. Study the following examples carefully.

EXAMPLE 1. $x + 12 = 19$ Given equation
    $x + 12 + {}^-12 = 19 + {}^-12$ Add $^-12$ to each member
    $x + 0 = 19 + {}^-12$ Property of additive inverses
    $x = 7$ Substitution principle

EXAMPLE 2. $5 = {}^-9 + x$ Given equation
    ${}^-9 + x = 5$ Symmetric property of equality
    $9 + {}^-9 + x = 5 + 9$ Add $^-9$ to each member
    $0 + x = 5 + 9$ Property of additive inverses
    $x = 14$ Substitution principle

Suppose that you wish to solve the equation $x + k = m$ for x. That is, you wish to write an equivalent equation such that x stands alone as one member of the equation. You would proceed as follows:

EXAMPLE 3. $x + k = m$ Given equation
    $x + k + (-k) = m + (-k)$ Add $-k$ to both members
    $x + 0 = m + (-k)$ Property of additive inverses
    $x = m + (-k)$ Substitution principle

Of course, the final equation can also be written as $x = m - k$.

ORAL EXERCISES

State each of the following in the form $x + a = b$.

SAMPLE: $k - 8 = {}^-10$ *What you say:* $k + {}^-8 = {}^-10$

1. $15 + n = 18$
2. $^-7 + x = 2$
3. $^-5 = w + 3$
4. $x - 9 = 20$
5. $b - \frac{2}{3} = \frac{7}{8}$
6. $t - 6 = 48$
7. $\frac{2}{3} + a = 1$
8. $^-\frac{1}{2} + y = 4$
9. $a + (3 + 7) = 6$
10. $\frac{3}{4} + y = \frac{1}{2}$
11. $17 = y + 2$
12. $13 = {}^-4 + k$
13. $h - \frac{2}{5} = \frac{5}{8}$
14. $n - 0.7 = 1.6$
15. $0.25 + m = 0.37$

Solving Equations and Inequalities 317

WRITTEN EXERCISES

Copy and complete the solution of each equation and write its solution set. Each replacement set is {the directed numbers}.

SAMPLE: $\quad x - 4 = 17 \qquad$ *Solution:* $\qquad x - 4 = 17$
$\qquad\qquad x + {}^-4 + 4 = 17 + 4 \qquad\qquad\qquad x + {}^-4 + 4 = 17 + 4$
$\qquad\qquad\quad x + \underline{\ ?\ } = 17 + 4 \qquad\qquad\qquad\quad x + 0 = 17 + 4$
$\qquad\qquad\qquad\qquad x = \underline{\ ?\ } \qquad\qquad\qquad\qquad\qquad x = 21 \quad \{21\}$

A

1. $\quad n + 6 = 19$
$\quad n + 6 + {}^-6 = 19 + {}^-6$
$\quad n + \underline{\ ?\ } = 19 + {}^-6$
$\quad n = \underline{\ ?\ }$

2. $\quad k + 12 = 5$
$\quad k + 12 + {}^-12 = 5 + {}^-12$
$\quad k + \underline{\ ?\ } = 5 + {}^-12$
$\quad k = \underline{\ ?\ }$

3. $\quad 4\frac{1}{2} + t = 7$
$\quad {}^-4\frac{1}{2} + 4\frac{1}{2} + t = 7 + \underline{\ ?\ }$
$\quad \underline{\ ?\ } + t = 7 + {}^-4\frac{1}{2}$
$\quad t = \underline{\ ?\ }$

4. $\quad {}^-3 + m = 12$
$\quad 3 + {}^-3 + m = 12 + \underline{\ ?\ }$
$\quad \underline{\ ?\ } + m = 12 + 3$
$\quad m = \underline{\ ?\ }$

Solve each equation. Check your solution, then write the solution set.

SAMPLE: $\quad r + {}^-8 = 27$

Solution: $\quad r + {}^-8 = 27 \qquad$ *Check:* $\quad r + {}^-8 = 27$
$\qquad\qquad r + {}^-8 + 8 = 27 + 8 \qquad\qquad\qquad \dfrac{35 + {}^-8 \ |\ 27}{27\ \ |\ 27} \quad \{35\}$
$\qquad\qquad\qquad r = 35$

5. $x + {}^-3 = 18$
6. $y + 7 = 5$
7. $14 + n = 25$
8. $^-3 + k = 17$
9. $w + \frac{1}{4} = 2\frac{1}{2}$

10. $a + {}^-\frac{2}{3} = 5$
11. $22 = t + 3$
12. $16 = m + {}^-5$
13. $^-5 = r + {}^-8$
14. $8 = 1 + s$

15. $y + 19 = 0$
16. $n + {}^-35 = 0$
17. $\frac{3}{4} + z = 6$
18. $a + 1.3 = 5.9$
19. $10.4 + k = 16.3$

B

20. $t - {}^-6 = 3\frac{1}{3}$
21. $a + {}^-\frac{4}{5} = {}^-8$
22. $^-3.5 + x = 4.6$

23. $y + \frac{3}{8} = {}^-\frac{3}{4}$
24. $n - {}^-\frac{1}{3} = \frac{1}{6}$
25. $^-\frac{1}{8} + m = {}^-\frac{1}{16}$

26. $^-\frac{3}{5} = {}^-\frac{10}{2} + p$
27. $^-4.9 = 2.1 + r$
28. $0.03 = {}^-2.6 + t$

Solve each equation for x.

29. $x + t = m$
30. $r + x = k$
31. $b = c + x$
32. $k = x + w$
33. $a + x = -s$
34. $x + -c = -t$

Solve each equation for the variable in red.

C

35. $s - a = r$
36. $b + (-n) = y$
37. $h = w - s$
38. $-z + x = r$
39. $p - (-q) = w$
40. $d - (-t) = g$

11-2 Equations of Type $ax = b$

Let us investigate equations of the general form $ax = b$. Do you agree that each of the following equations is of that type?

$$4y = 15 \qquad t \cdot \tfrac{2}{3} = {}^-6 \qquad {}^-2x = 8$$

$$\frac{3y}{2} = 7 \qquad {}^-\tfrac{3}{5}k = {}^-\tfrac{9}{10} \qquad \frac{x}{5} = 9$$

In general, equations of the $ax = b$ type can be solved with little difficulty by applying our knowledge of the **multiplication property** of **equality** and the **multiplication property** of **reciprocals**. The following examples illustrate these ideas.

EXAMPLE 1.

$6t = 5$	Given equation
$\tfrac{1}{6} \cdot 6 \cdot t = 5 \cdot \tfrac{1}{6}$	Multiply each member by $\tfrac{1}{6}$
$1 \cdot t = 5 \cdot \tfrac{1}{6}$	Multiplication property of reciprocals
$t = \tfrac{5}{6}$	Substitution principle

EXAMPLE 2.

$\dfrac{{}^-2m}{5} = 4$	Given equation
$\dfrac{5}{{}^-2} \cdot \dfrac{{}^-2}{5} \cdot m = 4 \cdot \dfrac{5}{{}^-2}$	Multiply each member by $\dfrac{5}{{}^-2}$
$1 \cdot m = 4 \cdot \dfrac{5}{{}^-2}$	Multiplication property of reciprocals
$m = {}^-10$	Substitution principle

The same methods used to solve the equations in Examples 1 and 2 can be applied to solving such an equation as $mx = k$ for x, provided $m \neq 0$. That is, we want to find an equivalent equation where x stands alone as one member.

EXAMPLE 3.

$mx = k$	Given equation
$\dfrac{1}{m} \cdot m \cdot x = k \cdot \dfrac{1}{m}$	Multiply each member by $\dfrac{1}{m}$
$1 \cdot x = k \cdot \dfrac{1}{m}$	Multiplication property of reciprocals
$x = \dfrac{k}{m}$	Substitution principle

ORAL EXERCISES

Name the reciprocal of the coefficient of the variable in each equation. Then state the solution set of the equation.

SAMPLE: $^-5n = 15$ *What you say:* $^-\frac{1}{5}$; $\{^-3\}$

1. $4x = 16$
2. $5n = 30$
3. $^-8t = 16$
4. $2k = \frac{1}{2}$
5. $\frac{m}{4} = 3$
6. $\frac{y}{5} = ^-2$
7. $\frac{s}{-2} = 3$
8. $\frac{2}{3}a = 6$
9. $\frac{1}{2}p = 4\frac{1}{2}$
10. $\frac{6n}{5} = 12$
11. $\frac{2s}{-3} = ^-6$
12. $-t = \frac{1}{2}$

Tell what number should replace each question mark to complete the equation:

SAMPLE: $? \cdot \frac{2m}{3} = m$ *What you say:* Since $\frac{3}{2} \cdot \frac{2}{3} \cdot m = 1m = m$, $\frac{3}{2}$ should replace the question mark.

13. $? \cdot 3r = r$
14. $? \cdot \frac{2}{5}k = k$
15. $? \cdot ^-\frac{2}{3}s = s$
16. $4b \cdot ? = b$
17. $^-9n \cdot ? = n$
18. $? \cdot \frac{3t}{5} = t$
19. $? \cdot \frac{7x}{8} = x$
20. $? \cdot \frac{-4m}{3} = m$
21. $? \cdot mx = x$, if $m \neq 0$

WRITTEN EXERCISES

Solve each equation. Check the solution, then write the solution set.

SAMPLE: $\frac{3t}{2} = 21$

Solution: $\frac{3t}{2} = 21$ Check: $\frac{3t}{2} = 21$

$\frac{2}{3} \cdot \frac{3}{2} \cdot t = 21 \cdot \frac{2}{3}$ $\frac{3 \cdot 14}{2}$ | 21

$t = 14$ 21 | 21 $\{14\}$

1. $8k = 56$
2. $9m = 45$
3. $12r = 30$
4. $\frac{1}{2}b = 21$
5. $^-3d = 12$
6. $\frac{2}{3}s = 10$
7. $x \cdot 7 = 28$
8. $y \cdot \frac{1}{3} = 14$
9. $^-4t = 20$

10. $n \cdot {}^-3 = 6$
11. $^-8r = {}^-24$
12. $5w = {}^-9$
13. $\frac{1}{8} \cdot s = {}^-10$
14. $-k = 3$
15. $^-3z = \frac{9}{10}$

Solve each equation for the variable in red. Assume that no divisor has the value 0.

16. $mv = a$
17. $mw = a$
18. $p = hd$
19. $p = hd$
20. $c = \pi d$
21. $v = at$
22. $at = v$
23. $d = rt$
24. $d = rt$

Solve each equation and write the solution set. Each replacement set is {the directed numbers}.

SAMPLE: $\dfrac{^-2t}{7} = 12$

Solution: $\dfrac{^-2t}{7} = 12$

$\dfrac{7}{-2} \cdot \dfrac{^-2}{7} \cdot t = 12 \cdot \dfrac{7}{-2}$

$1 \cdot t = \dfrac{84}{-2}$

$t = {}^-42 \quad \{^-42\}$

B 25. $3s = {}^-5$
26. $^-2n = 7$
27. $5t = \frac{1}{2}$
28. $\frac{2}{9}k = 4$
29. $y \cdot \frac{1}{5} = \frac{1}{2}$
30. $\frac{2}{3}h = {}^-10$
31. $\dfrac{4m}{5} = 2$
32. $\dfrac{5s}{-2} = 18$
33. $^-\frac{5}{9} \cdot w = 20$
34. $0.3r = 18$
35. $0.15q = {}^-1.2$
36. $^-1.5m = {}^-3$

Solve each formula for the indicated variable.

SAMPLE: Volume of a cone:

$V = \dfrac{1}{3} Bh$

Solve for h.

Solution: $V = \dfrac{1}{3} Bh$

$3 \cdot V = 3 \cdot \dfrac{1}{3} Bh$

$3V = Bh$

$\dfrac{1}{B} \cdot 3V = \dfrac{1}{B} \cdot Bh$

$\dfrac{3V}{B} = h$

37. Area of a rectangle:
$A = lw$
Solve for l.

38. Circumference of a circle:
$c = \pi d$
Solve for d.

39. Circumference of a circle:
$c = 2\pi r$
Solve for r.

40. Area of a triangle:
$A = \frac{1}{2}bh$
Solve for b.

Solve each equation for the variable in red. Then find the value of that variable for $a = \frac{2}{3}$, $b = ^-2$, $t = \frac{1}{3}$, and $k = ^-8$.

SAMPLE: $br = k$ Solution: $br = k$
$$\frac{1}{b} \cdot b \cdot r = k \cdot \frac{1}{b}$$
$$r = \frac{k}{b}; \quad r = \frac{^-8}{^-2} = 4$$

C
41. $ay = b$
42. $bw = t$
43. $na = b$
44. $k = gt$

45. $-b = pk$
46. $td = a$
47. $-tr = a$
48. $\frac{1}{b} \cdot s = k$

49. $an = bk$
50. $bw = -b$
51. $-tg = -k$
52. $-bq = -a$

11–3 Equations of Type $ax + bx = c$

Each of the equations below is of the $ax + bx = c$ type.

$3k + 7k = 48$ $^-5r + 2r = 19$ $^-7w = 48 + w$

$\frac{2}{3}t + \frac{1}{3}t = ^-6$ $\frac{3n}{2} + \frac{n}{4} = 1.7$ $\frac{b}{3} + \frac{b}{2} = ^-12$

You will probably recall that we can solve an equation of this type without much difficulty if we first combine **similar** terms. In the equation $\frac{5}{3}m + \frac{1}{3}m = ^-14$, the terms $\frac{5}{3}m$ and $\frac{1}{3}m$ are **similar**, since each contains the variable *m*. So we use the distributive property as follows:

$$\frac{5}{3}m + \frac{1}{3}m = (\frac{5}{3} + \frac{1}{3})m$$
$$= \frac{6}{3}m = 2m$$

The solution of $\frac{5}{3}m + \frac{1}{3}m = ^-14$ can now be found by applying the multiplication property of equality.

EXAMPLE 1.

$\frac{5}{3}m + \frac{1}{3}m = ^-14$	Given equation
$2m = ^-14$	Combine similar terms
$\frac{1}{2} \cdot 2 \cdot m = ^-14 \cdot \frac{1}{2}$	Multiply each member by $\frac{1}{2}$
$1 \cdot m = ^-14 \cdot \frac{1}{2}$	Multiplication property of reciprocals
$m = ^-7$	Substitution principle

Suppose you are to solve an equation like $^{-}6n = 18 + {}^{-}2n$. A good plan of attack is first to write an equivalent equation in the form $ax + bx = c$. Then the solution is easily found in the same manner as in Example 1.

EXAMPLE 2.

$^{-}6n = 18 + {}^{-}2n$	Given equation
$2n + {}^{-}6n = 18 + {}^{-}2n + 2n$	Add $2n$ to each member
$2n + {}^{-}6n = 18 + 0$	Property of additive inverses
$^{-}4n = 18$	Combine similar terms
$\frac{1}{^{-}4} \cdot {}^{-}4 \cdot n = 18 \cdot \frac{1}{^{-}4}$	Multiply each member by $\frac{1}{^{-}4}$
$1 \cdot n = 18 \cdot \frac{1}{^{-}4}$	Multiplication property of reciprocals
$n = {}^{-}4\frac{1}{2}$	Substitution principle

Let us consider the problem of solving the equation $tm + rm = k$ for m. Since the equation is of the $ax + bx = c$ type, we might arrange our work like this:

EXAMPLE 3.

$tm + rm = k$	Given equation
$m(t + r) = k$	Distributive property
$m(t + r) \cdot \frac{1}{t + r} = k \cdot \frac{1}{t + r}$	Multiplication property of equality
$m \cdot 1 = k \cdot \frac{1}{t + r}$	Multiplication property of reciprocals
$m = \frac{k}{t + r}$	Substitution principle

Of course, we must specify that $t \neq -r$. Why?

ORAL EXERCISES

Combine similar terms to give a simpler name for each expression.

1. $4t + (-t)$ **2.** $\frac{1}{5}q + \frac{2}{5}q$ **3.** $\frac{-3w}{5} + \frac{w}{5}$

Solving Equations and Inequalities   323

4. $^{-}3m + 8m + 5m$

5. $12x + {}^{-}15x$

6. $^{-}7n + {}^{-}5n$

7. $s + {}^{-}5s$

8. $\frac{1}{2}a + \frac{1}{4}a$

9. $^{-}\frac{3}{8}b + \frac{1}{2}b$

10. $^{-}\frac{2}{7}h + {}^{-}\frac{5}{7} + \frac{1}{7}h$

11. $\frac{2x}{5} + \frac{4x}{5}$

12. $\frac{2n}{-9} + \frac{3n}{-9}$

13. $^{-}5x + 2x + {}^{-}3y$

14. $\frac{3x}{4} + \left(-\frac{x}{4}\right)$

15. $\frac{2q}{3} + \frac{q}{2}$

Tell whether each statement is true or false.

16. $\frac{2w}{6} + \frac{3w}{6}$ is the same as $\frac{5w}{6}$.

17. $\frac{t}{4} + \frac{t}{2}$ is the same as $\frac{3}{4}t$.

18. $4n + {}^{-}2\frac{1}{2}$ is the same as $1\frac{1}{2}n$.

19. $^{-}3b + {}^{-}8b$ is the same as $^{-}11b$.

20. $\frac{x}{3} + \frac{x}{6}$ is the same as $\frac{1}{2}x$.

21. $2y + \frac{y}{2}$ is the same as $\frac{3y}{2}$.

22. $\frac{-3t}{8} + \frac{t}{4}$ is the same as $\frac{1}{8}t$.

WRITTEN EXERCISES

Copy and complete the solution of each equation. Write its solution set if the replacement set is {the directed numbers}.

SAMPLE: $^{-}5t + 2t = 24$
$? \cdot t = 24$
$? \cdot {}^{-}3 \cdot t = 24 \cdot ?$
$t = ?$

Solution: $^{-}5t + 2t = 24$
$^{-}3 \cdot t = 24$
$\frac{1}{-3} \cdot {}^{-}3t = 24 \cdot \frac{1}{-3}$
$t = {}^{-}8 \quad \{{}^{-}8\}$

1. $3s + 6s = 14$
$? \cdot s = 14$
$? \cdot 9s = 14 \cdot ?$
$s = ?$

2. $\frac{1}{8}m + \frac{3}{8}m = {}^{-}3$
$? \cdot m = {}^{-}3$
$? \cdot \frac{m}{2} = {}^{-}3 \cdot ?$
$m = ?$

3. $\frac{1}{2}k - \frac{1}{4}k = 9$
$? \cdot k = 9$
$? \cdot \frac{k}{4} = 9 \cdot ?$
$k = ?$

4. $^{-}2t + (-t) = 10$
$? \cdot t = 10$
$? \cdot {}^{-}3t = 10 \cdot ?$
$t = ?$

5. $\frac{3}{4}x + {}^{-}\frac{1}{4}x = {}^{-}5$
$? \cdot x = {}^{-}5$
$? \cdot \frac{x}{2} = {}^{-}5 \cdot ?$
$x = ?$

6. $\frac{2n}{3} + \frac{n}{2} = {}^{-}14$
$? \cdot n = {}^{-}14$
$? \cdot \frac{7n}{6} = {}^{-}14 \cdot ?$
$n = ?$

Solve each equation for the variable indicated, assuming that no variables have values which imply division by zero.

SAMPLE: Solve $ar + qr = m$ for r Solution: $ar + qr = m$
$$r(a + q) = m$$
$$r = \frac{m}{a + q}$$

7. Solve $tm + km = s$ for m.
8. Solve $ba + ca = z$ for a.
9. Solve $3k + nk = w$ for k.
10. Solve $st - ht = y$ for t.
11. Solve $dm + {}^-9m = 10$ for m.
12. Solve $h = rt + ru$ for r.
13. Solve $-t = mk + bk$ for k.
14. Solve $x = 3c - tc$ for c.

Solve each equation for the replacement set {the directed numbers}. Check, then write the solution set.

SAMPLE: $-3b + 9b = 15$

Solution: $-3b + 9b = 15$ Check: $-3b + 9b = 15$
$\qquad\qquad 6b = 15$ $-3(2\frac{1}{2}) + 9(2\frac{1}{2})$ | 15
$\qquad\qquad b = 2\frac{1}{2}$ $-7\frac{1}{2} + 22\frac{1}{2}$ | 15
$\qquad\qquad\qquad\qquad\qquad\qquad\qquad\qquad\qquad 15$ | $15 \quad \{2\frac{1}{2}\}$

15. $2t + 7t = 15$
16. $-10m + {}^-3m = 52$
17. $6k + k = {}^-3$
18. $4m - 5m = 27$
19. $3.2b + 1.7b = 9.8$
20. $35 = 10x + {}^-3x$
21. $4 = {}^-5s + {}^-4s$
22. $\frac{w}{3} + \frac{2w}{3} = \frac{-1}{2}$
23. $15 = 3y - 8y$
24. $28 = 13t - 6t$
25. $15a - 34a = {}^-38$
26. $\frac{3b}{7} + \frac{11b}{7} = 1$
27. $\frac{s}{2} + \frac{s}{2} = \frac{-3}{8}$
28. $4.1m - 1.6m = 10$
29. $0.34 = 0.04k + 0.13k$

SAMPLE: $13q = 5q + 4$

Solution: $\qquad 13q = 5q + 4$ Check: $13q = 5q + 4$
$\qquad 13q - 5q = 5q - 5q + 4$ $13(\frac{1}{2})$ | $5(\frac{1}{2}) + 4$
$\qquad\qquad 8q = 4$ $6\frac{1}{2}$ | $2\frac{1}{2} + 4$
$\qquad\qquad \frac{1}{8} \cdot 8q = 4 \cdot \frac{1}{8}$ $6\frac{1}{2}$ | $6\frac{1}{2} \quad \{\frac{1}{2}\}$
$\qquad\qquad q = \frac{1}{2}$

30. $7m = 2m + 10$
31. $18b = 3b + 45$
32. $-10k = 3k + 39$
33. $5a = 16 - a$
34. $3x = {}^-8x + 66$
35. $-5v = 3v - 64$

B
36. $-8y = 84 - y$
37. $\frac{3}{4}x = \frac{1}{4}x + 9\frac{1}{2}$
38. $\frac{2}{5}c = {}^-\frac{3}{5}c + 8$
39. $t = 5t + t + 3$
40. $0.6n = 0.2n + 1.6$
41. $w = 3 + 0.5w$
42. $t + 3t = 2t + 7$
43. $14z - 2z = z + 22$
44. $\frac{x}{2} + \frac{3x}{2} = {}^-3$

Solving Equations and Inequalities   325

C   45. $2(6a + a) = 10 + 18$
46. $3(x + 1) = 6(x - 2)$
47. $4(h + 5) + h = 35$
48. $\frac{1}{2}(8n - 10n) = {}^-15$

49. $4 + \frac{1}{3}(13y - y) = {}^-12$
50. $2k + \frac{1}{5}(10k + 5k) = 20 + 5$
51. $3(5b - 2b) - 6b = 4 + 11$
52. $13s - 6\left(\frac{s}{2} + \frac{s}{3}\right) = {}^-2$

Properties of Inequality

11–4   The Addition Property of Inequality

The properties that we have used to solve equations are similar to those for solving inequalities. First, however, let us recall the **order property** of numbers, which is important in dealing with inequalities. That is, for any two numbers *r* and *s*, *exactly one* of the following relationships holds:

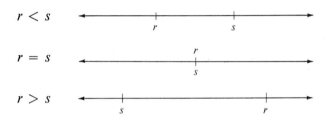

Do you recall the **transitive property** for equality? Let us look at a similar property for inequality. Consider the directed numbers *r*, *s*, and *t* as shown on the number line in Figure 11–1. Do you agree that $r < s$, $s < t$, and $r < t$?

Figure 11–1

Now consider the directed numbers *r*, *s*, and *t* on the number line in Figure 11–2. Do you see that in this case $r > s$, $s > t$, and $r > t$?

Figure 11–2

The number lines in Figures 11–1 and 11–2 illustrate the **transitive property** of **inequality**, which is summarized as follows:

For any directed numbers **r**, **s**, and **t**:

if $r < s$ and $s < t$, then $r < t$;
if $r > s$ and $s > t$, then $r > t$.

Let us investigate the possibility of an **addition property** for inequality. Study each statement to be sure you agree that it is true. In each, the numeral in red is added to each member of the first inequality.

Statement	True/False
6 > 2, and 6 + 5 > 2 + 5.	True, because 11 > 7.
5 > ⁻8, and 5 + 3 > ⁻8 + 3.	True, because 8 > ⁻5.
⁻3 > ⁻10, and ⁻3 + ⁻2 > ⁻10 + ⁻2.	True, because ⁻5 > ⁻12.
8 < 12, and 8 + 6 < 12 + 6.	True, because 14 < 18.
⁻7 < 3, and ⁻7 + 5 < 3 + 5.	True, because ⁻2 < 8.

Do you agree that adding the same number to each member of an inequality does not change the truth of the inequality? Thus it seems logical to assume the following **addition property** for **inequality**.

For any directed numbers *r*, *s*, and *t*:

if $r < s$, then $r + t < s + t$;
if $r > s$, then $r + t > s + t$.

Note that if 3 is **subtracted** from each member of the inequality 10 > 8, we have 10 − 3 > 8 − 3, which is true because 7 > 5. This outcome is not unexpected, since we arrive at the same result by adding ⁻3 to both members of 10 > 8. Thus we are, in fact, using the addition property of inequality, since **subtracting** a directed number is equivalent to **adding** its opposite.

To solve the inequality $x + 3 < 7$, we use the addition property of inequality to **transform** it into an **equivalent inequality** in which the variable stands alone as one member. (Inequalities that have the same solution set are **equivalent**.)

EXAMPLE.

$x + 3 < 7$	Given inequality
$x + 3 + ^-3 < 7 + ^-3$	Add ⁻3 to each member
$x + 0 < 7 + ^-3$	Property of additive inverses
$x < 4$	Substitution principle

Thus, if the replacement set is {the directed numbers} the solution set is {the directed numbers less than 4}.

The graph of the solution set looks like this:

ORAL EXERCISES

Tell whether each statement is true or false; explain your answer in terms of the number line.

SAMPLE: $^-8 < 2$ What you say: True; $^-8$ is to the left of 2 on the number line.

1. $6 > ^-3$
2. $0 < 12$
3. $0 < ^-8$
4. $\frac{3}{4} < \frac{7}{8}$
5. $^-\frac{1}{2} < \frac{1}{4}$
6. $^-\frac{1}{3} > ^-\frac{2}{3}$
7. $3\frac{1}{2} < ^-15$
8. $(3 + 4) > ^-10\frac{1}{2}$
9. $^-\frac{3}{5} < \frac{1}{5}$

Tell why each statement is true.

SAMPLE: If $n < ^-2$, then $n + 3 < ^-2 + 3$.

What you say: 3 is added to each member.

10. If $^-6 < m$, then $^-6 + 2 < m + 2$.
11. If $x < 15$, then $x + 5 < 15 + 5$.
12. If $19 > y$, then $19 + 6 > y + 6$.
13. If $a < 16$, then $a - 4 < 16 - 4$.
14. If $7 < b$, then $7 + ^-3 < b + ^-3$.

WRITTEN EXERCISES

Match each inequality in Column 1 with an equivalent inequality in Column 2.

COLUMN 1

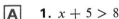
1. $x + 5 > 8$
2. $x - 6 < 4$
3. $x + 9 > 15$
4. $x + ^-7 \geq 8$
5. $x - 10 \geq ^-15$
6. $3 + x \leq 0$

COLUMN 2

A. $x > 6$
B. $x \leq ^-3$
C. $x < 10$
D. $x \geq ^-5$
E. $x > 3$
F. $x \leq 0$
G. $x \geq 15$

Copy and complete each solution, then graph the solution set. Each replacement set is {the directed numbers}.

SAMPLE: $t + 8 < 11$
$t + 8 + ? < 11 + ?$
$t < ?$

Solution: $t + 8 < 11$
$t + 8 + {}^-8 < 11 + {}^-8$
$t < 3$

7. $y + 5 > {}^-2$
$y + 5 + ? > {}^-2 + ?$
$y > ?$

8. $2 + m < 5$
$? + 2 + m < 5 + ?$
$m < ?$

9. $w - 2 \leq 0$
$w - 2 + ? \leq 0 + ?$
$w \leq ?$

10. ${}^-4 < k + {}^-8$
$? + {}^-4 < k + {}^-8 + ?$
$? < k$

11. $b + 3 \geq {}^-1$
$b + 3 + ? \geq {}^-1 + ?$
$b \geq ?$

12. ${}^-5 \leq a + {}^-4$
$? + {}^-5 \leq a + {}^-4 + ?$
$? \leq a$

Solve each inequality and write its solution set. Each replacement set is {the directed numbers}.

SAMPLE: $r + 2 < {}^-5$

Solution: $r + 2 < {}^-5$
$r + 2 + {}^-2 < {}^-5 + {}^-2$
$r < {}^-7$

{the directed numbers $< {}^-7$}

13. $t + 5 > 1$
14. $s + 2 < {}^-1$
15. $3 < m + 2$
16. $9 > n + {}^-8$

17. $9 + y > 4$
18. $5 + x \geq {}^-8$
19. $q - 5 < 7\frac{1}{2}$
20. $2 + s \leq 11$

21. $5 < t + 17$
22. ${}^-3 \leq k + {}^-3$
23. $z + 7 \geq 0$
24. $p - 3 > {}^-1$

B 25. $23 \geq {}^-10 + n$
26. $0 \leq r - 14$
27. $\frac{1}{2}(2m + 6) > 4$
28. $(3c + 9)\frac{1}{3} < 12$
29. $\frac{{}^-10 + 2m}{2} \geq {}^-5$

30. $2\left({}^-3 + \frac{a}{2}\right) \leq {}^-4$
31. ${}^-12 \leq \frac{1}{4}({}^-24 + 4k)$
32. $0.5(8 + 2c) \geq 2$
33. $3y - 1 > 2y$
34. $\frac{3m - 6}{3} < 0$

C 35. $5\left(\frac{x}{5} + 3\right) \geq 2(3 - 6)$
36. $2(z + {}^-3) < z - 8$
37. $\frac{{}^-18 + 3z}{2} \geq \frac{z}{2} + {}^-1$

38. $\frac{2}{3}\left(\frac{3r}{2} + 6\right) > 7\frac{1}{2}$
39. $3({}^-2m + m) + 4m \leq \frac{{}^-3}{2}$
40. $\frac{{}^-18 + 6y}{3} + 5y \geq {}^-3.2$

Solving Equations and Inequalities 329

11-5 The Multiplication Property of Inequality

Do you suppose there is also a **multiplication property** for inequality? Study these sample statements to help you arrive at a decision. The numerals in red indicate the number by which each member of the original inequality is multiplied. Notice that each multiplier is a **positive** number.

Statements	True/False
$5 > 2$, and $5 \cdot 3 > 2 \cdot 3$.	True, because $15 > 6$.
$^-6 < 2$, and $^-6 \cdot 5 < 2 \cdot 5$.	True, because $^-30 < 10$.
$10 > ^-3$ and $10 \cdot \frac{2}{5} > ^-3 \cdot \frac{2}{5}$.	True, because $4 > ^-\frac{6}{5}$.

Do you see that multiplying both members of an inequality by a **positive** number does not change the truth of the inequality?

For any directed numbers *r*, *s*, and *t*, with $t > 0$,

if $r < s$, then $rt < st$;
if $r > s$, then $rt > st$.

Of course the result of multiplying both members of an inequality by **0** results in an equation, since the product of 0 and any number is 0. For example, do you see that the following statements are true?

$^-3 < 7$ and $^-3 \cdot 0 = 7 \cdot 0$; $8 > ^-5$ and $8 \cdot 0 = ^-5 \cdot 0$.

For any directed numbers *r*, *s*, and *t*, with $t = 0$,

if $r < s$, then $rt = st$;
if $r > s$, then $rt = st$.

Now let us consider the outcome of multiplying each member of an inequality by a **negative** number. Study these two examples.

EXAMPLE 1.

Given the inequality $^-3 < ^-1$,
if we multiply each member by $^-2$, $^-3 \cdot ^-2$? $^-1 \cdot ^-2$
  or 6 ? 2.
Since we know that $6 > 2$,
we see that $^-3 \cdot ^-2 > ^-1 \cdot ^-2$.

EXAMPLE 2.

Given the inequality $\quad 5 > {}^-3,$
if we multiply each member by ${}^-4,\quad 5 \cdot {}^-4\ ?\ {}^-3 \cdot {}^-4$
or $\quad {}^-20\ ?\ 12.$
Since we know that $\quad {}^-20 < 12,$
we see that $\quad 5 \cdot {}^-4 < {}^-3 \cdot {}^-4.$

Do you agree that multiplying each member of an inequality by a **negative** number still gives us an inequality, but that the **sense** (greater than or less than) has been **reversed**?

For any directed numbers *r, s,* and *t,* with $t < 0$,

if $r < s$, then $rt > st$;
if $r > s$, then $rt < st$.

Application of the addition and multiplication properties of inequality can help us solve an inequality such as $3x + {}^-5 > 13$. The plan of attack is to write an **equivalent** inequality with the variable standing alone as one member. Follow each step carefully.

EXAMPLE.		
	$3x + {}^-5 < 13$	Given inequality
	$3x + {}^-5 + 5 < 13 + 5$	Add 5 to both members
	$3x + 0 < 13 + 5$	Property of additive inverses
	$3x < 18$	Substitution principle
	$\frac{1}{3} \cdot 3x < 18 \cdot \frac{1}{3}$	Multiply each member by $\frac{1}{3}$
	$x < 6$	Substitution principle

Thus the solution set is {the directed numbers less than six}. The graph of the solution set is:

ORAL EXERCISES

Tell how to complete each statement so that it is true.

SAMPLE: ${}^-3 < 2$, and ${}^-3 \cdot 4 < 2 \cdot 4$ since ___?___.

What you say: ${}^-3 < 2$, and ${}^-3 \cdot 4 < 2 \cdot 4$ since ${}^-12 < 8$.

1. $12 > {}^-15$, and $12 \cdot 3 > {}^-15 \cdot 3$ since ___?___.
2. ${}^-7 < 4$, and ${}^-7 \cdot 8 < 4 \cdot 8$ since ___?___.
3. $5 > {}^-1$, and $5 \cdot {}^-2 < {}^-1 \cdot {}^-2$ since ___?___.
4. ${}^-8 < {}^-3$, and ${}^-8 \cdot 0 = {}^-3 \cdot 0$ since ___?___.
5. $3 < 15$, and $3 \cdot 8 < 15 \cdot 8$ since ___?___.
6. ${}^-8 < {}^-5$, and ${}^-8 \cdot {}^-3 > {}^-5 \cdot {}^-3$ since ___?___.

Tell whether the symbol $>$, $<$, or $=$ should replace each question mark to make a true statement.

SAMPLE: $3 \; ? \; 10$ and $3 \cdot 2 \; ? \; 10 \cdot 2$

What you say: $3 < 10$ and $3 \cdot 2 < 10 \cdot 2$

7. ${}^-5 > {}^-8$ and ${}^-5 \cdot 2 \; ? \; {}^-8 \cdot 2$
8. ${}^-10 \; ? \; 6$ and ${}^-10 \cdot 3 \; ? \; 6 \cdot 3$
9. $4 > {}^-7$ and $4 \cdot 9 \; ? \; {}^-7 \cdot 9$
10. ${}^-9 \; ? \; 4$ and ${}^-9 \cdot 0 \; ? \; 4 \cdot 0$
11. ${}^-3 \; ? \; {}^-8$ and ${}^-3 \cdot 0 \; ? \; {}^-8 \cdot 0$
12. $4 < 9$ and $4 \cdot {}^-2 \; ? \; 9 \cdot {}^-2$
13. ${}^-8 > {}^-9$ and ${}^-8 \cdot {}^-5 \; ? \; {}^-9 \cdot {}^-5$
14. ${}^-2 < 6$ and ${}^-2 \cdot {}^-3 \; ? \; 6 \cdot {}^-3$

WRITTEN EXERCISES

Show how you use the multiplication property of inequality to transform each first inequality into the second inequality.

SAMPLE: $8 > {}^-10$; $24 > {}^-30$ *Solution:* $8 > {}^-10$
$8 \cdot 3 > {}^-10 \cdot 3$
$24 > {}^-30$

A
1. $12 < 25$; $48 < 100$
2. ${}^-10 < 4$; ${}^-20 < 8$
3. $11 > {}^-12$; $55 > {}^-60$
4. $16 > 4$; $8 > 2$
5. $\frac{1}{2} > {}^-3$; $1 > {}^-6$
6. ${}^-\frac{1}{4} < \frac{1}{3}$; ${}^-1 < \frac{4}{3}$
7. $4 > {}^-2$; ${}^-8 < 4$
8. ${}^-3 > {}^-5$; $12 < 20$
9. ${}^-9 < 10$; $18 > {}^-20$
10. $3t > 12$; $t > 4$
11. $-m < 4$; $m > {}^-4$
12. $\frac{k}{4} \geq 3$; $k \geq 12$

Solve each inequality. Then graph the solution set.

SAMPLE: $6a < 15$ *Solution:* $6a < 15$
$\frac{1}{6} \cdot 6 \cdot a < 15 \cdot \frac{1}{6}$
$a < 2\frac{1}{2}$

13. $3x > 9$
14. $21y \leq 42$
15. $\frac{2}{3}s \geq 2$

16. $5b < 10$
17. $2k > {}^-7$
18. $6n \geq 9$
19. $12t < 4$
20. $7c \leq {}^-14$
21. $9m \leq {}^-30$
22. $3w > 0$
23. $\frac{1}{2}d < {}^-1$
24. $15 > 3x$
25. ${}^-8 > 2z$
26. $0 \leq 2n$
27. $0.5r > {}^-\frac{1}{4}$

Solve each inequality and write the solution set. Remember that multiplying both members of an inequality by a **negative** number **reverses** the sense of the inequality.

SAMPLE: ${}^-3h \leq 12$

Solution:
$${}^-3h \leq 12$$
$$-\tfrac{1}{3} \cdot {}^-3 \cdot h \geq 12 \cdot -\tfrac{1}{3}$$
$$h \geq {}^-4$$

Solution set: {the directed numbers greater than or equal to ${}^-4$}

28. $-t > 30$
29. ${}^-2b < {}^-5$
30. ${}^-5w \geq 35$
31. ${}^-3k \leq {}^-10$
32. $16 > -r$
33. ${}^-20 \leq {}^-3t$
34. $-\dfrac{n}{2} > \dfrac{{}^-1}{4}$
35. $-\dfrac{b}{5} < 3$
36. $-\dfrac{w}{3} \geq 0$
37. $2t + 1 \leq 11$
38. $4c + 11 \leq 31$
39. $3 + 2s > 13$

B
40. $-\tfrac{2}{3}p \leq {}^-3$
41. $20 \leq \dfrac{{}^-4n}{5}$
42. $-\dfrac{r}{6} > 5$
43. ${}^-0.25x < \tfrac{1}{2}$
44. $0 \geq {}^-3m$
45. $-\dfrac{3b}{4} < 9$
46. ${}^-5a - 2 < 18$
47. $\dfrac{n}{4} + 7 \geq 0$
48. ${}^-14 \leq 3t - 2$
49. $w + 5 > 3w - 5$
50. $\tfrac{1}{2}s + 2 < 7$
51. ${}^-6c + 3 + c < 15 + c$
52. $3k + 2(4 - k) \geq 0$
53. $-b + 3(9 + b) < b + {}^-3$
54. ${}^-2(3m + {}^-6) < 2m$

C
55. $6\left(\dfrac{t}{2} - \dfrac{1}{3}\right) < 2t$
56. $12\left(\dfrac{1}{4} + \dfrac{y}{6}\right) > {}^-2y$
57. $-s + 3(s - 9) \leq 0$
58. $(h - 18){}^-\tfrac{1}{3} - 2 \geq h + 4$

Solving Equations and Inequalities   333

Formulas

11–6 Using Formulas to Solve Problems

Many practical problems can be solved by applying appropriate formulas. For example, to find the volume of the box shown here we can apply the formula $V = lwh$. If we know the length is 9 inches, the width is 4 inches, and the height is 5 inches, the solution is carried out like this.

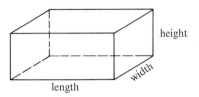

$$V = \text{length} \cdot \text{width} \cdot \text{height}$$
$$= lwh$$
$$= 9 \cdot 4 \cdot 5$$
$$= 180 \qquad \text{The volume is 180 cubic inches.}$$

Suppose that we know the volume, length, and width of the box pictured here, and wish to find its height. Again we can use $V = lwh$. However, if we were faced with several problems asking for the heights of various boxes, we would be wise to begin by solving the formula for h, the unknown dimension, as follows:

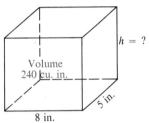

$V = lwh$	Given formula
$V \cdot \dfrac{1}{lw} = \dfrac{1}{lw} \cdot lwh$	Multiply both members by $\dfrac{1}{lw}$
$V \cdot \dfrac{1}{lw} = 1 \cdot h$	Multiplication property of reciprocals
$\dfrac{V}{lw} = h$	Substitution principle

Now the given information can be used to replace all the variables, except h, in $\dfrac{V}{lw} = h$.

$$\dfrac{240}{5 \cdot 8} = h$$
$$6 = h \qquad \text{The height is 6 inches.}$$

In general, when a particular formula is to be used to solve several problems for the same variable, it is often more efficient to transform the formula into one in which the variable representing the unknown quantity stands alone as one member.

PROBLEMS

Use the formula $A = lw$ (Area = length × width) to solve each problem. In each case write the formula so that the variable that represents the unknown dimension stands alone as one member.

1. The length of a rectangular shaped room is 17 feet. The area of the room is 138 square feet. Find the width of the room.

2. A roll of plastic film is $4\frac{1}{2}$ feet wide. How long a piece must be cut from the roll so that its area will be 50 square feet?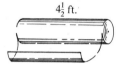

3. A sheet of plywood is 4 feet wide. How long must the sheet of plywood be if its area is more than 27 square feet? *Hint:* $lw = A$, and $A > 27$, so $lw > 27$.

4. Mr. Cater purchased $35\frac{1}{2}$ square yards of carpeting from a roll that was 3 feet wide. How many yards long was the piece of carpeting?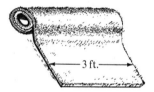

Use the formula $V = lwh$ (Volume = length × width × height) to solve each problem. In each case, first solve the formula for the variable that represents the unknown dimension.

5. Use the figure at the right and the given information to find the width. The volume is 324 cubic inches.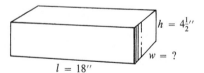

6. A metal tank is to be built with a bottom that measures 30 cm. in length and 14 cm. in width. How deep must the tank be if it is to hold 4200 cubic centimeters of water?

7. An air conditioning unit is designed to cool a room whose volume is no greater than 3200 cubic feet. Show whether or not the air conditioning unit should operate properly in a room that is 20 feet long and 18 feet wide, if the ceiling is 8 feet high.

8. A box is designed to hold 240 wooden blocks that measure 1 inch on each edge. The box is 12 inches long. Find the width of the box if it is 2 inches deep.

Use the formula $I = prt$ (Interest = principal × rate × time) to solve each problem. In each case solve the formula for the variable that represents the unknown quantity.

SAMPLE: Mr. Johnson borrows $1800 from the bank at 9% interest. If he keeps the money for $1\frac{1}{2}$ years, how much interest must he pay?

Solution: $I = prt$
$= 1800 \cdot \frac{9}{100} \cdot \frac{3}{2}$
$= 243$ He must pay $243.00 interest.

9. A bank makes a loan to a business man in the amount of $25,000 for 1 year. How much interest must he pay if the rate of interest is $7\frac{1}{2}\%$?

10. Mr. Carlson borrowed $2400 from the bank at an 8% rate of interest. The bank charged him $384 interest. For how long a period of time did he borrow the money?

11. A loan was made by a bank at the rate of $8\frac{1}{2}\%$ for a period of 2 years. The interest charged was $255. What was the amount of the loan?

12. Mr. McFarland loaned $2000 to Mr. Norton for a period of $1\frac{1}{2}$ years. At the end of this time Mr. Norton paid $210 interest. What rate of interest did Mr. McFarland charge?

CHAPTER SUMMARY

Inventory of Structure and Concepts

1. Equations of the type $x + a = b$ are solved by using the addition property of equality and the property of additive inverses.

2. Equations of the type $ax = b$ are solved by using the multiplication property of equality and the multiplication property of reciprocals.

3. Equations of the type $ax + bx = c$ are solved by combining similar terms and then using the multiplication property of equality.

4. The **transitive property** of **inequality** states that for any directed numbers r, s, and t:
   if $r < s$ and $s < t$, then $r < t$;
   if $r > s$ and $s > t$, then $r > t$.

5. The **addition property** of **inequality** states that for any directed numbers r, s, and t:

$$\text{if } r < s, \text{ then } r + t < s + t;$$
$$\text{if } r > s, \text{ then } r + t > s + t.$$

6. Inequalities that have the same truth set are **equivalent**.

7. The **multiplication property** for **inequality** states that, for any numbers r, s, and t:

$$\text{when } t > 0, \text{ if } r < s, \text{ then } rt < st,$$
$$\text{and if } r > s, \text{ then } rt > st.$$
$$\text{when } t = 0, \text{ if } r < s, \text{ then } rt = st = 0,$$
$$\text{and if } r > s, \text{ then } rt = st = 0.$$
$$\text{when } t < 0, \text{ if } r < s, \text{ then } rt > st,$$
$$\text{and if } r > s, \text{ then } rt < st.$$

Vocabulary and Spelling

property of additive inverses (p. 316)
equivalent equation (p. 316)
addition property of equality (p. 316)
multiplication property of equality (p. 318)
multiplication property of reciprocals (p. 318)
similar terms (p. 321)
order property of numbers (p. 325)
transitive property of inequality (p. 325)
addition property of inequality (p. 326)
transform (p. 326)
equivalent inequality (p. 326)
multiplication property of inequality (p. 329)

Chapter Test

Determine the solution set of each of the following equations. The replacement set is {the directed numbers}.

1. $x + {}^-7 = {}^-2$
2. $2x = {}^-12$
3. $3t + 5t = 4$
4. $15 = {}^-13 + x$
5. ${}^-4x = 14$
6. $t - 4t = {}^-6$

7. $\frac{x}{4} - 2 = 3$

8. $\frac{1}{2}x + \frac{3}{4}x = 15$

9. $-\frac{3k}{4} = 9$

10. $1.5x - {}^-2 = 1$

Solve each equation for x, assuming that no divisor has the value zero.

11. $x - b = a$

12. $\frac{x}{a} = b$

13. $ax - c = b$

14. $ax + cx = a$

Graph the solution set of each inequality, if the replacement set is {the directed numbers}.

15. $2x > {}^-6$

16. $^-1.5 \geq {}^-\frac{1}{2}x$

17. $^-2 < 3 + x$

18. $x + {}^-2x \leq 1$

19. $^-3k < 12$

20. $x - 5 < 1$

21. $2x - 5 \geq {}^-9$

22. What is the additive inverse of $^-3$?

23. Is the statement $3 > 2$ equivalent to the statement $2 < 3$?

24. Using the formula $V = lwh$ (Volume = length × width × height), find the height of the figure at the right if its volume is 156 cubic inches.

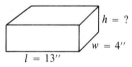

$l = 13''$, $w = 4''$, $h = ?$

25. A room is to be 12 feet wide. How long will you make it, if you want its floor area to be 24 square yards?

Chapter Review

Unless otherwise specified, assume that each replacement set is {the directed numbers}.

11–1 Equations of Type $x + a = b$

Solve each of the following equations for x.

1. $x + {}^-4 = 11$

2. $x - 9 = {}^-2$

3. $x + \frac{3}{4} = {}^-\frac{3}{4}$

4. $^-12 = x + 8$

5. $^-16 = {}^-4 + x$

6. $^-1.3 + x = {}^-2.4$

7. $a + x = c$

11–2 Equations of Type $ax = b$

Solve each equation for x.

8. $^-5x = {}^-30$
9. $3x = {}^-10$
10. $\dfrac{x}{6} = 17$
11. $^-\tfrac{1}{2}x = \tfrac{3}{4}$
12. $1.2x = 48$
13. $^-\tfrac{3}{8}x = 30$
14. $bx = c,\ b \neq 0$
15. $ax + b = c,\ a \neq 0$
16. $\dfrac{a}{b}x = c,\ a \neq 0,\ b \neq 0$

11–3 Equations of Type $ax + bx = c$

Solve each equation for x.

17. $3x + 9x = {}^-60$
18. $15 = \tfrac{3}{4}x + \tfrac{1}{2}x$
19. $15 = \dfrac{x}{3} - \dfrac{x}{4}$
20. $^-3x = x + {}^-15$
21. $\tfrac{1}{8}x - \tfrac{3}{8}x = 4$
22. $4x + 5x = 21$
23. $5x = 7x - 10$
24. $4.1x - 2.3x + 0.2x = 5$
25. $ax = bx + c,\ a \neq b$
26. $ax + bx = c,\ a \neq -b$

11–4 The Addition Property of Inequality

For each of the following, tell what property of inequality is illustrated.

27. $2 < 5$ and $2 + {}^-5 < 5 + {}^-5$
28. $^-3 < {}^-2$ and $^-2 < 0$, so $^-3 < 0$

Graph the solution set of each of the following inequalities, if the replacement set is {the directed numbers}.

29. $6 + y < {}^-1$
30. $4 < {}^-5 + x$
31. $\tfrac{1}{2}(6 + 2k) < 3$
32. $w - 5 \geq 0$
33. $p + {}^-4 \leq {}^-7$
34. $\dfrac{3x - 12}{3} \geq {}^-3$

11–5 The Multiplication Property of Inequality

35. What true inequality results when each member of $3 > {}^-2$ is multiplied by $^-5$?

Graph the solution set of each equation, if the replacement set is {the directed numbers}.

36. $3x < 15$
37. $\frac{x}{2} \geq {}^-1$
38. $3t + 2 < -7$
39. ${}^-2k < {}^-4$
40. ${}^-15m > 0$
41. $-x + 10 \leq {}^-6x$
42. ${}^-8 < 4y$
43. $\frac{3n}{-4} > 3$
44. $\frac{3x}{2} - 5 < {}^-7$

11–6 Using Formulas to Solve Problems

45. Using the formula $A = lw$ (Area = length × width), determine the length of a piece of sheet metal that is 3.5 inches wide and has an area of 70 square inches.

46. For how long a time did a man borrow $8000 at a rate of 5% per year, if the bank charged him $200 interest? Use the formula $I = prt$ (Interest = principal × rate × time).

Review of Skills

Write each of the following as a power of 10.

SAMPLE: 1000   Solution: $1000 = 10 \cdot 10 \cdot 10 = 10^3$

1. 100
2. 10,000
3. 10
4. 1,000,000

Write each of the following in expanded form.

SAMPLE: 5263

Solution: $5000 + 200 + 60 + 3 = 5 \cdot 10^3 + 2 \cdot 10^2 + 6 \cdot 10 + 3$

5. 543
6. 947
7. 1379
8. 10,354

Simplify each expression.

9. $3x + 2x$
10. $5y - 2y$
11. $3k + 15k$
12. $3 \cdot 4 + 3 \cdot 6$
13. $3(4 + 6)$
14. $12 \cdot 3 + 12 \cdot 7$
15. $23 \cdot 2 + 23 \cdot 8$
16. $24 \cdot 14 + 24 \cdot 6$
17. $15 + {}^-2$
18. $15 - 2$
19. $8 + {}^-8$
20. $8 - 8$
21. $35 + 23$
22. $51 + 32$
23. $32 + 51$
24. $(15 + 4) + 6$
25. $15 + (4 + 6)$
26. $27 + (3 + 15)$

Chapter 11

27. 12(3 + 7)	29. ⁻15 + 15	31. (27 + 3) + 15
28. 23(2 + 8)	30. 23 + 35	

State the additive inverse of each of the following.

32. ⁺54	34. ⁻16	36. 3
33. ⁺$\frac{3}{4}$	35. ⁻$\frac{1}{2}$	37. ⁻5

Solve each equation and write its solution set.

38. $3n - 5 = 10$	41. $5n = 3n + 14$
39. $n - 12 = 2n - 13$	42. $3n - 5n = 24$
40. $12 - n = 3n - 8$	43. $⁻24 - n = 3n + 4$

What number, multiplied by itself, gives:

44. 4	45. 25	46. 100

Find the value of $f(x)$, using the given replacement for x.

47. $f(x) = 3x - 5$; 3	49. $f(x) = 2x + 9$; 4
48. $f(x) = 10 + x$; 6	50. $f(x) = 12 - 3x$; 7

■ ■

CHECK POINT FOR EXPERTS

Pairs of Directed Numbers

You have used a lattice of points for graphing number pairs of whole numbers. In order to graph pairs of **integers** (whole numbers and their opposites), the lattice must be extended as shown in Figure 1. Notice that the x-scale and the $f(x)$-scale divide the lattice into four regions, which are called **quadrants**. As before, the point of intersection of the two scales is called the **origin**, and is labeled **0**.

The point marked A in the first quadrant of the lattice corresponds to the number pair **(4, 2)**. The point marked B is in the second quadrant, and corresponds to the number pair (⁻3, 4). Recall that the first number in the pair

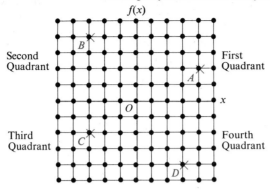

Figure 1

is the number on the *x*-scale, and the second number is the number on the *f(x)*-scale. Do you see that point **C** is in the third quadrant and corresponds to the pair (⁻3, ⁻2)? And that point **D** in the fourth quadrant corresponds to (3, ⁻4)?

Questions

1. In which quadrant does each triangle lie, and what set of number pairs corresponds to its vertices?

    a. Triangle *ABC*
    b. Triangle *DEF*
    c. Triangle *MNP*
    d. Triangle *QRS*

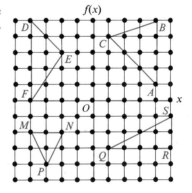

2. Which word, **positive** or **negative**, should replace each blank to make a true statement?

    a. If a number pair corresponds to a point in the first quadrant, both numbers in the pair must be __?__.
    b. If a number pair corresponds to a point in the second quadrant, the first number must be __?__, and the second number must be __?__.
    c. If a number pair corresponds to a point in the third quadrant, both numbers in the pair must be __?__.
    d. If a number pair corresponds to a point in the fourth quadrant, the first number must be __?__ and the second number must be __?__.

3. a. What set of ordered pairs of numbers corresponds to the points marked with black ×'s in the lattice? Is this set of number pairs a function? Explain.

b. What set of ordered pairs of numbers corresponds to the points marked with red ×'s? Is this set of number pairs a function? Explain.

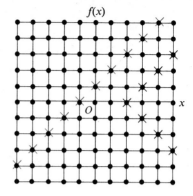

4. Which of the given sets of number pairs is a function and which is not? Make a lattice like the one in Question 3 and graph each set of number pairs.

a. {(4, 0), (3, 1), (2, 2), (1, 3), (0, 4), (⁻1, 5)}.

b. {(⁻3, 1), (⁻2, 1), (⁻2, 0), (⁻2, ⁻1), (⁻1, ⁻1), (0, ⁻1), (1, ⁻1)}.

THE HUMAN ELEMENT

Leonhard Euler

One of the most remarkable mathematicians of all time was Leonhard Euler (1707–1783). Over his life span of seventy-six years, he is said to have produced about 600 important articles on mathematics. His early studies also included theology, medicine, oriental languages, astronomy, and physics. This wide interest brought him recognition as one of the outstanding scholars of his time.

Euler found, in solving the classical Königsburg Bridge Problem of the 18th century, an important scientific principle concealed in the puzzle. The sketch in Figure 1 represents the city of Königsburg and the seven bridges that connected parts of the city. The problem was to determine if it was possible to walk a continuous path, crossing all seven bridges, but without crossing any bridge more than once.

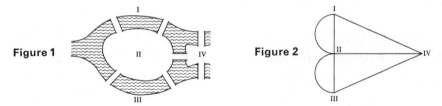

Figure 1

Figure 2

Euler reduced the problem to the line drawing in Figure 2, called a *network*. Each land area is replaced by a point and each bridge by a line connecting two points. The points are called vertices, and each vertex is called **odd** or **even** according to whether the number of lines leading from the vertex is odd or even. Thus in the network, each of vertices I, II, III, and IV is odd. The problem of crossing the bridges becomes that of tracing the network without lifting the pencil from the paper, and without going over any line a second time. Euler discovered that this can be done, ending where you started, if all of the vertices are even. Since lines in the network represent bridges in the Bridge Problem, the walk cannot be made without crossing some bridge twice.

The solution of the bridge problem was not important in itself. What was significant was the fact that Euler's method of solution led to the development of a new branch of mathematics called topology.

One of the earliest paved roads . . .

Modern superhighways . . .

Addition and Subtraction of Polynomials

Although in early times most roads developed from cart paths, the ancient Romans often constructed well-built roads, of which the most famous was the Appian Way, named for the official Appius Claudius Caecus, under whose direction it was built. This road was about eighteen feet wide, and was paved with blocks of lava, laid on a bed of broken stone and cemented with lime. Parts of the Appian Way, for which construction began in 312 B.C., are still in use today. Descendants of the Roman road builders are the Italian engineers who design and build Italy's modern superhighways. Shown is an intersection of two such highways, north of Bologna, Italy.

Adding Polynomials

12-1 Introduction to the Set of Polynomials

In the study of algebra, terms such as $16n^2$, n^2, $5k$, $0.03y^2$, $\frac{2}{3}$, and $8ab$ are called **monomials**. That is, a term may be a numeral, a variable, or a product of a numeral and one or more variables. The expression $2x + 5$ consists of two terms and is called a **binomial**. The expression $2c^2 + 3c + 4$ consists of three terms, so it is called a **trinomial**.

Monomials	Binomials	Trinomials
$10y^2$	$b + 10$	$5m^2 - 3mn + 2$
$3r^4$	$3 + t^4$	$7 + x + 5y^2$
$\dfrac{15m}{4}$	$5x + 3y$	$a^2 + 2ab + b^2$
$2r^3s^2t^6$	$ax + bx$	$2r^2 + r - 3$

The expressions listed as monomials, binomials, and trinomials are all called **polynomials**. A polynomial may have more than three terms. If so, it is not usually given a special name, but is simply called a polynomial.

In general, when *a*, *b*, *c*, and *d* represent directed numbers and *m*, *n*, and *p* are different positive integers, a polynomial in one variable takes the form

$$ax^m + bx^n + cx^p + \ldots + d.$$

As a matter of keeping our work uniform, let us agree to write polynomials in a standard form. To illustrate, let us consider the polynomial $x^3 + 10x^2 + 3x - 5$. Notice that the terms are arranged so that the first term has the largest exponent, the second term has the next largest exponent, and so on. The **degree** of a polynomial in one variable is indicated by the largest-valued exponent of any of its terms. In $x^3 + 10x^2 + 3x - 5$ the term x^3 has the largest exponent, 3, so the degree of the polynomial is 3.

The polynomial $x^4 + 2x^2 - 2$ seems to have some "holes" in it. We may fill such "holes" by supplying the missing terms as shown here:

$$x^4 + 0x^3 + 2x^2 + 0x - 2$$

Do you agree that putting in the missing terms, $0x^3$ and $0x$, has not changed the value of the original polynomial for any value of x? What properties of zero apply?

Now let us consider polynomials in more than one variable. Notice that the variables x and y both appear in $2x^3 + 5x^2y + xy$. It is a generally accepted procedure, when writing a polynomial in more than one variable, to order the terms according to values of the exponents of one of the variables.

Thus we would write $7x^2y + x^3 + 3xy^2 - 8y^3$ as $x^3 + 7x^2y + 3xy^2 - 8y^3$.

ORAL EXERCISES

Tell whether each of the following is a monomial, a binomial, or a trinomial.

1. $3r$
2. $k + n + 4$
3. $5r + 8r - 3r$
4. $x^2 + 17y - x^2$
5. $2a + 3b$
6. $5xyz$
7. $7a^2bc^3$
8. $a^2 - b^2$
9. $m + n$
10. $5^2 + 3ax$
11. $45abcd$
12. $a^2 + \dfrac{a}{2}$

Tell whether the right member of each formula is a monomial, a binomial, or a trinomial.

13. $V = lwh$
14. $a = \frac{1}{2}bh$
15. $p = 2l + 2w$
16. $d = rt$
17. $F = \frac{9}{5}C + 32$
18. $c^2 = a^2 + b^2$
19. $0 = F_1 + F_2 + F_3$
20. $C = \pi d$
21. $A = 2\pi r^2 + \pi dh$

WRITTEN EXERCISES

Give an example of each of the following, using some combination of the symbols m^3, n^2, m, and 4.

SAMPLE: A polynomial with five terms

Solution: $m^3n^2 + mn^2 + m - n^2 + 4$

1. A trinomial
2. A binomial
3. A monomial
4. A polynomial with five terms
5. A polynomial with four terms
6. A polynomial with six terms

Write each of these polynomials in standard form.

SAMPLE: $4 + 8y^3 + y$ Solution: $8y^3 + y + 4$

7. $2w + w^3$
8. $3 + a^2$
9. $6 + 5y + 3y^2$
10. $19x^2 + x^3$
11. $k^3 - 2k + k^2$
12. $3c - 5c^2 + c$
13. $8a + 7a^5 + a^2$
14. $t^3 + 5 + t^2$
15. $4d^3 + 3d^4 - d^6$
16. $2r^2 - 7r^4 + 3r^5 + 6$

Write each of these polynomials in standard form.

SAMPLE: $3b^2 + a^2 + 6ab$ Solution: $a^2 + 6ab + 3b^2$

B 17. $mn + m^2$
18. $xy^2 + y^4 - x^2$
19. $2r^2s^2 - 7r^4s + 12r^5$
20. $3pq^2 + p^3 - 2q^2 + 5$
21. $x^2 + y^2 + xy$
22. $5 + 3a^2 + b^2 + 4ab$
23. $x^2y + y^2 + x^3y + x$
24. $19c + 5a - a^3 + 7a^5$

Write each polynomial in standard form, inserting any "missing" terms.

SAMPLE: $3m^4 + m + 2$ Solution: $3m^4 + 0m^3 + 0m^2 + m + 2$

25. $2x^3 + 3$
26. $3x^5 - x^3 + 8$
27. $t^4 + 5t + 1$
28. $m^2 + 3m^5 - 2$
29. $y^4 + 8 - y^3$
30. $10 - k^2 + 5k^3$
31. $h - 2h^4 + h^6 + 25$
32. $7r^4 - 1$

12–2 Addition of Polynomials

Suppose we are to add the polynomials $3x + 5$ and $5x + 7$. Applying ideas we have already learned about simplifying expressions, we might proceed like this:

$$(3x + 5) + (5x + 7) = 3x + 5x + 5 + 7$$
$$= (3 + 5)x + 12$$
$$= 8x + 12$$

As you can see we can rearrange the terms of the two polynomials and combine similar terms. This is done by applying the commutative, associative, and distributive properties.

It is sometimes more convenient to complete the addition of polynomials by arranging them vertically as shown in these examples.

EXAMPLE 1. Add $10x^3 + 4xy + 3$ and $8x^3 + 2xy + 4$.

$$\begin{array}{r} 10x^3 + 4xy + 3 \\ 8x^3 + 2xy + 4 \\ \hline 18x^3 + 6xy + 7 \end{array}$$

EXAMPLE 2. Add $5b^4 + 6b^3 + 4$ and $3b^4 + b^2 + 5$.

$$\begin{array}{r} 5b^4 + 6b^3 + 0b^2 + 4 \\ 3b^4 + 0b^3 + b^2 + 5 \\ \hline 8b^4 + 6b^3 + b^2 + 9 \end{array}$$

You will recall from an earlier chapter that we have used the small raised minus sign to indicate a **negative** number, and the lowered minus sign to indicate the **opposite** of a number. For example

$$^-3 \text{ is read } \textbf{negative 3}$$
$$-3 \text{ is read } \textbf{the opposite of 3}$$

However, since the opposite of any positive number is the corresponding negative number, it is true that $-3 = {^-3}$. Hence from now on, whenever there is no possibility of misunderstanding, we shall use the lowered minus sign to indicate either "the opposite of" or "negative."

Since we have seen that subtracting a directed number is the same as adding the opposite of the number, we can write

$$5 - 7 \text{ as either } 5 + (-7) \text{ or } 5 + {^-7}, \text{ and}$$
$$-t - 3 \text{ as either } -t + (-3) \text{ or } -t + {^-3}, \text{ but}$$
$$3 - k \text{ only as } 3 + (-k).$$

Addition and Subtraction of Polynomials

Do you see why $3 - k$ cannot be written as $3 + {}^-k$? How do we know that ${}^-k$ represents a negative number? It depends upon what value k has, of course.

If $k = 5$, then $-k = {}^-5$, but
if $k = 0$, then $-k = 0$, and
if $k = {}^-5$, then $-k = 5$.

That is why, with a variable, we *always* use the symbol for "opposite of," rather than the one for "negative."

To add the polynomials $9x - 3$ and $6x + 7$, we can arrange them vertically, rewrite $9x - 3$ as a sum, then add:

$$\begin{array}{rl} 9x - 3 = & 9x + (-3) \\ 6x + 7 = & \underline{6x + 7} \\ & 15x + 4 \end{array}$$

Study the following examples:

EXAMPLE 1.
$$\begin{array}{rl} 10x^2 - 2xy = & 10x^2 + (-2xy) \\ 2x^2 - 4xy = & \underline{2x^2 + (-4xy)} \\ & 12x^2 + (-6xy) = 12x^2 - 6xy \end{array}$$

EXAMPLE 2.
$$\begin{array}{rl} 7x^2 - 8y^2 + 9 = & 7x^2 + (-8y^2) + 9 \\ -3x^2 + 2y^2 - 5 = & \underline{-3x^2 + 2y^2 + (-5)} \\ & 4x^2 + (-6y^2) + 4 \\ = & 4x^2 - 6y^2 + 4 \end{array}$$

To find the sum of polynomials $12 - 8y^2 + 6x^2$ and $12y^2 - 2x^2 - 5$ it is best to rearrange terms so that each is in standard form:

$$\begin{array}{rl} 12 - 8y^2 + 6x^2 = & 6x^2 + (-8y^2) + 12 \\ \underline{12y^2 - 2x^2 - 5} = & \underline{-2x^2 + 12y^2 + (-5)} \\ & 4x^2 + 4y^2 + 7 \end{array}$$

ORAL EXERCISES

Give a meaning in terms of "opposite" for each of the following.

SAMPLE 1: -18 *What you say:* the opposite of 18

SAMPLE 2: $-3 + k$ *What you say:* the opposite of 3, plus k

350 Chapter 12

SAMPLE 3: $(8m - n) - 3m$ *What you say:* $8m$ plus the opposite of n, plus the opposite of $3m$.

1. $-m$
2. $5 - x$
3. $y - 7$
4. $t - 35$
5. $6 - t$
6. $8 - a$
7. $m + n - 3$
8. $\frac{3}{4} - d$
9. $(a - 3b) + b$
10. $y - 12$
11. $-3 + m$
12. $-10 + 2k$
13. $-5 - n$
14. $-y - 6$
15. $x - x$

Simplify each of the following.

SAMPLE 1: $(5m + 2n) + 3m$ *What you say:* $8m + 2n$
SAMPLE 2: $(8x - 6y) + 2y$ *What you say:* $8x - 4y$

16. $(2x + 4y) + 3x$
17. $10k + (3h + 7k)$
18. $(8a + 4) + 5a$
19. $(4r + 5s) + 9s$
20. $\frac{1}{2}a + (2a + b)$
21. $(\frac{1}{4}w + 2y) + \frac{1}{2}w$
22. $(4m - 3n) + 2n$
23. $(2r + s) + (r + s)$
24. $(3x^2 - 2y^2) - 9y^2$
25. $(4.2a + 3.6b) + 3.4a$
26. $0.5s + (1.3r - 0.7s)$
27. $0.03p + (0.06p + 0.2q)$
28. $2rs + (r^2 + s^2)$
29. $(x + y) + (x - y)$

WRITTEN EXERCISES

Find the sum indicated in each exercise.

SAMPLE 1: $4k + 6$
           $8k + 1$

Solution: $12k + 7$

SAMPLE 2: $s - 3t^2$
           $2s + 5t^2$

Solution: $3s + 2t^2$

1. $6z + 3$
   $7z + 4$

2. $9r + 5s$
   $3r + s$

3. $2x^2 + 3$
   $3x^2 + 1$

4. $7a - 3b$
      $8b$

5. $x^3 + 5y^2$
   $x^3 - 5y^2$

6. $2m + 4$
   $3m - 1$

7. $y^2 + 3$
   $y^2 - 3$

8. $2x^2 + y^2 + z^2$
   $2x^2 + 3y^2 + 2z^2$

9. $6r^2 + 8rs + s^2$
   $-2r^2 - 3rs + 3s^2$

Find each sum.

10. $5t + 7$
     $3t - 2$

11. $r^2 - s^2 + t^2$
     $2r^2 - 4s^2 - 2t^2$

12. $3w^3 + 4z$
      $2w + 3z$

Addition and Subtraction of Polynomials

13. $8k - 10$
    $\underline{2k + 2}$

14. $2x - 3$
    $\underline{3x - 1}$

15. $2x - y^2$
    $\underline{x - y^2}$

16. $2a - 3b$
    $\underline{-2a - 3b}$

17. $2a^2 - 5b + 7$
    $\underline{a^2 - b - 8}$

18. $4c^2 + 9d + 5$
    $\underline{c^2 + 2d}$

19. $p^2 + q - r$
    $\underline{p^2 - r}$

20. $2n^2 - 5s + 6$
    $\underline{n^2 - s}$

21. $8.5r + 4.2s + 7$
    $\underline{1.6r - 1.7s + 2}$

22. $0.06x + 0.07y - 0.9$
    $\underline{0.27x + 0.7}$

23. $3k + 4\tfrac{1}{2}n - 6\tfrac{3}{4}$
    $\underline{-k - 2\tfrac{1}{4}n}$

24. $x^2 - 2x + 14$
    $\underline{-3x^2 + 6x - 10}$

25. $(5r + 3s) + (6r + 4s)$
26. $(4w - 7) + (w + 2)$
27. $(5x - 6y) + (5x - 6y)$
28. $(3z^2 + b) + (8z^2 - b)$
29. $(6x - 4y^2) + (10x - y^2)$
30. $(c^2 + 5c - 9) + (3c^2 - 10c + 3)$
31. $(a^2 - 2ab + b^2) + (a^2 + 2ab + b^2)$
32. $(^-2m^3 + 5m) + (m^4 - 5m^2 + 4)$
33. $(4t^4 - 10t^2) + (10t^2 + 20)$
34. $(h^2 - 3h + 7) + (2h^2 + 5h - 10)$

Rewrite each polynomial in standard form.

SAMPLE 1: $5t + 15t^2 - 6 + t^3$ Solution: $t^3 + 15t^2 + 5t - 6$
SAMPLE 2: $a^3 + b^3 + a^2b - 2ab^2$ Solution: $a^3 + a^2b - 2ab^2 + b^3$

35. $2y + 3y^2 - 2 + y^3$
36. $6m^2 + 4 + 4m$
37. $25x^2 + 3 + 10x$
38. $34 + d^2 - 15d$
39. $65 + z^2 - 20z$
40. $2t^4 + 3t + 5t^3 + 8t^2$
41. $2k^3 - 3 - 5k^2 + k$
42. $4n^3 + 3n - 5 - 2n^2 + n^4$

B 43. $x^2 + 2y^2 + 3xy$
44. $4w^2 + x^2 - 4wx$
45. $a^3 + b^3 + 5a^2b + 5ab^2$
46. $3r^2s + r^3 - s^3 + 3rs^2$

Complete each addition. Check for $x = 2$, $y = 3$, $r = 4$, and $s = 5$.

SAMPLE: $3x - 8$ Solution: $3x - 8$ Check: $3(2) - 8 = -2$
$\underline{x + 2}$ $\underline{x + 2}$ $\underline{1(2) + 2 = 4}$
$4x - 6$ $4(2) - 6 = 2$

47. $4s - 8$
    $2s + 1$
    $\underline{s - 3}$

48. $5y + 7$
    $\underline{2y - 10}$

49. $5x - y + 3$
    $x - y - 5$
    $\underline{8x + 2y - 10}$

50. $r^2 + r + 6$
    $\underline{3r^2 - r - 2}$

51. $5r - s + 7$
    $r + s + 3$
    $\underline{3r - 2}$

52. $2x + 10 - s$
    $3x - 12 - s$
    $\underline{x + s}$

352 *Chapter 12*

53. $3r + 8$
    $5r - 2$

54. $s^2 - s + 5$
    $2s^2 - 2s - 5$

55. $x^2 + 5x - 1$
    $ -x - 8$

Find each sum.

56. $-6.1a + 3.2b$
    $-3.4a + 1.6b$

57. $2.5x - 1.8w$
    $-0.5x - 1.4w$

58. $1\frac{1}{2}n^3 - m^3$
    $\frac{1}{2}n^3 + \frac{1}{2}m^3$

59. $ab^2 + ac^2$
    $3ab^2 - 4ac^2$

60. $0.5xy - 0.70$
    $0.3xy + 0.85$
    $2.4xy + 0.05$

61. $r^3 - s^3$
    $\frac{1}{3}r^3 - \frac{1}{2}s^3$

62. $\frac{5}{8}x - y$
    $\frac{1}{4}x - y$

63. $-2k + \frac{1}{2}m$
    $k - \frac{1}{2}m$

64. $4cd - 16$
    $-4cd + 20$

65. $-3rs + 4$
    $0.4rs - 2$
    $5rs + 2$

66. $5yz - 6$
    $-8yz + 14$

67. $5ab - 9$
    $-15ab + 17$

68. $h + 5$
    $5h + 3$
    $6h - 4$

69. $3a - 10$
    $a - 8$
    $5a + 20$

C 70. $m + n + 1.7$
    $-m - n + 3.2$

71. $9b^2 - 2.4bc + 2c^2$
    $4b^2 - 6bc - 3.4c^2$

72. $8t^3 - 35t^2 + 45t$
    $ 20t^2 - 30t - 10$

73. $2k^3 - 6k^2 + 5k$
    $-k^3 + k - 8$

74. $(12y^3 + 3y^2 - 5y - 14) + (5y + 10 - y^2 + 2y^3) + (y + 3)$

75. $(2t^2 - 3t^4 - 1 + t) + (6t + t^2 + t^5 + 8) + (-3t^2)$

12–3 Polynomials and Addition Properties

Do you agree that the work you have already completed with polynomials seems to suggest that when two polynomials are added the result is always going to be a unique polynomial? That is, the set of polynomials is **closed** under **addition**.

Now let us see if the **commutative** property holds for addition of polynomials. To help verify this property consider the sum of $(5x + 5)$ and $(3x - 2)$.

EXAMPLE 1. $(5x + 5) + (3x - 2) = 5x + 5 + 3x - 2$
$ = 5x + 3x + 5 - 2$
$ = 8x + 3$

EXAMPLE 2. $(3x - 2) + (5x + 5) = 3x - 2 + 5x + 5$
$= 3x + 5x - 2 + 5$
$= 8x + 3$

Do you agree that, regardless of the order in which we add the polynomials $5x + 5$ and $3x - 2$, the result is the same? If you were to repeat this process again and again, using different polynomials, you would observe a similar outcome.

Now let us think about the **associative** property, and polynomials. If we are to add the polynomials $(10x - 6)$, $(4x + 1)$, and $(x + 8)$ do you suppose that the way they are grouped will affect the result? Study the examples below to help verify your conclusion. Notice the use of brackets to show how the polynomials are grouped in each case.

EXAMPLE 1. $(10x - 6) + (4x + 1) + (x + 8)$
$= (10x - 6) + [(4x + 1) + (x + 8)]$
$= (10x - 6) + [4x + 1 + x + 8]$
$= (10x - 6) + [5x + 9]$
$= 10x - 6 + 5x + 9$
$= 15x + 3$

EXAMPLE 2. $(10x - 6) + (4x + 1) + (x + 8)$
$= [(10x - 6) + (4x + 1)] + (x + 8)$
$= [10x - 6 + 4x + 1] + (x + 8)$
$= [14x - 5] + (x + 8)$
$= 14x - 5 + x + 8$
$= 15x + 3$

Thus we can list the following properties of the **addition** of **polynomials**:

The set of polynomials is **closed under addition**.
Addition of polynomials is **commutative**.
Addition of polynomials is **associative**.

It should not seem surprising that polynomials have the same properties as numbers, for a polynomial is really a number. When the variables are replaced by numbers chosen from their replacement sets, each polynomial becomes a particular number, so it is only natural that addition of polynomials should turn out to be both commutative and associative.

ORAL EXERCISES

Tell which property of addition for polynomials justifies each statement.

SAMPLE: $(3x^2 - 1) + (x + 10) = (x + 10) + (3x^2 - 1)$

What you say: The commutative property of addition

1. $3t + (8 - t) = (8 - t) + 3t$
2. $(5 - x^2) + (2 + x^2) = (2 + x^2) + (5 - x^2)$
3. $(k + 4) + (k + 9) = (k + 9) + (k + 4)$
4. $m + [3m^2 + (2m + 1)] = [m + 3m^2] + (2m + 1)$
5. $(a^2 - 1) + 5a = 5a + (a^2 - 1)$
6. $2b + [b^2 + (2b + 1)] = [b^2 + (2b + 1)] + 2b$
7. $[(ab + 2) + (a^2 + b)] + (a - b) = (ab + 2) + [(a^2 + b) + (a - b)]$
8. $(y^2 + 3) + [(y + 1) + (2y - 5)] = [(y^2 + 3) + (y + 1)] + (2y - 5)$
9. $(3r + s) + [(r^2 - s^2) + (2 + s)] = [(r^2 - s^2) + (2 + s)] + (3r + s)$
10. $[(3 - x) + 4x^2] + (1 + x) = (3 - x) + [4x^2 + (1 + x)]$

WRITTEN EXERCISES

Simplify both members of each equation to show that they are equal.

SAMPLE: $(4x - 2) + (3 + x) = (3 + x) + (4x - 2)$

Solution: $(4x - 2) + (3 + x) = (3 + x) + (4x - 2)$

$4x - 2 + 3 + x$	$3 + x + 4x - 2$
$4x + x - 2 + 3$	$x + 4x - 2 + 3$
$5x + 1$	$5x + 1$

1. $(8 + y) + (3 + 2y) = (3 + 2y) + (8 + y)$
2. $3t^2 + (t^2 + 5t + 6) = (t^2 + 5t + 6) + 3t^2$
3. $(4k + 1) + (6 - 4k) = (6 - 4k) + (4k + 1)$
4. $(x - y) + (3x + y) = (3x + y) + (x - y)$
5. $(3r - 6) + [(2r + 1) + (r + 3)] = [(3r - 6) + (2r + 1)] + (r + 3)$
6. $(4y - 3y^2) + (8y^2 - 14y + 5) = (8y^2 - 14y + 5) + (4y - 3y^2)$
7. $[(x - y - 7) + (x + y + 9)] + (y - 3)$
    $= (x - y - 7) + [(x + y + 9) + (y - 3)]$
8. $(x^2 + 6) + [(3x^2 + x - 8) + (-2x^2)]$
    $= [(x^2 + 6) + (-2x^2)] + (3x^2 + x - 8)$

Addition and Subtraction of Polynomials

9. $(-5a^2 + 2ab + b^2) + (8a^2 - 8b^2)$
   $= (8a^2 - 8b^2) + (-5a^2 + 2ab + b^2)$
10. $(-2m^3 + 5m) + (m^4 - 5m^2 + m)$
    $= (m^4 - 5m^2 + m) + (-2m^3 + 5m)$

Add the polynomials in each problem pair. Compare the answers.

11. $6x - 5$ $x + 2$
    $x + 2$ $6x - 5$

12. $3t + 7$ $2t - 45$
    $2t - 45$ $3t + 7$

13. $3r - 8$ $-r + 1$
    $-2r + 3$ $-2r + 3$
    $-r + 1$ $3r - 8$

14. $5x - y + 4$ $-x + y + 3$
    $-x + y + 3$ $5x - y + 4$

15. $r + s + 9$ $r + s + 9$
    $5r - s + 7$ $-s + 3$
    $-s + 3$ $5r - s + 7$

16. $x^2 + 5x - 1$ $3x^2 - x - 8$
    $-x^2 \quad\quad + 1$ $-x^2 \quad\quad + 1$
    $3x^2 - x - 8$ $x^2 + 5x - 1$

Find the value of each polynomial for $x = 2$, $y = 3$, $a = 0$, and $b = -1$.

SAMPLE 1: $3x^2 + a - 5$ *Solution:* $3(4) + 0 - 5 = 12 - 5$
$= 7$

SAMPLE 2: $5b^2 + 3b + 4$ *Solution:* $5(1) + 3(-1) + 4 = 5 - 3 + 4$
$= 6$

17. $2x^2 + 2x + 7$
18. $5y^2 - 5y - 8$
19. $3a^2 + 4ab - 10$
20. $1.8a^2 - 0.3a + 6$
21. $y^3 - y^2 + 5y$
22. $2 - 3b + 2b^2$
23. $0.3x^2 + 5x + 0$
24. $a^2 - 9ab - 0.2$
25. $(x + y) + (-x - y)$
26. $ax + ay + ab$
27. $b^2 + bx + by$
28. $\dfrac{4xy}{6} + x - 3y$

Show that each equation results in a true statement when the variable is replaced by the suggested value.

SAMPLE: $(7y + 3) + (8y - 16) = (8y - 16) + (7y + 3)$; let $y = 5$.

Solution:

$(7y + 3) + (8y - 16) = (8y - 16) + (7y + 3)$	
$(7 \cdot 5 + 3) + (8 \cdot 5 - 16)$	$(8 \cdot 5 - 16) + (7 \cdot 5 + 3)$
$38 + 24$	$24 + 38$
62	62

29. $12a + (a + 21) = (a + 21) + 12a$; let $a = 2$.
30. $7n + (n^2 - 9n + 1) = (n^2 - 9n + 1) + 7n$; let $n = 3$.
31. $(4x^2 - 7) + (5x^2 - 3x + 8) = (5x^2 - 3x + 8) + (4x^2 - 7)$;
    let $x = 10$.
32. $3b + [5b + (6b - 2)] = (3b + 5b) + (6b - 2)$; let $b = 5$.

B 33. $(x^3 + 2x^2) + (x^2 + 3x - 4) = (x^3 + 2x^2 + x^2) + (3x - 4)$; let $x = 4$.
34. $(5n + 6) + (3n - 2) = (3n - 2) + (5n + 6)$; let $n = -1$.
35. $(r^2 - r) + (7r^2 - r + 3) = (7r^2 - r + 3) + (r^2 - r)$; let $r = 0$.
36. $9k + (1 - 2k + k^2) = (9k + 1) + (-2k + k^2)$; let $k = 6$.
37. $(4b^2 + 10b - 2) + (b^3 + b^2) = (b^3 + b^2) + (4b^2 + 10b - 2)$; let $b = \frac{1}{2}$.

Subtracting Polynomials

12–4 Polynomials, Additive Identity, and Additive Inverses

Do you suppose that zero is still the **additive identity element** when we are working with polynomials? If your answer to this question is "yes" you are thinking correctly. These true statements help verify this idea.

$$(x + 3) + 0 = 0 + (x + 3) = (x + 3)$$
$$(x^2 + xy - 6) + 0 = 0 + (x^2 + xy - 6) = (x^2 + xy - 6)$$
$$(-4x^2 + 3x) + 0 = 0 + (-4x^2 + 3x) = (-4x^2 + 3x)$$

An important idea to consider in the study of polynomials is that of the **opposite**, or the **additive inverse**. In our earlier work with numbers we found that the **sum** of **any number** and its **opposite** is **zero**. For example:

$$3 + (-3) = 0 \qquad -15 + 15 = 0 \qquad 8x + (-8x) = 0$$

It may already have occurred to you, then, that the **opposite** of a given polynomial should be another polynomial whose **sum** with the given polynomial is **zero**. Thus

$$(k - 5) + [-(k - 5)] = 0.$$

However, note also that

$$(k - 5) + (-k + 5) = k + (-5) + (-k) + 5$$
$$= k + (-k) + (-5) + 5$$
$$= 0 + 0 = 0.$$

Thus we see that $(-k + 5)$ is the additive inverse of $(k - 5)$, and conclude that

$$-(k - 5) = -k + 5.$$

Let's try a few more examples:

EXAMPLE 1. $(x + 2) + [-(x + 2)] = 0$,
and $(x + 2) + [-x + (-2)] = 0$;
hence $-(x + 2) = -x + (-2)$.

EXAMPLE 2. $(-5a - 7) + [-(-5a - 7)] = 0$,
and $(-5a - 7) + (5a + 7) = 0$;
hence $-(-5a - 7) = 5a + 7$.

Does it become apparent to you that the opposite of a polynomial may be written as the sum of the opposites of the terms of the polynomial?

For any polynomial $a + b + c$,

$$-(a + b + c) = -a + (-b) + (-c).$$

The sum of $(x - 2)$ and $-(x - 2)$ is $(x - 2) + (-x + 2) = 0$.
The sum of $(a - 3ab + 10)$ and $-(a - 3ab + 10)$ is

$$(a - 3ab + 10) + (-a + 3ab - 10) = 0.$$

Study this summary of ideas about polynomials:

Zero is the additive identity for polynomials.
Every polynomial has an opposite (or additive inverse).
The sum of any polynomial and its opposite is zero.
The opposite of a given polynomial is a polynomial each of whose terms is the opposite of the corresponding term of the given polynomial.

ORAL EXERCISES

Give the sum of the two polynomials and include a reason for your answer.

SAMPLE 1: $(3ab + b - 3)$ and $-(3ab + b - 3)$

What you say: 0; the sum of any polynomial and its opposite is zero.

SAMPLE 2: $(3x^2 + 5)$ and $(x - x)$

What you say: $3x^2 + 5$; zero is the additive identity for polynomials.

1. $(8k^2 + 10)$ and $-(8k^2 + 10)$
2. $(3ab - b)$ and $(-3ab + b)$

3. $(10r^2s^2 + rs - s^2)$ and $(s^2 - rs - 10r^2s^2)$
4. $(15a^2b^2 - 15a^2b^2)$ and $(4a^2 + 10ab - 5b^2)$
5. $-(x^2 - x - 3)$ and $(x^2 - x - 3)$
6. $(3x^2 - 3x^2)$ and $(10y^2 - 1)$

Match each polynomial in Column 1 with its additive inverse in Column 2.

COLUMN 1

7. $(3r^2 - 2s + 2)$
8. $(x^2 + 4xy + y^2)$
9. $(r^2 - 3s^2 - 3)$
10. $(-r^2 - 3s^2 + 3)$
11. $(3r^2 + 2s + 2)$
12. $-(x^2 + 4xy - y^2)$

COLUMN 2

A. $(-r^2 + 3s^2 + 3)$
B. $-(-x^2 - 4xy + y^2)$
C. $(x^2 - 4xy - y)$
D. $-(x^2 + 4xy + y^2)$
E. $(-3r^2 - 2s - 2)$
F. $(r^2 + 3s^2 - 3)$
G. $(-3r^2 + 2s - 2)$

WRITTEN EXERCISES

Give the opposite of each polynomial. Write your answer in standard form.

SAMPLE: $x^2 - x + 4$ Solution: $-x^2 + x - 4$

A
1. $3x^2 - x + 10$
2. $-5x^2 - 3x + 4$
3. $4x^5 + 2x^3 - x + 5$
4. $-\frac{x^2}{3} + \frac{x}{5} + 6$
5. $15x^7 + 10x^6 + x + 3$
6. $10x^2 - 6x^3 - 4x - 10$
7. $-(25x^4 + 10x^3 - 7x^2 + x - 19)$
8. $\frac{a}{5} + \frac{b}{7} - \frac{ab}{6} - \frac{2}{3}$
9. $-1 + 2x - x^2 + 3x^5 - x^7$
10. $-[3x + (2x^2 - x + 2)]$

Write each polynomial in standard form without using parentheses.

SAMPLE: $-(4x^5 + 2x^3 + 7x - 3)$ Solution: $-4x^5 - 2x^3 - 7x + 3$

11. $-(2b^2 + 5b + 6)$
12. $-(5x^2 - 3x + 8)$
13. $-(2 - 3m + 2m^2)$
14. $-\left(\frac{t^2}{3} - \frac{t}{2} + \frac{1}{5}\right)$
15. $-(-a^2 + a - 1)$
16. $-(10y^3 + 3y^5 - 8y^4 - y^3 + y - 1)$
17. $-(3.7w^4 - 2.5w^2 + 0.3w + 10)$
18. $-\left(-\frac{a^2}{2} - \frac{ab}{4} - \frac{b}{2} + \frac{a}{7}\right)$
19. $-(-x^3 + 6x^2 - x + 10)$
20. $-(1.8n^3 - 0.5n^2 + 1.2n - 1.6)$

Find the sum of each pair of polynomials. First arrange your work in vertical form as shown in the sample.

SAMPLE: $(3x^2 - 5x + 2)$ and $-(x^2 - 5x - 7)$

Solution:
$$(3x^2 - 5x + 2) = 3x^2 - 5x + 2$$
$$-(x^2 - 5x + 7) = -x^2 + 5x + 7$$
$$\overline{ 2x^2 + 0x + 9} = 2x^2 + 9$$

21. $(10a^2 - 6)$ and $-(3a^2 + 10)$
22. $(x + y)$ and $-(x - y)$
23. $(2a + b)$ and $-(2a + b)$
24. $-(2a + b)$ and $(3a - b)$
25. $(x + y)$ and $-(x + y)$
26. $-(x^2 - 3)$ and $(3 - x^2)$

B 27. $(5a^2 - 6b + c - 3)$ and $-(-3a^2 + b - 5)$
28. $(1.5x^2 + 0.3xy - 2.4y^2)$ and $-(-1.5x^2 - 0.3xy + 2.4y^2)$
29. $-(\frac{1}{2}x^2 + \frac{1}{4}x + \frac{5}{6})$ and $(\frac{3}{2}x^2 + \frac{1}{2}x + 1)$
30. $(10k^3 + 3k^2 + k - 6)$ and $-(10k^3 + 3k^2 + k - 6)$

Write each polynomial in standard form without using grouping symbols.

SAMPLE 1: $-[3x^2 + (5y^2 + 2)]$

Solution:
$\ -[3x^2 + (5y^2 + 2)]$
$= -[3x^2 + 5y^2 + 2]$
$= -3x^2 - 5y^2 - 2$

SAMPLE 2: $-[4t^3 - (t + 3t^2 - 8)]$

Solution:
$\ -[4t^3 - (t + 3t^2 - 8)]$
$= -[4t^3 - t - 3t^2 + 8]$
$= -4t^3 + 3t^2 + t - 8$

31. $-[(4t^7 + t^6) + (3t^2 - 3)]$
32. $-[(x^5 - 3) + (x^3 - x)]$
33. $-[(-a^4 - a^3) + (-a + 5)]$
34. $[-(m^3 + m) + (m^2 - 2)]$
35. $[(b^3 - b^2) + (3b + 8)]$
36. $-[-(w^3 + 3) + (w^2 + 2w)]$
37. $-[-(5x^6 + x^5) + (x^2 - 3)]$
38. $-[-(t^3 + t^2) - (t - 1)]$

12–5 Subtraction of Polynomials

You previously learned that subtracting a directed number is the same as adding the opposite, or additive inverse, of the number. We will use this same procedure in subtracting polynomials. That is, we will **add the opposite** of the polynomial to be subtracted. Study these examples.

EXAMPLE 1. To subtract $\begin{array}{r} 10x - 8 \\ 7x + 5 \\ \hline \end{array}$ we add $\begin{array}{r} 10x - 8 \\ -7x - 5 \\ \hline 3x - 13 \end{array}$

EXAMPLE 2. To subtract $12x^2 - 5x + 8$
$\phantom{\text{To subtract }}2x^2 + 2x - 3$

we add $12x^2 - 5x + 8$
$\phantom{\text{we add }}-2x^2 - 2x + 3$
$\phantom{\text{we add }}\overline{10x^2 - 7x + 11}$

Subtracting one polynomial from another is a little more tricky when arranged horizontally. Follow this example carefully.

EXAMPLE. $(7x + 6) - (2x - 1) = 7x + 6 + (-2x) + 1$
$= 7x - 2x + 6 + 1$
$= 5x + 7$

The same problem arranged in vertical form is shown below. Subtraction should always be checked by comparing the first polynomial with the sum of the addends, as is indicated.

 Subtract: $7x + 6$ Check: $2x - 1$
$2x - 1$ $5x + 7$
$\overline{5x + 7}$ $\overline{7x + 6}$

ORAL EXERCISES

State the opposite of each polynomial.

SAMPLE: $3x - 7$ What you say: $-3x + 7$

1. $5a$
2. $3x^2$
3. $y + 5$
4. $a + b$
5. $-t + 5$
6. $-3k - 4$
7. $2x^2 - 3x + 5$
8. $-3m^2 + m - 3$

Subtract the second polynomial from the first.

SAMPLE 1: $3b$ SAMPLE 2: $5x - 2$
$\phantom{\text{SAMPLE 1: }}\underline{-b}$ $\underline{-x + 3}$

What you say: $4b$ What you say: $6x - 5$

9. $-5x$ **10.** $5t$ **11.** $\frac{1}{2}m$ **12.** $-0.6ac$
$\phantom{\textbf{9. }}\underline{7x}$ $\underline{-9t}$ $\underline{-\frac{1}{4}m}$ $\underline{0.2ac}$

Addition and Subtraction of Polynomials   361

13. 9ab
    −5ab

14. 14rs
    −10rs

15. −15h
      3h

16. −y
      2y

17. −10w
      3w

18. −6ab²
    −2ab²

19. −18k
     −3k

20. 15xy
    −10xy

21.   0.75m
    −0.23m

22. 5a + 6
     2a + 4

23. 5n − 2
     n + 3

24. −7x + 6
     −x − 3

25. $(2a + 3) - (a)$
26. $(7r + 10s) - (-2r)$

27. $(3x + y) - (x - y)$
28. $(5a + b) - (3b)$

WRITTEN EXERCISES

Subtract the second polynomial from the first. Check by addition.

SAMPLE: $7a + b$ Solution: $7a + b$ Check: $3a + 5b$
          $3a + 5b$ $3a + 5b$ $4a - 4b$
                                 $4a - 4b$ $7a + b$

1. $4m + 2n$
    $2m − 8n$

2. $5a − 6$
    $2a + 1$

3. $3x + 7$
    $2x − 15$

4. $a^2 + 5$
    $2a^2 + 2$

5. $10 − 3r$
    $1 − 2r$

6. $5a − 1$
     $− 9$

7. $4y + 5$
    $10y + 7$

8. $10x^2 − 3x$
    $16x^2 + x − 3$

9. $6k + 3$
    $−5k + 2$

10. $9r^2 + 6r$
    $4r^2$

11. $5b^2 − 3c$
    $−4b^2$

12. $8a^2 + 2a$
    $−6a^2\quad + 5$

Write each of these expressions without parentheses. Do not combine similar terms.

SAMPLE: $(5w - 13) - (-3w + 4)$ Solution: $5w - 13 + 3w - 4$

13. $(x - 16) - (2x + 5)$
14. $(7y - 2) - (12y - 6)$
15. $(x^2 + 5x) - (3x^2 - 10x)$
16. $(4m + 3n) - (m + 6)$

17. $(7a^2 + 8a) - (-4a^2 + a)$
18. $(-3n^2 + 2n) - (4n + 7)$
19. $(a^2 - a + 5) - (2a^2 + 2a - 3)$
20. $(b + 15) - (-b^2 + 5 - 2b)$

Simplify each of the following by removing parentheses and combining similar terms. Write each answer in standard form.

SAMPLE 1: $(4a + 3) - (-a - 5)$ Solution: $(4a + 3) - (-a - 5)$
                                              $= 4a + 3 + a + 5$
                                              $= 5a + 8$

SAMPLE 2: $(6k + 8) - (k - 2)$ **Solution:** $(6k + 8) - (k - 2)$
$= 6k + 8 - k + 2$
$= 5k + 10$

21. $(9x - 12) - (x + 2)$
22. $(3m + 4) - (m - 10)$
23. $(4r^2 - 8) - (3r^2 + 5r)$
24. $(12m - r) - (m - r)$
25. $(10s - 9t) - (3s + 5)$
26. $(2a + b) - (2a + b)$

B 27. $(5 + 10x) - (3 - x)$
28. $(7.3x + 2y) - (1.5x + 6y)$
29. $(7b + 4a) - (-5b + a)$
30. $(x - y) - (x - y)$
31. $(1.8k^2 - 4.6) - (0.6k + 3.2)$
32. $(4a^2 - 7) - (5a^2 - 3a - 8)$

Subtract the second polynomial from the first. Check by addition.

SAMPLE: $3b^2 + b - 7$
$b^2 + 3b + 9$

Solution: $3b^2 + b - 7$
$b^2 + 3b + 9$
$\overline{2b^2 - 2b - 16}$

Check: $b^2 + 3b + 9$
$2b^2 - 2b - 16$
$\overline{3b^2 + b - 7}$

33. $15x^2 + 7x + 10$
$x^2 + 6x - 5$

34. $8k^2 + k - 1$
$3k^2 + 6k + 6$

35. $m^2 - 3m - 9$
$4m^2 + m + 6$

36. $4y^2 + y - 1$
$-y^2 + 5y + 10$

37. $5w^2 - 6$
$w^2 - 3w + 8$

38. $18n^2 + 4n - 1$
$n + 4$

39. $6 - 3b + 8b^2$
$9 + 6b + b^2$

40. $4y - 19$
$10y^2 - 3y - 6$

41. $2z^2 + z$
$z + 1$

42. $t^2 + 3t$
$t - 1$

43. $m^2 - 3$
$m - 1$

44. $0.6a^2 + 10a - 1.5$
$a^2 - 2a - 2$

Rewrite each of the following, removing all parentheses and brackets. Do not combine similar terms.

SAMPLE 1: $2x + [5x - (3x + 7)]$ **Solution:** $2x + [5x - (3x + 7)]$
$= 2x + [5x - 3x - 7]$
$= 2x + 5x - 3x - 7$

SAMPLE 2: $5a - [2a - (4a + 8)]$ **Solution:** $5a - [2a - (4a + 8)]$
$= 5a - [2a - 4a - 8]$
$= 5a - 2a + 4a + 8$

C 45. $5t + [8 - (2t + 3)]$
46. $7a + [a - (2a + 5)]$
47. $-[x + (2x - 8)] + 3x$
48. $2m - [7m - (6m + 2)]$
49. $(n + 4) + [3n - (n + 5)]$
50. $-(xy + 3) - [xy + (10 + 3xy)]$
51. $10k - [2k - (k + 5) + k^2 + 6]$
52. $-[-(5 - 3t) - 8] - (4t + 1)$

Simplify each of the following by removing all grouping symbols and combining similar terms. Write each answer in standard form.

53. $15a - [2a - (a + 7) + 3a^2 - 10] + 5a$
54. $7n - [n - (3n + 8) + 10 - n^2] + (6n + 4)$
55. $-[-3x - (6 - x) + 5] - (7x - 4)$
56. $[3k - (6 + k)] - [3 - (k + 2)] + 10$
57. $2x - [6x - (x - 2)] - [-5x - (4 - x) + 9]$

Using Polynomials

12–6 Problems Using Polynomials

Let us consider the solution of problems where given information is expressed as polynomials. For example, consider the two triangles formed by the diagonal drawn across the quadrilateral in Figure 12–1. If each polynomial represents the area of a triangle, the area of the quadrilateral can be expressed as a polynomial which is the sum of the given polynomials.

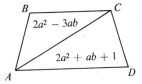

Figure 12–1

$$\begin{array}{ll} 2a^2 - 3ab & \text{(Area of } \triangle ABC\text{)} \\ \underline{2a^2 + ab + 1} & \text{(Area of } \triangle ACD\text{)} \\ 4a^2 - 2ab + 1 & \text{(Area of quadrilateral } ABCD\text{)} \end{array}$$

Now suppose we know that the area of $\triangle ACD$ is greater than the area of $\triangle ABC$. To find out how much greater, we subtract.

$$\begin{array}{ll} 2a^2 + ab + 1 & \text{(Area of } \triangle ACD\text{)} \\ \underline{2a^2 - 3ab} & \text{(Area of } \triangle ABC\text{)} \\ 4ab + 1 & \text{(The amount by which the area of } \triangle ACD \\ & \text{is greater than the area of } \triangle ABC\text{)} \end{array}$$

PROBLEMS

Write each answer as a polynomial in standard form.

 1. The lengths of the sides of a triangle are indicated by polynomials as shown in the figure. Find the perimeter of the triangle.

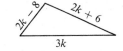

364   Chapter 12

2. The lengths of the sides of a right triangle are indicated by polynomials as shown in the figure. Which polynomial represents the length of the hypotenuse? Find the sum of the lengths of the hypotenuse and the longer leg. Find the sum of the lengths of the two legs.

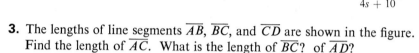

3. The lengths of line segments $\overline{AB}$, $\overline{BC}$, and $\overline{CD}$ are shown in the figure. Find the length of $\overline{AC}$. What is the length of $\overline{BC}$? of $\overline{AD}$?

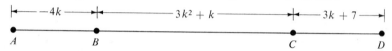

4. Use the information given on triangle ABC to answer each of the following. How much longer is side $\overline{AC}$ than side $\overline{BC}$? How much longer is side $\overline{BC}$ than side $\overline{AB}$? How much longer is side $\overline{AC}$ than side $\overline{AB}$?

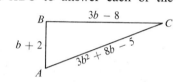

5. The volume of each of three water tanks is indicated in the accompanying figure. Find the combined volume of tanks A and B; of tanks B and C; of tanks A, B, and C. How much greater is the volume of tank B than of tank A? How much greater is the volume of tank B than the combined volume of tanks A and C?

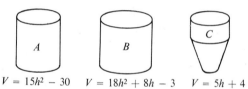

$V = 15h^2 - 30$ $V = 18h^2 + 8h - 3$ $V = 5h + 4$

B   6. Use the following figure to answer each question. How much longer is $\overline{QS}$ than $\overline{PR}$? How long is $\overline{PS}$? How long is $\overline{QR}$?

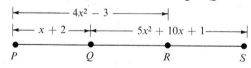

7. Two rockets are launched at the same moment. After a certain length of time the velocity of rocket A is represented by the polynomial $6at^2 + g$ and the velocity of rocket B is given by $2at^2 - 3g - 5$. How much greater is the velocity of rocket A than that of rocket B?

8. The area of the larger circular region in the figure is given by the polynomial $3m^2 + 18mn - 5n^2$. The area of the smaller circular region is $2m^2 - 3mn + n^2$. What is the area of the shaded portion of the figure?

9. The area of the regions in the accompanying figure are indicated by the given polynomials. How much larger is region III than region II? What is the combined area of the three regions? How much larger is region II than region I?

I	II
$\frac{x^2}{2} + 3x + 4\frac{1}{2}$	$x^2 + 6x + 9$

III
$x^2 + 8x + 15$

12–7 Equations and Polynomials

Many of the equations we need to solve in algebra require that we combine the ideas we learned some time ago about solving equations with our new knowledge of polynomials. Study the methods used to solve each of these equations and to check the solution.

EXAMPLE 1.
$$2y + (8 - y) = 12$$
$$2y + 8 - y = 12$$
$$y + 8 = 12$$
$$y = 4$$

Check: $2y + (8 - y) \stackrel{?}{=} 12$

$2(4) + (8 - 4)$	12
$8 + 4$	12
12	12

The solution set is $\{4\}$.

EXAMPLE 2.
$$(8t + 4) - (t - 7) = 17 + (4 - 3t)$$
$$8t + 4 - t + 7 = 17 + 4 - 3t$$
$$7t + 11 = 21 - 3t$$
$$10t = 10$$
$$t = 1$$

Check: $(8t + 4) - (t - 7) \stackrel{?}{=} 17 + (4 - 3t)$

$(8 \cdot 1 + 4) - (1 - 7)$	$17 + (4 - 3 \cdot 1)$
$12 - (-6)$	$17 + 1$
$12 + 6$	18
18	18

The solution set is $\{1\}$.

WRITTEN EXERCISES

Solve each equation. Check your solution, then write the solution set.

A
1. $2t + (15 - t) = 18$
2. $3x + (2x + 3) = 13$
3. $10a + (4 - 6a) = 52$
4. $(5 - y) + 2y = 9$
5. $42 = (2w + 3) + 11w$
6. $5w - (2w + 4) = 12$
7. $13y - (y + 21) = 21$
8. $3x - (5 - 2x) = 35$
9. $7m + (8m + 54) = 9$
10. $(7k + 67) + 9k = 3$
11. $30 = 5x + (2 - 7x)$
12. $9n - (3n - 6) = 42$
13. $(3x + 4) + (x + 8) = 20$
14. $(3b + 4) - (b + 2) = 12$

Solve each equation and write the solution set.

B
15. $(6c + 4) + (c - 7) = (10 - 3c) + 7$
16. $(2x + 4) - (x - 7) = (4 - 3x) + (3x + 1)$
17. $7n - (n^2 - n - 9) = 17 - n^2$
18. $19x - (1 - 2x - x^2) = x^2 + 34$
19. $12 - (z + 8) - (5z - 7) = -6z + 16 - (-6z + 7)$
20. $x - (3x + 1) + (7x + 2) = 5x - (3x - 13)$
21. $0.5n + (n - 1) - (0.2n + 10) = 0.2n - (0.2n - 28)$
22. $w - (0.5w + 1) - (0.3w + 2) = 0.5w - (0.5w - 3)$
23. $(2x - 6) - (8x + 4) = -x - 25$
24. $(5b + 3) + (2b - 4) = (b + 3) - (b + 3)$
25. $(n^2 + 2) - (n^2 + 2) = 3n - (5n + 3)$

C
26. $2x - [5x - (6x + 2)] = 10 - x$
27. $7a - [a - (2a + 8)] = 3a - 2$
28. $20t - [2t - (t + 2) + t^2 - 6] = 28 - t - t^2$
29. $r^2 - r - 5 = 5r - [r - (2r + 8) + 3 - r^2]$
30. $-[-(7 - 2k) - 8 + 3k] - (3k + 1) = -3k - 1$

12-8 Functions and Polynomials

Now that we are accustomed to working with polynomials let us see what happens when we use polynomials as function rules in our function machine.

Addition and Subtraction of Polynomials

The function machine in Figure 12-2 uses the polynomial $2x - 3$ as its rule. The replacement set is $\{-3, -2, -1, 0, 1, 2, 3\}$. Each value for x listed in the first column of the table is a member of the replacement set. Each $f(x)$ value is found by replacing x in the right member of the function equation $f(x) = 2x - 3$ with the corresponding value of x in the first column.

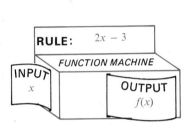

Figure 12-2

x	$f(x)$	Number Pair $(x, f(x))$
-3	-9	$(-3, -9)$
-2	-7	$(-2, -7)$
-1	-5	$(-1, -5)$
0	-3	$(0, -3)$
1	-1	$(1, -1)$
2	1	$(2, 1)$
3	3	$(3, 3)$

Do you agree that the set of number pairs $\{(-3, -9), (-2, -7), (-1, -5), (0, -3), (1, -1), (2, 1), (3, 3)\}$ is a function? If we were to continue using different polynomials as rules for our function machine it would soon become clear that any polynomial can be used as a rule to write a set of number pairs that is a function.

ORAL EXERCISES

Tell how to complete the table according to the pictured function machine.

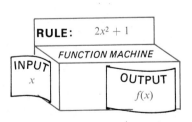

	x	$f(x)$	$(x, f(x))$
	-6	73	$(-6, 73)$
1.	-4	33	?
2.	-2	9	?
3.	0	?	?
4.	2	?	?
5.	4	?	?
6.	5	?	?
7.	6	?	?

368 *Chapter 12*

Tell how to complete each table if the given polynomial is used as the function rule.

Rule: $2x^2 + x - 8$

	x	f(x)	(x, f(x))
	4	28	(4, 28)
8.	2	?	?
9.	0	?	?
10.	−2	?	?
11.	−4	?	?
12.	−6	?	?

Rule: $(t + 2) - 3$

	t	f(t)	(t, f(t))
	−3	−4	(−3, −4)
13.	−2	?	?
14.	−1	?	?
15.	0	?	?
16.	1	?	?
17.	2	?	?

WRITTEN EXERCISES

Find the value of f(x) for the given replacement for x.

SAMPLE: $f(x) = x^2 + 2x - 1$
Let $x = -2$.

Solution: $f(-2) = (-2)^2 + 2(-2) - 1$
$= 4 + (-4) - 1$
$= -1$

A

1. $f(x) = x^2 - 10x + 6$
Let $x = 0$.

2. $f(x) = 6x^2 - 2x + 3$
Let $x = 1$.

3. $f(x) = 4x^2 + x - 1$
Let $x = -2$.

4. $f(x) = 6x + 10$
Let $x = -\frac{1}{2}$.

5. $f(x) = 4x^2 + 8x - 5$
Let $x = \frac{1}{2}$.

6. $f(x) = x - (2x + 8)$
Let $x = -2$.

7. $f(x) = 9 - (-5x + 3)$
Let $x = 0.6$.

8. $f(x) = (17 - 2) + 7x$
Let $x = -1$.

9. $f(x) = (x^2 + 2) - 3$
Let $x = 4$.

10. $f(x) = -(x - 10)$
Let $x = -3$.

11. $f(x) = 2 + x^2 + (3x + 1)$
Let $x = -5$.

12. $x^2 + 5x - 7 = f(x)$
Let $x = 0$.

13. $32x^2 + 12x - 2 = f(x)$
Let $x = \frac{1}{4}$.

14. $f(x) = (5x + 7) - (x + 2)$
Let $x = -5$.

15. $f(x) = x - (2x + 8)$
Let $x = -10$.

Use the function machine and the rule expressed by the polynomial $x^2 - 3x - 4$ to complete the table.

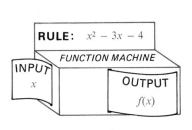

	x	$f(x)$	$(x, f(x))$
16.	-6	50	?
17.	-4	24	?
18.	-2	?	?
19.	0	?	?
20.	1	?	?
21.	3	?	?
22.	5	?	?

Use the given function equation and replacement set to write the number pairs that represent the function.

SAMPLE: $f(b) = 3b^2 + b - 1$; $\{-3, -1, 0, 1, 3\}$

Solution: $\{(-3, 23), (-1, 1), (0, -1), (1, 3), (3, 29)\}$

B 23. $f(m) = (m + 2) + m^2$; $\{-3, -2, -1, 0, 1, 2, 3\}$
24. $f(n) = 5n - (n + 3)$; $\{-4, -2, 0, 2, 4\}$
25. $f(t) = t^2 + 5t - 1$; $\{-4, -3, -2, -1, 0\}$
26. $f(x) = 9x^2 - 6x - 8$; $\{-1, -\frac{1}{3}, 0, \frac{1}{3}, 1\}$
27. $f(h) = 4h^2 + 4h - 9$; $\{-1, -\frac{1}{2}, 0, \frac{1}{2}, 1\}$
28. $f(n) = n^3 + 3n + 4$; $\{-2, -1, 0, 1, 2\}$
29. $f(x) = x^2 + \frac{1}{8}$; $\{-\frac{1}{2}, -\frac{1}{4}, 0, \frac{1}{4}, \frac{1}{2}\}$
30. $f(m) = 0.5m^2 + m + 2$; $\{-3, -2, -1, 0, 1, 2\}$

C 31. $f(t) = 2t - 5t - (6t + 1)$; $\{-2, -1, 0, 1, 2\}$
32. $f(x) = 7x - x - (2x + 8)$; $\{-10, -5, 0, 5, 10\}$
33. $f(h) = 2h - h - (h + 2) + h^2$; $\{-3, -2, -1, 0, 1, 2, 3\}$

CHAPTER SUMMARY

Inventory of Structure and Concepts

1. A term of a **polynomial** may be a numeral, a variable, or a product of a numeral and one or more variables.

2. A **monomial** is a polynomial consisting of just one term.

3. A **binomial** is a polynomial consisting of two terms.

4. A **trinomial** is a polynomial consisting of three terms.
5. A **polynomial in one variable**, where a, b, c, and d are directed numbers, x is the variable, and m, n, and p are positive integers such that $m > n > p$, takes the standard form $ax^m + bx^n + cx^p + \ldots + d$.
6. The **degree** of a polynomial in one variable is the number indicated by the largest-valued exponent of any term that contains a variable.
7. A polynomial in more than one variable is written in standard form by ordering the terms according to the values of exponents of one of the variables.
8. The sum of any two polynomials is a unique polynomial.
9. Addition of polynomials is **commutative**.
10. Addition of polynomials is **associative**.
11. Zero is the **additive identity** for polynomials.
12. Every polynomial has an **opposite (additive inverse)**.
13. The **opposite** of a given polynomial is a polynomial each of whose terms is the opposite of the corresponding term of the given polynomial.
14. The sum of any polynomial and its opposite is **zero**.
15. **Subtracting** a polynomial is equivalent to **adding its opposite**.
16. Any polynomial can be used as a function rule to write a set of number pairs that is a function.

Vocabulary and Spelling

polynomial (p. 345)
monomial (p. 345)
binomial (p. 345)
trinomial (p. 345)
degree of a polynomial (p. 346)
commutative property (p. 352)

associative property (p. 353)
additive identity (p. 356)
opposite (p. 356)
additive inverse (p. 356)
function rules (p. 366)

Chapter Test

1. Indicate whether each polynomial is a monomial, a binomial, or a trinomial.
    a. 2
    b. $2 - a$
    c. x^2
    d. $2ab$
    e. $2x^3 - 3x + 1$

2. Write each polynomial in standard form.
   a. $x^2 - 5 + x$ b. $25 - x^2$ c. $2ab^2 - 3a^2b + 5b^3$
3. For each of the following, add or subtract as indicated.

   a. Add:

   $6x^2 - 5x + 7$
   $-x^2 + 4x - 3$

   b. Subtract:

   $3x^2 - 5x + 3$
   $-x^2 + 3x + 1$

   c. Subtract:

   $2ab - b^2$
   $5ab + b^2$

4. Find the value of each polynomial when $x = 3$ and $y = 1$.
   a. $2x^2 - 5xy + y^2$ b. $25 - 6x - x^2$
5. Simplify each of the following by removing parentheses and brackets and combining similar terms.
   a. $(15 - 2x) + (x - 7)$ c. $[x - (2 - 3x)] + (x - 3)$
   b. $(2a - 3b) - (3a - 2b)$ d. $(x^2 - 5x - 3) - (2x^2 - x + 5)$
6. Write each of the following as a polynomial in standard form.
   a. The sum of the lengths of legs $\overline{AB}$ and $\overline{BC}$.
   b. The difference between the sum of the lengths of the legs and the length of the hypotenuse.

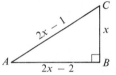

7. Solve each equation and check your solution.
   a. $t - (14 - 3t) = 2$ c. $12 - (x - 5) = 11$
   b. $(2t - 3) = 12 - (t - 3)$ d. $(6 + k^2) = k^2 - 2k + 5$
8. Find the value of $f(x)$ for the given replacement for x.
   a. $f(x) = 2x^3 - 3x + 1$; 2   b. $f(x) = 12 - 3x - x^2$; 3
9. Use the given function equation and replacement set to write the number pairs that represent the function.
   a. $f(k) = 4k^2 - 12k + 9$ $\{-2, -1, 0, 1, 2\}$
   b. $f(t) = 2t - t^2$ $\{-\frac{1}{2}, 0, \frac{1}{2}, 1, 2\}$

Chapter Review

12–1 Introduction to the System of Polynomials

1. Give an example of a binomial.
2. Give an example of a monomial.

Write each polynomial in standard form.

3. $5 - 2x$ 4. $10x^2 - 5 + x$

5. $2x^2 - 5x^4 + 3x$
6. $2 - y^3$
7. $3a^2b - 2ab^2 + b^3 - a^3$
8. $b^2 + 3ab - a^2$

12–2 Addition of Polynomials

Write each of the following as a polynomial in standard form.

9. $(2x^2 - 5) + (x^2 - x + 3)$
10. $(2 - 3x) + (7x - 10)$
11. $(1 - x^2) + (x^2 - 1)$

Find the sum of the polynomials in each question.

12. $7x - 5$
    $2x + 3$

13. $\frac{3}{4}x - 9$
    $\frac{3}{8}x + 10$

14. $6t^2 - 3t + 5$
    $-t^2 - 4t - 7$

15. $3t^2 - 0t + 5$
    $-t^2 + 3t - 2$

16. $4y^2 - y - 6$
    $3y^2 + 6y$

12–3 Polynomials and Addition Properties

Match each expression in Column 1 with an equivalent expression from Column 2.

COLUMN 1

17. $(8 - x) + (3 + 2x)$
18. $(8 + x) + 3x$
19. $(8 + x) - 3x$
20. $3x + (8 - x)$
21. $(8 + x) + (2x + 3)$

COLUMN 2

A. $(8 - x) + 3x$
B. $(x + 8) - 3x$
C. $(3 + 2x) + (8 - x)$
D. $(x + 8) + (2x + 3)$
E. $8 + (x + 3x)$

Find the value of each polynomial when $x = 2$ and $y = 3$.

22. $2x^2 - 3x + 7$
23. $x^2 - 3xy + 2y^2$
24. $y^3 - 5y + 8$

12–4 Polynomials, Additive Identity, and Additive Inverse

Give the opposite of each polynomial.

25. $2 - 5x$
26. $3x^2 - 5x$
27. $-x^2 + 5x + 12$

Write each polynomial in standard form.

28. $-(2x^3 - 2x + 5)$
29. $-(-2 + 3x - 5x^2)$
30. $-(28 - 3x)$

Addition and Subtraction of Polynomials

Write each of the following as a polynomial in standard form.

31. $(10t^2 - 5t + 6) + [-(10t^2 - 5t + 6)]$
32. $(x - y) + [-(x + y)]$
33. $(5a^2 + 3a + 3) + [-(3 - 3a + 5a^2)]$

12–5 Subtraction of Polynomials

For each of the following, subtract the second polynomial from the first.

34. $3k - 5$
$k + 3$

35. $10 - x$
$-5 + x$

36. $2t^2 - 5t + 6$
$3t^2 + t - 7$

Simplify each of the following by removing parentheses and brackets and combining similar terms. Write each answer in standard form.

37. $(5x - 4) - (3x + 7)$
38. $(2a - b) - (a - 2b)$
39. $(2x - 7) - [(3 - x) - 5]$
40. $(r^2 - 3r + 8) - (r^2 + 7)$
41. $(a - 2b) - (a + 2b)$
42. $[x - (3 - x)] - (2x - 3)$

12–6 Problems Using Polynomials

Write the perimeter of each figure as a polynomial in standard form.

43.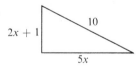

44.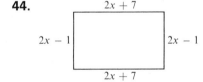

12–7 Equations and Polynomials

Solve each equation and check your solution.

45. $t + (15 + t) = 12$
46. $3x - (x - 3) = 24$
47. $x - (3x - 5) + (3x + 5) = 4$
48. $12 - (3 - t) = 4$
49. $18 - (12 + 2x) = 0$
50. $(n^2 - n - 1) - n^2 = -1$

12–8 Functions and Polynomials

Find the value of $f(x)$ for the given replacement for x.

51. $f(x) = 3x^2 - 5x + 2;\ 1$
52. $f(x) = 7 - 3x - 2x^2;\ 2$
53. $f(x) = 5 - 3x^2;\ -1$
54. $f(x) = 2x - 4x^2;\ \frac{1}{2}$

Use the given function equation and replacement set to write the number pairs that represent the function.

55. $f(x) = x^2 - 3x - 1$; $\{-2, -1, 0, 1, 2\}$
56. $f(t) = 25 - 10t + t^2$; $\{-3, -1, 1, 3\}$
57. $f(n) = n^3 + 3n^2 + 3n + 1$; $\{-3, 0, 1, 2\}$
58. $f(y) = 1 - y^3$; $\{-2, -1, 0, 1, 2\}$

Review of Skills

Match each expressions in Column 1 with a corresponding expression from Column 2.

COLUMN 1

1. $2 \cdot 2 \cdot 2 \cdot 2$
2. $3 \cdot 3 \cdot 5 \cdot 5 \cdot 5$
3. $x \cdot x \cdot x \cdot y$
4. $a^2 \cdot a^3$
5. $2 + 2 + 2 + 2$
6. $3 + 3 + 5 + 5 + 5$
7. $x + x + x + y$
8. $(2h)(2h)^4$

COLUMN 2

A. $3^2 \cdot 5^3$
B. $4 \cdot 2$
C. a^5
D. $2 \cdot 3 + 3 \cdot 5$
E. $x^3 \cdot y$
F. $3x + y$
G. $(2h)^5$
H. 2^4

Use exponents to write a simpler form for each expression.

SAMPLE: $2^4 \cdot 2^2$ **Solution:** $2^4 \cdot 2^2 = 2^6$

9. $r^2 \cdot r^1$
10. $k \cdot k^3$
11. $t^2 \cdot t^5$
12. $(2t)(2t)(2t)$

Simplify each expression.

13. 2^3
14. $(3)(-5)$
15. $(-2)(-x)$
16. $2 \cdot 3$
17. $(-2)(-2)(-2)$
18. $(2)^3$
19. 3^2
20. $3 \cdot 2$
21. $(2^3)(2^3)$

Use the distributive property to rewrite each expression.

SAMPLE: $a(b + c)$ **Solution:** $a(b + c) = ab + ac$

22. $-3(h - k)$
23. $(y - 2k)(-2)$
24. $x(2 - y)$
25. $(x - y + z)(5)$
26. $(s + t)(k)$
27. $t(m + n - p)$

Addition and Subtraction of Polynomials

For each statement, tell which of the following properties is illustrated: Commutative Property; Associative Property; Multiplicative Property of One; Property of Reciprocals.

28. $2 \cdot 3 = 3 \cdot 2$
29. $3 \cdot (5 \cdot 7) = (3 \cdot 5) \cdot 7$
30. $5 \cdot \tfrac{1}{5} = 1$
31. $xy = yx$
32. $5 \cdot 1 = 5$
33. $1 \cdot y = y$

Simplify each of the following.

34. $\dfrac{5 \cdot 5 \cdot 5 \cdot 5 \cdot 5}{5 \cdot 5 \cdot 5}$
36. $\dfrac{5^5}{5^3}$
38. $\dfrac{3^5}{3^3}$

35. $\dfrac{2 \cdot 2 \cdot 2 \cdot 2 \cdot 2 \cdot 2 \cdot 2}{2 \cdot 2 \cdot 2 \cdot 2}$
37. $\dfrac{2^7}{2^4}$
39. $\dfrac{3^7}{3^4}$

Determine the value of *n* that makes the statement true.

40. $4n = 12$
42. $4 \cdot n = 8$
44. $2^2 \cdot n = 2^3$
41. $3^4 \cdot n = 3^5$
43. $2^4 \cdot n = 2^6$

Simplify each of the following.

45. $\dfrac{5 \cdot 4 + 5 \cdot 7}{5}$
47. $\dfrac{14 \cdot 13 - 14 \cdot 8}{7}$

46. $\dfrac{24 \cdot 6 - 30 \cdot 5}{6}$
48. $\dfrac{6x - 4}{2}$

■ ■

CHECK POINT FOR EXPERTS

Absolute Value

Study the number line pictured below. How many units from the zero point is the point corresponding to the number 4? How many units from the zero point is the point corresponding to the number ⁻4? Do you see that each point is **four** units from **0**? The distance from 0 of a point on the number line is called the **absolute value** of the number corresponding to the point.

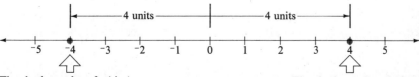

The absolute value of ⁻4 is 4.   The absolute value of 4 is 4.

The symbol $|x|$ stands for the absolute value of the directed number x. The symbol is read "the absolute value of x." Study these examples.

$$|10| = 10; \quad |\tfrac{2}{3}| = \tfrac{2}{3}; \quad |35| = 35$$
$$|{}^-10| = 10; \quad |{}^-\tfrac{2}{3}| = \tfrac{2}{3}; \quad |{}^-35| = 35$$

We define absolute value for directed numbers as follows:

When $r > 0$, $|r| = r$;
when $r = 0$, $|r| = r = 0$;
when $r < 0$, $|r| = -r$.

Notice the meaning of the statement "When $r < 0$, $|r| = -r$." If r is a **negative** number, its absolute value is its additive inverse, a **positive** number. For example, $|{}^-2|$ is the additive inverse of ${}^-2$, or 2. Thus the **absolute value** of *any* directed number is *always* a **positive** number or **zero**.

Questions

What is the simplest name for each of the following?

SAMPLE 1: $|12|$ Solution: 12
SAMPLE 2: $|{}^-5|$ Solution: 5

1. $|4|$
2. $|9|$
3. $|{}^-8|$
4. $|{}^-32|$
5. $|{}^-16|$
6. $|{}^-0.25|$
7. $-|{}^-\tfrac{1}{2}| + 0$
8. $|\tfrac{3}{5}| + 1$
9. $|0| + |{}^-2|$

Which sentences are true and which are false?

10. $|{}^-2| = |2|$
11. $|4| > |{}^-2|$
12. $|{}^-6| \neq |6|$
13. $|{}^-8| < |{}^-2|$
14. $|4| - |{}^-2| = |2|$
15. $|{}^-8| + |8| = 0$

What is the solution set for each sentence? The replacement set is the set of directed numbers.

SAMPLE: $|t| = 5$ Solution: $\{{}^-5, 5\}$

16. $|m| = 6$
17. $|k| = 25$
18. $|r| = 0$
19. $|x| + 4 = 8$
20. $|a| + 2 = 3$
21. $-|n| = {}^-9$
22. $7 + |t| = 12$
23. $2 + (-|y|) = {}^-5$
24. $-|k| + 3 = 0$

What is the value of each expression?

25. $|3 + 6|$
26. $-|{}^-2 + {}^-5|$
27. $|4 - 3|$
28. $|2 - 10|$
29. $-|2 + 3| + |1|$
30. $|{}^-3 + 9| + 5$

THE HUMAN ELEMENT

The Egyptians

The ancient civilization of Egypt, dating back as far as 5000 years, left behind records that included descriptions of work in mathematics.

The planning and construction of the pyramids, built as tombs for the Pharaohs, or kings, required many applications of mathematics. The largest of the pyramids, called the Great Pyramid, is over 480 feet high and has a base that is very nearly an exact square, 755 feet on each side. The faces of this pyramid are in the shape of isosceles triangles. The Egyptians quarried nearly $2\frac{1}{3}$ million stone blocks, weighing about $2\frac{1}{2}$ tons each, in order to construct the Great Pyramid. Each block had to be transported and placed in position. The pyramids were built without mortar, so the blocks had to fit together perfectly.

The pyramid planners had the pyramids constructed so they faced almost exactly in a north-south, east-west direction. Since the compass had not yet been invented it is interesting to speculate on how the direc-

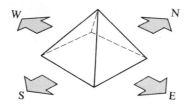

tions were determined. It seems likely that this was accomplished by solar observation.

The records left behind by the Egyptians show that they knew a good deal about geometry. They probably gained much of this knowledge by solving the practical problems presented by their agricultural system. Farmers tilled the best land on the banks of the Nile River. Each spring when the Nile flooded over its banks, it erased the markers identifying parcels of land. It was necessary to re-survey the land each spring, and to allot to each farmer his proper share. The scribes who did the surveying eventually picked up some knowledge of geometry, which helped them to do the job better and faster.

For weaving cloth by hand . . .

A modern weave room . . .

Multiplication and Division of Polynomials

The textile industry has benefitted from modern research and development. Before power looms were developed, cloth was woven by hand, on looms like the one from the Smithsonian Institution. Notice the shuttles resting on the finished part of the work, at the front. There are three of them, which indicates that the pattern calls for three colors. The picture of the weave room of a modern textile plant typifies the progress that has been made. These looms weave fabric at a high speed. They are of the "shuttleless" variety, which makes for greater weaving speeds and increased economies in the cost of both weave room labor and maintenance.

Multiplication

13–1 Polynomials and Exponents

The previous discussion of exponents in this book led to a definition of exponents that implied that:

$$5^4 = 5 \cdot 5 \cdot 5 \cdot 5 \quad \text{(5 is a factor 4 times)}$$
$$k^3 = k \cdot k \cdot k \quad \text{(k is a factor 3 times)}$$

We have also worked with expressions like the following:

EXAMPLE 1. $x^2 \cdot x^3 = (x \cdot x) \cdot (x \cdot x \cdot x)$
$= x^5$ (x is a factor 5 times)

EXAMPLE 2. $a^2 \cdot a \cdot a^3 = (a \cdot a) \cdot (a) \cdot (a \cdot a \cdot a)$
$= a^6$ (a is a factor 6 times)

EXAMPLE 3. $(xy)(xy)^3 = (xy) \cdot (xy \cdot xy \cdot xy)$
$= (xy)^4$ (xy is a factor 4 times)

From the examples above it appears that when powers of the same base are multiplied, the product can be written as the base raised to a power equal to the sum of the exponents.

For any directed number *a*, and positive integers *p* and *q*,

$$a^p \cdot a^q = a^{p+q}.$$

Can you simplify the expression $r^2 \cdot s^4$? Notice that two **different** variables appear as bases, so our rule of exponents does not apply in this case.

$$r^2 \cdot s^4 = (r \cdot r) \cdot (s \cdot s \cdot s \cdot s)$$
$$= r^2 s^4$$

We cannot further simplify this expression.

ORAL EXERCISES

Use the rule of exponents for multiplication to find a simpler name for each of the following.

1. $t \cdot t \cdot t$
2. $m \cdot m \cdot m \cdot m$
3. $\frac{1}{2} \cdot \frac{1}{2} \cdot \frac{1}{2}$
4. $(xy)(xy)$
5. $(ab)(ab)(ab)$
6. $9 \cdot b \cdot b \cdot b \cdot b$
7. $(x \cdot x) \cdot (y \cdot y \cdot y)$
8. $m \cdot m \cdot m \cdot n \cdot n \cdot n$
9. $7 \cdot r \cdot r \cdot r \cdot s \cdot s$
10. $(a \cdot a)(b \cdot b)(c \cdot c)$
11. $a \cdot b \cdot a \cdot b \cdot a$
12. $3 \cdot m \cdot m \cdot n \cdot n \cdot n$
13. $\frac{a}{b} \cdot \frac{a}{b} \cdot \frac{a}{b}$

Answer each question "yes" or "no."

14. Is 2^4 the same as $2 \cdot 4$?
15. Is 3^2 the same as 2^3?
16. Is 1^5 equal to 1^{10}?
17. Is a^3 always equal to $3a$?
18. Is m^2 always equal to $m \cdot m$?
19. Is x^3 always equal to $x + x + x$?

WRITTEN EXERCISES

Give the value of each expression for $x = 2$ and $y = 3$.

A 1. x^3
2. $12x^2$
3. $(x \cdot x)(y \cdot y)$

Multiplication and Division of Polynomials

4. y^2
5. $x \cdot x \cdot x \cdot x$
6. $y \cdot y \cdot y$
7. $5 \cdot x \cdot x \cdot x$
8. $(xy)^2$
9. $(xy)(xy)$
10. $10y^3$
11. $7x^3$
12. $x^2 y^3$

13. the third power of y
14. twice the cube of x
15. twice the square of y
16. the fourth power of x
17. the square of xy
18. ten times the third power of y

Copy and complete each of the following.

SAMPLE: $5^2 \cdot 5^4 = 5^{2+4}$
$= ?$

Solution: $5^2 \cdot 5^4 = 5^{2+4}$
$= 5^6$

19. $10^3 \cdot 10^5 = 10^{3+5}$
$= ?$
20. $a^4 \cdot a = a^{4+1}$
$= ?$
21. $m \cdot m \cdot m^2 = m^{1+1+2}$
$= ?$
22. $(ab)^2 \cdot (ab)^2 = (ab)^{2+2}$
$= ?$

23. $(r)(r^6) = r^{1+6}$
$= ?$
24. $k^5 \cdot k^4 = k^{5+4}$
$= ?$
25. $-3(t)(t^2) = -3t^{1+2}$
$= ?$
26. $(x^2)(x^3)(-1) = -x^{2+3}$
$= ?$

Simplify each of the following by using exponential notation.

SAMPLE 1: $(x - y)(x - y)$ Solution: $(x - y)^2$
SAMPLE 2: $(4 \cdot 7)(z \cdot z)$ Solution: $28z^2$

27. $(3 \cdot 7)(k \cdot k \cdot k)$
28. $(5 \cdot 11)(t \cdot t)$
29. $(a + b)(a + b)$
30. $(n + 1)(n + 1)$
31. $(-3 \cdot 5)(x \cdot x \cdot x)$
32. $(-1)(k \cdot k)$
33. $(-2 \cdot 7)(w \cdot w)$
34. $(5)(3)(x \cdot x \cdot x)$
35. $(-5)(-7)(t \cdot t)$

36. $(3)(-10)(a \cdot a \cdot a)$
37. $(4 \cdot 5)(x^2 \cdot x)$
38. $(3)(x)(-2)(x)$
39. $(11)(3)(a^2)(a^3)$
40. $(7 \cdot 2)(m^2 \cdot m^3)$
41. $(-2)(5)(b^2 \cdot b)$
42. $(6)(-7)(t^3)(t^2)$
43. $(x)(x)(x)(-x)$
44. $(-1)(x \cdot x \cdot x \cdot x \cdot x)$

Find the correct replacement for the variable x in each equation.

SAMPLE: $2^5 = 2^2 \cdot 2^x$ Solution: $2^5 = 2^2 \cdot 2^3$

45. $5^2 \cdot 5^4 = 5^x$
46. $(n \cdot n) \cdot n = n^x$
47. $a^6 = a^4 \cdot a^x$
48. $k^7 = k^2 \cdot k^2 \cdot k^x$

49. $m^2 \cdot m^x \cdot m^5 = m^{10}$
50. $(s + t)^3 = (s + t)(s + t)^x$
51. $(a - b)^5 = (a - b)^2(a - b)^x$
52. $(ab)^5 = (ab)(ab)^x$

13–2 Multiplication of Monomials

You will probably recall that earlier we simplified indicated products of monomials. For example, $(5a)(6b) = 30ab$ can be justified by the commutative and associative properties of multiplication. Now let us consider how to simplify such products as $(5x^2y^3)(3xy^4)$ and $(2ab^3)(-4a^2b^2)$. Again we will use the commutative and associative properties, as well as the rule of exponents for multiplication.

EXAMPLE 1.
$$\begin{aligned}(5x^2y^3)(3xy^4) &= (5 \cdot 3)(x^2 \cdot x)(y^3 \cdot y^4) \\ &= (15)(x^{2+1})(y^{3+4}) \\ &= 15 \cdot x^3 \cdot y^7 \\ &= 15x^3y^7\end{aligned}$$

EXAMPLE 2.
$$\begin{aligned}(2ab^3)(-4a^2b^2) &= (2 \cdot -4)(a \cdot a^2)(b^3 \cdot b^2) \\ &= (-8)(a^{1+2})(b^{3+2}) \\ &= -8 \cdot a^3 \cdot b^5 \\ &= -8a^3b^5\end{aligned}$$

Do you see that exponents have been added only when the bases of the powers are the same?

ORAL EXERCISES

State each of the following expressions in a simpler form.

SAMPLE 1: $(-6)(2x)$ *What you say:* $-12x$

SAMPLE 2: $(-5)(-3)(a^2 \cdot a)(b^2)$ *What you say:* $15a^3b^2$

1. $(3x)(-5)$
2. $(-t)(t)$
3. $(x)(-x^2)$
4. $(-m^3)(-m^2)$
5. $(5x^3)(-x)$
6. $(-3)(m^2 \cdot m^2)(n^3)$
7. $(3 \cdot 4)(a^2 \cdot a^3)(b \cdot b^2)$
8. $(5 \cdot 8)(t^4 \cdot t^3)(w^2 \cdot w^5)$
9. $(-1)(b \cdot b^3)(d^2 \cdot d^4)$
10. $(3 \cdot -5)(x \cdot x^2)(y^3 \cdot y^4)$
11. $(-2 \cdot -3)(a^3 \cdot a^2)$
12. $(-5)(r \cdot r^2)(s \cdot s)$

WRITTEN EXERCISES

Copy and complete each of the following.

SAMPLE: $(-3ab)(5ab^2)$ *Solution:* $(-3ab)(5ab^2)$
$= (-3 \cdot 5)(a \cdot a)(b \cdot b^2)$ $= (-3 \cdot 5)(a \cdot a)(b \cdot b^2)$
$= (\ ?\ )(\ ?\ )(\ ?\ )$ $= (-15)(a^{1+1})(b^{1+2})$
$= ?$ $= -15a^2b^3$

Multiplication and Division of Polynomials 383

A 1. $(6xy^2)(8x^2y^2)$
$= (6 \cdot 8)(x \cdot x^2)(y^2 \cdot y^2)$
$= (\ ? \ )(\ ? \ )(\ ? \ )$
$= ?$

3. $(-a^4b^2)(4a^3b^3)$
$= (-1 \cdot 4)(a^4 \cdot a^3)(b^2 \cdot b^3)$
$= (\ ? \ )(\ ? \ )(\ ? \ )$
$= ?$

2. $(-2r^2s^3)(-8rs^2)$
$= (-2 \cdot -8)(r^2 \cdot r)(s^3 \cdot s^2)$
$= (\ ? \ )(\ ? \ )(\ ? \ )$
$= ?$

Simplify each indicated product.

4. $(3rs)(5rs^2)$
5. $(6xy)(8x^2y)$
6. $(-3b)(-3b)(3b)$
7. $(-2ab)(5ab^2)b^3$
8. $(-4a)(3a)(5ab)$
9. $(10r^2)(-8r^3)$
10. $(-5x)(-6y)(z)$
11. $(6x)(2xy)(-3y)$
12. $a^2(-3ab)(2b)$
13. $-x(-3xy)(-y)$
14. $5t(rt)(-t^2)$
15. $(-4ab)(2bc)(-2b)$
16. $(-7a)(-b)(-3c)$
17. $(\tfrac{1}{2}ab)(\tfrac{1}{4}ab)$
18. $(rs)(-r)(-rs)$

B 19. $-rt(r^2t)(-rt)$
20. $(-3y)(4x^3y^2)(-y^4x^2)$
21. $-5a(2a^2b^5)(-ab^2)$
22. $(0.2r^3s^5)(0.4rs^4)$
23. $(3^2xy^5)(-10x^3y^4)$
24. $(-12ab)(-7a^2b^2)(-2c^3)$
25. $(-a)^2(a^2b)b^2$
26. $(2m)^3(mn)(3m)^2$

Simplify each of the following. (Find the indicated products, then combine similar terms.)

SAMPLE: $(9a^2)(2b)(-b) + (4ab)(-2ab)$

Solution: $(9a^2)(2b)(-b) + (4ab)(-2ab)$
$= (-18a^2b^2) + (-8a^2b^2) = -26a^2b^2$

C 27. $(5p^2)(-3q) + (7pq)(p)$
28. $(-3rs)(4r^2s) - (6s^2)(-r^3)$
29. $(9x^2y^3)(-2xy) + (xy^2)(x^2y^2)$
30. $(-qp^2)(5p^2)(-3q) - (7qp^3)(4pq)$
31. $(-6s^2)(2rt^2)(-r^2) + (-3rst)(4r^2st)$
32. $(-v^2wz^3)(-2vw^2z) + (-3z^2)(-6v^3w^2)(wz^2)$

13–3 Power of a Product

Do you think the expressions $5x^3$ and $(5x)^3$ have the same meaning or different meanings? Check your idea against this explanation.

$5x^3 = 5 \cdot x^3$
$= 5 \cdot x \cdot x \cdot x$

$(5x)^3 = 5x \cdot 5x \cdot 5x$
$= 5^3 \cdot x^3$
$= 125x^3$

Thus it appears that $5x^3$ and $(5x)^3$ are not the same.

We can say that $(ab)^m$, where m is a positive integer, has this meaning:

$$(ab)^m = ab \cdot ab \cdot ab \ldots ab \qquad (ab \text{ is a factor } m \text{ times.})$$

Also, $(ab)^m = (a \cdot a \cdot a \ldots a)(b \cdot b \cdot b \ldots b)$
$\qquad\qquad\qquad (a \text{ is a factor } m \text{ times and } b \text{ is a factor } m \text{ times.})$

Thus we state the following general rule of exponents for a power of a product:

For all directed numbers **a** and **b**, and every positive integer **p**,

$$(ab)^p = a^p \cdot b^p.$$

Now let us consider the meaning of expressions like $(2^3)^2$ and $(x^2)^4$. Look for a common pattern.

$$(2^3)^2 = 2^3 \cdot 2^3 \qquad\qquad (x^2)^4 = x^2 \cdot x^2 \cdot x^2 \cdot x^2$$
$$ = 2^{3+3} \qquad\qquad = x^{2+2+2+2}$$
$$ = 2^6 = 2^{3 \cdot 2} \qquad = x^8 = x^{2 \cdot 4}$$

We can summarize these ideas as a general rule of exponents for a power of a power:

For every directed number **a** and all positive integers **p** and **q**,

$$(a^p)^q = a^{pq}.$$

ORAL EXERCISES

Answer "yes" or "no" to each question.

1. Is $(5x)^3$ the same as $5x \cdot 5x \cdot 5x$?
2. Is $(t^4)^3$ the same as $t^4 \cdot t^4 \cdot t^4$?
3. Is $(2n^3)^3$ the same as $8 \cdot n^{3+3+3}$?
4. Is $(3a^3)^2$ the same as $(3)^2(a^3)^2$?
5. Is $(2^3)^2$ the same as $(2^2)^3$?
6. Is $4 \cdot k^{12}$ the same as $(2k^4)^3$?
7. Is $(-2^3)^2$ the same as $(2^3)^2$?
8. Is $(-2^2)^3$ the same as $(2^2)^3$?

Simplify each of the following:

SAMPLE: $(5ab)^2 \qquad\qquad$ *What you say:* $25a^2b^2$

9. $(5xy)^3$
10. $(10ab)^4$
11. $(-10st)^2$
12. $(-10ab)^3$
13. $(-x^2y)^3$
14. $(-4y^5)^2$
15. $(k^3)^3$
16. $(-m^3)^3$

Multiplication and Division of Polynomials 385

17. $(2abc)^5$ 19. $(12x^2y^2)^2$ 21. $(-ab)^5$ 23. $(4m^2n)^1$
18. $(-3xy)^2$ 20. $(-3ab^2)^4$ 22. $(-xyz)^2$ 24. $(-cd^2)^4$

WRITTEN EXERCISES

Find the value of each expression. Use the given values for variables.

SAMPLE 1: $(2^2)^3$ Solution: $(2^2)^3 = 4^3 = 64$
SAMPLE 2: $(2n^3)^2$; $n = 2$ Solution: $(2 \cdot 2^3)^2 = (2 \cdot 8)^2$
$= 16^2 = 256$

A
1. $(5^2)^2$
2. $(10^3)^2$
3. $(-10^3)^2$
4. $(3x^2)^2$; $x = 3$
5. $(t^4)^4$; $t = 1$
6. $(-2m^2)^3$; $m = 2$
7. $(2a)^4$; $a = 5$
8. $(-7y^3)^2$; $y = 10$

9. $(24t^3)^2$; $t = \frac{1}{2}$
10. $(3m^2)^3$; $m = -2$
11. $(-18b^5)^3$; $b = 0$
12. $(-cd^2)^5$; $c = 4$, $d = \frac{1}{2}$
13. $(0.3x^2)^2$; $x = 10$
14. $(-0.05t^3)^2$; $t = 10$
15. $(-16r^2s)^2$; $r = \frac{1}{4}$, $s = 5$

Find the square of each expression.

SAMPLE: $-5x^2y$ Solution: $(-5x^2y)^2 = 25x^4y^2$

16. $3xy^2$
17. $-6a^2b$
18. $2ab^2$

19. $-0.3rs^2$
20. $-0.2vw^3$
21. $6b^3c^2$

22. $-5r^2s^3t$
23. $-8xyz$
24. $0.5abc^3$

Simplify each expression.
SAMPLE: $(4xy)^2 + (5x)(-4xy)(y)$ Solution: $(4xy)^2 + (5x)(-4xy)(y)$
$= 16x^2y^2 + (-20x^2y^2)$
$= -4x^2y^2$

B
25. $(2ab)^3 + (3a)^2(-5ab^2)(b)$
26. $(3x)^2(-2xy)^2 + (2x)^2(xy)^2(y)^2$
27. $(-2c)(-3cd)^3 + (3c)^2(cd)^2(-6d)$
28. $(a^2)a + (2a^2)(-2a) + (2a)^3$
29. $(2x^2)(-5x^3) + (-4x)(2x^2)^2 + 10x^5$
30. $(mx)^5(2mx^2) + (m^2x)^3 + (2x^2)^2$

C
31. $(-zt)^3(0.2z^2t) + (t^2z)^2(0.3z^3) + (t^2z)(t^2z)^2$
32. $(-mk)^5(0.3mk^2) + (m^2k)^3(0.2k^2)^2 + (-m^3k)^2(-k)^4k$
33. $(2b)^2(-25b^4) + (-4b)(-b)^2 + (-10b^3)^2$

13–4 Multiplying a Polynomial by a Binomial

You will probably recall that in our earlier work we made use of the distributive property to complete indicated multiplications like those in Examples 1 and 2. This can be done by arranging the work either in vertical form or horizontal form, as shown.

(vertical)   (horizontal)

EXAMPLE 1.
$$\begin{array}{r} 56 \\ \times 8 \\ \hline 48 \\ 400 \\ \hline 448 \end{array}$$
$$\begin{aligned} 8 \cdot 56 &= 8(50 + 6) \\ &= (8 \cdot 50) + (8 \cdot 6) \\ &= 400 + 48 \\ &= 448 \end{aligned}$$

EXAMPLE 2.
$$\begin{array}{r} 5x + 3 \\ 7 \\ \hline 35x + 21 \end{array}$$
$$\begin{aligned} 7(5x + 3) &= (7 \cdot 5x) + (7 \cdot 3) \\ &= 35x + 21 \end{aligned}$$

By applying the rules of exponents we have developed for multiplication we can multiply any polynomial by a monomial. Again either the vertical form or horizontal form may be used, as shown in Examples 3 and 4.

EXAMPLE 3.
$$\begin{array}{r} 3x + 4 \\ 6x \\ \hline 18x^2 + 24x \end{array}$$
$$\begin{aligned} 6x(3x + 4) &= (6x \cdot 3x) + (6x \cdot 4) \\ &= 18x^2 + 24x \end{aligned}$$

EXAMPLE 4.
$$\begin{array}{r} 3n^2 - 2n + 5 \\ 4n \\ \hline 12n^3 - 8n^2 + 20n \end{array}$$

$$\begin{aligned} 4n(3n^2 - 2n + 5) &= (4 \cdot 3n^2) - (4n \cdot 2n) + (4n \cdot 5) \\ &= 12n^3 - 8n^2 + 20n \end{aligned}$$

ORAL EXERCISES

Give the product for each indicated multiplication.

SAMPLE: $3(2 - x)$ *What you say:* $6 - 3x$

1. $5(x + 4)$
2. $7(4a - 3b)$
3. $-6(x^2 - 3)$
4. $(5b - 2c)(-c)$
5. $-5a^5(-3 + a)$
6. $-1(8 + 2b)$

7. $(a - 2)8$
8. $(2x - 1)10$
9. $7x(x + 3)$
10. $(2y - 4)8y$
11. $st(5s + 7)$
12. $6m(3m^2 - 4)$
13. $4qr(q + 3r + 5)$
14. $a^2(a^2 + 2a - 3)$
15. $-3x^4(-2 + x)$
16. $-4b(5b^2 - 3bc)$
17. $(5r - s)(-1)$
18. $-3a(-2a^2 - 4a)$
19. $-1(3 + 6k - 10k^2)$
20. $x^2(2 + x - x^2)$
21. $-3x(-3x^3 - 1)$
22. $(-5 - 3r)(-7r)$
23. $(-4w - 6)(\frac{1}{2}w)$
24. $2m(-m^2 + 3m + 2)$

WRITTEN EXERCISES

Find each product.

A
1. $-6(x^2 - 2x + 1)$
2. $-5(2a^2 - 3a + 7)$
3. $a(a^2 - 2ac + c^2)$
4. $x(x^2 - 2xy + y^2)$
5. $b^2(-a - b - c)$
6. $r^2(r - s - t)$
7. $-x^2(5x^3 - x^2 + 2x + 3)$
8. $(7 - 3y - y^3)(-y^2)$
9. $(2v^2 - 3v - 4)(-7v^3)$
10. $-3x^3(x^2 - 6x + 9)$

11. $3x^2y(5 - 2xy^4 + 3x^3y^2 - y^5)$
12. $5rs^2(8 + 6r^5s - 2r^3s^4 - 7rs^5)$
13. $-2a^3b^2(a^4b - a^3b^2 + 2a^2b^4 - b^5)$
14. $-3c^4d^2(cd^5 + 9c^3d^4 - c^2d^4 - 2d^5)$
15. $7a^2c^2(-3a^2 + 4ac - 5c^2)$

16. $-3x^3 + 9x - 1$
    $\underline{-5x}$
17. $-y^2 + 2y + 5$
    $\underline{-4y}$
18. $-b^2 + 3c + 2d$
    $\underline{b^3}$
19. $-m^2 - 3n^2 + 8$
    $\underline{-4.5m}$
20. $4x + 6y + z$
    $\underline{-2.5x^2}$
21. $12a^2 - 8ab - 11$
    $\underline{-ab}$

Simplify each of the following expressions.

22. $5a(a - 2b) + 4b(3a + b)$
23. $3r(r + s) + 5r(r + s)$
24. $3c(2c + d) + c(3c - 2d)$
25. $7n(m - n) + m(-m - n)$
26. $3p(q - p) + p(q - p)$
27. $r(r + 3t) + 3(-rt)$
28. $4x(x + 5y) - 2x(x + 5y)$
29. $2st(s + 3t) - 4s(2st - 3t^2)$
30. $2w^2(3x - 4w) + (wx + w^2)(7w)$
31. $y(-3y - 2z) - y(y + 3z)$
32. $ab(a^2 - ab^2) + (3b)(2a^3 + a^2b^2)$
33. $r^2s(r - s^2) + (3r^2s)(s^2 - 2r)$

Solve each equation.

B
34. $3n + 5(n - 3) = 9$
35. $8a + 6(-a - 2) = 16$
36. $1 = (5m + 3) - 3(m - 2)$
37. $(8b + 7) - 5(b - 8) = 44$

38. $7t + 0.8(7 - t) = 18$
39. $-6x - 0.5(10 - x) = 17$
40. $12(x + 1) = 18 - 3(4x - 2)$
41. $7(8 - r) - 96 = -5(2 + 2r)$

C 42. $2x - 2(x + 21) = 3x - 3(7 - x)$
43. $5k - (k - 8) = 14k - 7(2k + 8)$
44. $0.2(3a - 12) - (6a - 5) = 13 + 0.5(3a - 7)$
45. $2(5 - 3n) = 3(2n) - 5$

PROBLEMS

Express each answer as a polynomial in simplest form.

A 1. What is the area of the pictured rectangle? What is its perimeter?

2. Find the perimeter of a regular hexagon if the length of one of its sides is $x^2 - 3x - 2$.

3. A rectangle is divided into four rectangles as shown below. What is the perimeter of the large rectangle? What is the area of one small rectangle? What is the area of the large rectangle?

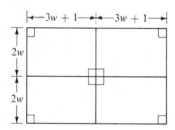

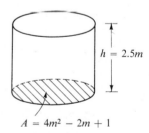

4. The volume of a cylinder is given by formula $V = Ah$ (Volume = area of base × height). Find the volume of the cylinder in the illustration.

B 5. The total volume of the figure at the right is given by the formula $V = \frac{1}{2} \times$ height $\times$ area of the base. Find the area of the base. What is the total volume? What is the volume of part I of the figure? of part II?

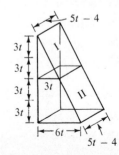

6. Find the area of each of the four small rectangles in this figure. What is the total area? What is the perimeter of the large rectangle?

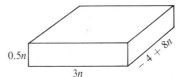

7. Suppose that the average speed of a supersonic aircraft is given by the expression $m^3 - 2m^2 + 10m$. What distance does the aircraft travel if it maintains that average speed for a period of time represented by $5m^2$? (Use: distance = rate × time.)

8. Find the volume of the pictured figure if each of its faces is a rectangular region. (Use: Volume = length × width × height.)

C 9. The amount of water in the cylinder-shaped tank tipped as shown is given by $1.9r^3 + 6.5r^2 - 0.04$. What is the total volume of the tank?

10. A rectangular piece of sheet metal has six circular holes punched in it. The area of each circle is $0.4c^2 + 2$. Find the area of the remaining piece of metal (the shaded part of the figure).

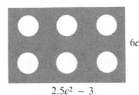

13–5 Multiplying a Polynomial by a Polynomial

You have learned from long experience how to complete such multiplications as 27 × 35, but have you understood that you are really making use of the distributive property? We can see just what happens by first writing 27 and 35 as binomials and then completing the work in the usual vertical form.

$$\begin{array}{c}27\\ \times 35\\ \hline\end{array} \text{ is the same as } \begin{array}{r}20 + 7\\ \times 30 + 5\\ \hline 100 + 35 = (20 + 7)\cdot 5\\ 600 + 210 = (20 + 7)\cdot 30\\ \hline 600 + 310 + 35 = 945\end{array}$$

Since in algebra we often arrange our work in horizontal form let us multiply 27×35 horizontally. Be sure that you understand each step.

$$27 \times 35 = (20 + 7)(30 + 5)$$
$$= [(20 + 7) \cdot 30] + [(20 + 7) \cdot 5]$$
$$= [(20 \cdot 30) + (7 \cdot 30)] + [(20 \cdot 5) + (7 \cdot 5)]$$
$$= 600 + 210 + 100 + 35$$
$$= 945$$

Now let us apply the same idea in finding the product of $(x + 5)$ and $(2x + 3)$.

$$(x + 5)(2x + 3) = [(x + 5) \cdot 2x] + [(x + 5) \cdot 3]$$
$$= [(x \cdot 2x) + (5 \cdot 2x)] + [(x \cdot 3) + (5 \cdot 3)]$$
$$= 2x^2 + 10x + 3x + 15$$
$$= 2x^2 + 13x + 15$$

Thus to find the product of two polynomials we multiply each term of one polynomial by every term of the other, and then add like terms. Although we illustrated this for a product of binomials, the method applies to finding the product of any two polynomials.

Study the following examples:

EXAMPLE 1. $(x^2 + 3x - 2)(x + 4)$

Vertical form:

$$\begin{array}{r} x^2 + 3x - 2 \\ x + 4 \\ \hline 4x^2 + 12x - 8 \\ x^3 + 3x^2 - 2x \\ \hline x^3 + 7x^2 + 10x - 8 \end{array}$$

Horizontal form:

$$(x^2 + 3x - 2)(x + 4)$$
$$= [(x^2 + 3x - 2) \cdot x] + [(x^2 + 3x - 2) \cdot 4]$$
$$= [x^3 + 3x^2 - 2x] + [4x^2 + 12x - 8]$$
$$= x^3 + 3x^2 - 2x + 4x^2 + 12x - 8$$
$$= x^3 + 7x^2 + 10x - 8$$

EXAMPLE 2. $(2n^2 - 3n + 4)(n - 2)$

Vertical form:
$$\begin{array}{r} 2n^2 - 3n + 4 \\ n - 2 \\ \hline -4n^2 + 6n - 8 \\ 2n^3 - 3n^2 + 4n \\ \hline 2n^3 - 7n^2 + 10n - 8 \end{array}$$

Horizontal form:

$(2n^2 - 3n + 4)(n - 2)$
$= [(2n^2 - 3n + 4) \cdot n] + [(2n^2 - 3n + 4) \cdot (-2)]$
$= [2n^3 - 3n^2 + 4n] + [-4n^2 + 6n - 8]$
$= 2n^3 - 3n^2 + 4n - 4n^2 + 6n - 8$
$= 2n^3 - 7n^2 + 10n - 8$

ORAL EXERCISES

For each question mark, name the replacement that will complete the sentence.

SAMPLE: $(x + 8)(5x + 1) = [(x + 8) \cdot ?] + [(x + 8) \cdot ?]$

What you say: 5x and 1

1. $(2x + 3)(x + 1) = [(2x + 3) \cdot ?] + [(2x + 3) \cdot ?]$
2. $(x + 2)(x + 5) = [(x + 2) \cdot ?] + [(x + 2) \cdot ?]$
3. $(t - 3)(t + 5) = [(t - 3) \cdot ?] + [(t - 3) \cdot ?]$
4. $(14n - 6)(2n + 8) = [(14n - 6) \cdot ?] + [(14n - 6) \cdot ?]$
5. $(2b + 7)(b - 3) = [(2b + 7) \cdot ?] + [(2b + 7) \cdot ?]$
6. $(5r - 1)(r - 3) = [(5r - 1) \cdot ?] + [(5r - 1) \cdot ?]$
7. $(n^2 + 2n + 3)(n + 1) = [(n^2 + 2n + 3) \cdot ?] + [(n^2 + 2n + 3) \cdot ?]$

WRITTEN EXERCISES

Find each product by arranging your work in *both* the vertical and the horizontal form.

SAMPLE: $(4y - 3)(2y + 1)$

Solution:
$$\begin{array}{r} 4y - 3 \\ 2y + 1 \\ \hline 4y - 3 \\ 8y^2 - 6y \\ \hline 8y^2 - 2y - 3 \end{array}$$

$(4y - 3)(2y + 1)$
$= (4y - 3) \cdot 2y + (4y - 3) \cdot 1$
$= 8y^2 - 6y + 4y - 3$
$= 8y^2 - 2y - 3$

A
1. $(x + 7)(x + 2)$
2. $(y + 3)(y + 5)$
3. $(3n + 2)(n + 6)$
4. $(4k - 6)(k + 3)$
5. $(2a + 5)(a - 3)$
6. $(b - 9)(b + 5)$
7. $(m - 9)(m + 4)$
8. $(x - 7)(x - 6)$
9. $(y + 3)(y - 8)$
10. $(2x + 3)(5x + 1)$
11. $(7n + 4)(8n + 9)$
12. $(7t - 3)(5t + 2)$
13. $(3y - 4)(4y - 6)$
14. $(12c + 5)(5c - 2)$
15. $(a + b)(a + b)$

Find each product. If you can, complete the work mentally, using the horizontal form. Check your work with paper and pencil.

SAMPLE: $(4a + 2)(a + 3)$

Solution: $4a^2 + 14a + 6$ Check:
$$\begin{array}{r} 4a + 2 \\ a + 3 \\ \hline 12a + 6 \\ 4a^2 + 2a \\ \hline 4a^2 + 14a + 6 \end{array}$$

16. $(3y + 2)(2y + 5)$
17. $(2n + 6)(n + 1)$
18. $(2a + 3)(4a + 1)$
19. $(2x - 5)(x + 4)$
20. $(4t - 5)(t + 2)$
21. $(x + 3)(x - 2)$
22. $(2m + 3)(m - 1)$
23. $(x + y)(x + y)$
24. $(4k - 1)(k - 5)$
25. $(r - s)(r + s)$
26. $(3n - 5)(n - 1)$
27. $(r + t)(r + t)$

Find each product in its simplest form.

B
28. $(8w + 5)(5w - 2)$
29. $(3a + b)(3a + b)$
30. $(4m - n)(4m + n)$
31. $(2a - 0.3b)(a + 5b)$
32. $(2m - 0.5n)(m + 3n)$
33. $(3m^2 - 3m + 4)(m + 2)$
34. $(x^2 + 2x + 5)(x + 3)$
35. $(a^2 + ab + b^2)(a - b)$
36. $(s^2 - st + t^2)(2s + t)$
37. $(n^2 - 6n - 1)(2n - 3)$

C
38. $(a^2 - 8)(a^2 + 1)$
39. $(y^2 + 10)(y^2 - 3)$
40. $(ab - 1)(ab - 1)$
41. $(3 + ab)(1 - ab)$
42. $(5 + 2t)(7 - 3t - t^2)$
43. $(2a^2 + a - 1)(a^2 + 3a + 4)$
44. $(2a - 4b + 3c)(4a + 4b - 5c)$
45. $(x - 1)(x^4 + x^3 + x^2 + x + 1)$

PROBLEMS

Express each answer as a polynomial in simplest form.

 1. What is the area of the pictured rectangle? What is its perimeter?

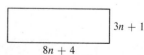

2. Find the area of a square if the length of one side is $t + 4$.

Multiplication and Division of Polynomials

3. Find the area of each small rectangle. Find the area of the entire region by finding a product, then check by comparing the result with the sum of the areas of the two small rectangles.

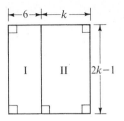

4. Find the area of each small rectangle. Find the area of the entire region by finding a product of two binomials. Check by comparing the result with the sum of the areas of the four small rectangles.

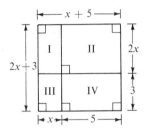

B 5. The dimensions of a box-shaped figure are given in the illustration. What is the volume of the figure? What is its total surface area?

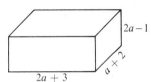

For each problem, write an equation representing the given information. Express each member of the equation as a polynomial in simplest form.

6. One piece of paper is in the shape of a square. A second piece of paper is a rectangle 3 inches longer than a side of the square. The width of the rectangle is 2 inches less than a side of the square. The areas of the two pieces of paper are equal.

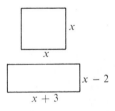

7. One opening cut into a piece of metal is in the shape of a square and another is in the shape of a rectangle. The length of the rectangle is 1 inch longer than a side of the square and its width is 3 inches less than a side of the square. The area of the rectangle is 45 square inches less than the area of the square.

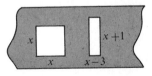

8. The distance from home plate to first base in softball is 30 feet less than the corresponding distance in baseball. The area of the square which is the softball diamond is 4500 square feet less than the area of a baseball diamond.

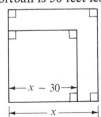

Special Products; Properties of Multiplication

13-6 Special Products of Polynomials

Let us consider the **square of a binomial**. As you recall, we use exponential notation to show how many times a polynomial is to be used as a factor. Thus the expression $(3x - 2)^2$ means $(3x - 2)(3x - 2)$. When we perform the multiplication we find that

$$(3x - 2)^2 = 9x^2 - 12x + 4$$

and we say that we have **expanded** the expression $(3x - 2)^2$.

It is easy to expand the square of a binomial once you discover the pattern to follow. Study these examples to find the pattern.

EXAMPLE 1.

$$\begin{array}{r} 2x + 3 \\ 2x + 3 \\ \hline 6x + 9 \\ 4x^2 + 6x \\ \hline 4x^2 + 12x + 9 \end{array}$$

EXAMPLE 2.

$$\begin{array}{r} 3n - 2 \\ 3n - 2 \\ \hline -6n + 4 \\ 9n^2 - 6n \\ \hline 9n^2 - 12n + 4 \end{array}$$

EXAMPLE 3.

$$\begin{array}{r} 2c - b \\ 2c - b \\ \hline -2cb + b^2 \\ 4c^2 - 2bc \\ \hline 4c^2 - 4bc + b^2 \end{array}$$

Did you discover the pattern? Did you find that the square of a binomial is this sum?

$$(\text{first term})^2 + (\text{twice product of terms}) + (\text{second term})^2$$

Thus,
$$(3t + 4)^2 = (3t)^2 + 2(3t \cdot 4) + 4^2$$
$$= 9t^2 + 24t + 16$$

We can summarize the pattern for the square of a binomial:

If **a** and **b** are any directed numbers,

$$(a + b)^2 = a^2 + 2ab + b^2.$$

Another special product involving binomials is illustrated below. Can you find a pattern?

EXAMPLE 1.

$$\begin{array}{r} x + 4 \\ x - 4 \\ \hline -4x - 16 \\ x^2 + 4x \\ \hline x^2 - 16 \end{array}$$

EXAMPLE 2.

$$\begin{array}{r} n + 8 \\ n - 8 \\ \hline -8n - 64 \\ n^2 + 8n \\ \hline n^2 - 64 \end{array}$$

EXAMPLE 3.

$$\begin{array}{r} a + 3c \\ a - 3c \\ \hline -3ac - 9c^2 \\ a^2 + 3ac \\ \hline a^2 - 9c^2 \end{array}$$

Did you notice that, in each example, one factor is the **sum** of two terms and the other factor is the **difference** of the same two terms, and that the product is the **difference** of the **squares** of the two terms? This pattern can be summarized:

If *a* and *b* are any directed numbers,

$$(a + b)(a - b) = a^2 - b^2.$$

ORAL EXERCISES

Name the missing term(s) in each expansion.

1. $(y + 4)^2 = (y + 4)(y + 4)$
   $= y^2 + ? + 16$
2. $(x + 3)^2 = (x + 3)(x + 3)$
   $= x^2 + ? + 9$
3. $(5 - n)^2 = (5 - n)(5 - n)$
   $= 25 + ? + n^2$
4. $(2y - 3)^2 = (2y - 3)(2y - 3)$
   $= ? - 12y + ?$
5. $(2k - 4)(2k + 4) = 4k^2 - 8k + 8k - 16$
   $= ? - ?$

WRITTEN EXERCISES

Find each product mentally. Then check your work with paper and pencil.

SAMPLE: $(2x - 5)^2$

Solution: $4x^2 - 20x + 25$

Check:
$$\begin{array}{r} 2x - 5 \\ 2x - 5 \\ \hline -10x + 25 \\ 4x^2 - 10x \\ \hline 4x^2 - 20x + 25 \end{array}$$

1. $(x + 1)^2$
2. $(y - 5)^2$
3. $(x - 2)(x + 2)$
4. $(x + 6)(x - 6)$
5. $(x + 6)^2$

6. $(r - s)(r + s)$
7. $(s - t)^2$
8. $(4m + 10)^2$
9. $(10x - 12)^2$
10. $(8x + 3)(8x - 3)$

11. $(9 - 2k)(9 + 2k)$
12. $(4a + 3b)^2$
13. $(3n - 5m)(3n + 5m)$
14. $(3a - b)(3a + b)$
15. $(5c + 6d)^2$

Expand each of the following.

16. $(x^2 + 12)^2$
17. $(y^2 - 11)^2$
18. $(xy + 3)^2$

19. $(8 - xy)^2$
20. $(6 + rs^2)^2$
21. $(c^2 - 3d)^2$

22. $(6b + \frac{1}{2})^2$
23. $(5k - \frac{1}{5})^2$
24. $(0.3ab - 0.5)^2$

25. $(a^2b + 5)^2$
26. $(9 + rs)^2$
27. $(x + 0.5)^2$

28. $(a^2 - b^2)^2$
29. $(a^2 + cd)^2$
30. $(0.8t^2 - 0.9t)^2$

31. $(1.6 + k)^2$
32. $(8 - rs^2)^2$
33. $(0.02s - 3t)^2$

C 34. $(t - 0.03)^2$
35. $(2y + 0.8)^2$
36. $(0.7 + 3y)^2$

37. $(0.3y^2 + y)^2$
38. $(\frac{1}{2}a + \frac{1}{4}b)^2$
39. $(1.5n^2 + 0.4)^2$

40. $(0.3y^2 + 0.2)^2$
41. $(18ab + 10c)^2$
42. $(9w^2 - 15xy)^2$

PROBLEMS

Write each answer as a polynomial in simplest form.

A 1. What is the area of a square if the length of one side is $(4x - 10)$ inches?

2. What is the area of a rectangle whose length is $(4n + 1.8)$ feet and whose width is $(n + 6)$ feet?

Solve each of the following problems.

3. The length of one side of a square piece of glass is 3 inches more than the length of one side of another square piece of glass. Their areas differ by 81 square inches. Find the dimensions of each piece of glass.

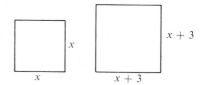

4. Of two square pieces of plastic, one has a side 10 inches shorter than a side of the other. The area of the smaller is 1100 square inches less than the area of the larger. What is the length of the side of the smaller piece?

5. The square of an integer is 103 less than the square of its successor. What are the two integers?

6. The square of an integer and the square of its successor differ by 95. Find the integers.

7. The product of an integer and its successor exceeds the square of the smaller integer by 13. Find the integers.

8. A square picture is in a frame 1 inch wide. The area of the frame is 24 square inches. What is the width of the picture?

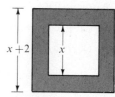

9. The product of an integer and its successor is 17 less than the square of the larger integer. Find the integers.

13-7 Multiplication Properties for Polynomials

Let us consider some further properties of operations with polynomials. If in our work with polynomials you observed carefully the results of multiplying one polynomial by another it seems clear that every product of two polynomials is a unique polynomial. Thus we see that the set of polynomials is **closed** under multiplication.

Do you suppose that multiplication of polynomials is both commutative and associative? Study the following examples:

EXAMPLE 1.

$$\begin{array}{cc} 12xy^2 & 3x^3y \\ \underline{3x^3y} & \underline{12xy^2} \\ 36x^4y^3 & 36x^4y^3 \end{array}$$

EXAMPLE 2.

$$\begin{array}{cc} 2n + 4 & n + 3 \\ \underline{n + 3} & \underline{2n + 4} \\ 6n + 12 & 4n + 12 \\ \underline{2n^2 + 4n} & \underline{2n^2 + 6n} \\ 2n^2 + 10n + 12 & 2n^2 + 10n + 12 \end{array}$$

Does multiplication of polynomials appear to be **commutative**?

Consider these examples:

EXAMPLE 1.

$$[(4ab)(2ab^2)]\,5abc \qquad 4ab\,[(2ab^2)(5abc)]$$
$$= [8a^2b^3]\,5abc \qquad\quad = 4ab\,[10a^2b^3c]$$
$$= 40a^3b^4c \qquad\qquad\quad = 40a^3b^4c$$

EXAMPLE 2.

$$[(x + 3)(2x + 1)](x + 2) \qquad (x + 3)[(2x + 1)(x + 2)]$$
$$= [2x^2 + 7x + 3](x + 2) \qquad = (x + 3)[2x^2 + 5x + 2]$$
$$= 2x^3 + 11x^2 + 17x + 6 \qquad = 2x^3 + 11x^2 + 17x + 6$$

Do you think that multiplication of polynomials is **associative**?

What about the distributive property? Is there a multiplicative identity for polynomials? You may recall that earlier we used the distributive property of multiplication over addition to simplify such expressions as $(x + 5)(x + 7)$ and $3x^2(x + 6)$. Since we know that the product of a given polynomial and 1 is the given polynomial, we see that the number 1, which is a polynomial, is the **multiplicative identity**.

All of the properties of operations with polynomials that have been discussed are similar to those presented earlier for directed numbers. This seems logical, since directed numbers such as $3\frac{1}{2}$, -4.05 and $\frac{5}{8}$ are polynomials according to our definition. Also, we noted that an expression like $x^2 + 3x - 5$, or like $n^3 + 4$, would represent a directed number if the variables were replaced by members of the set of directed numbers.

Now let us consider the idea of **multiplicative inverses** (**reciprocals**) for polynomials. For example, to find the multiplicative inverse of $3x^2$ we think "The product of $3x^2$ and what number is equal to 1?" Obviously, if $x = 0$ there is no reciprocal. If $x \neq 0$, we have

$$3x^2 \cdot \ ? \ = 1$$

$$3x^2 \cdot \frac{1}{3x^2} = 1$$

Thus $\frac{1}{3x^2}$ is the reciprocal, or multiplicative inverse, of $3x^2$, when $x \neq 0$. But $\frac{1}{3x^2}$ is not a polynomial. In fact, we shall find that the reciprocals of most polynomials are not polynomials, just as the reciprocals of most whole numbers are not whole numbers.

Study the following summary of the ideas of this section.

Multiplication of polynomials is **commutative**.
Multiplication of polynomials is **associative**.
The **multiplicative identity** for polynomials is **1**.
Multiplication of polynomials is **distributive** over addition.
The **reciprocal** of a polynomial is usually not a polynomial.

ORAL EXERCISES

Match each polynomial in Column 1 with its multiplicative inverse (reciprocal) in Column 2. Assume that replacements for the variables are such that no denominator of a fraction equals zero.

COLUMN 1

1. x^2y
2. $x + y$
3. $3x^2 + y$
4. $\frac{2xy^2}{3}$
5. $\frac{x + y}{x - y}$
6. $\frac{x^2 + y^2}{x - y}$

COLUMN 2

A. $\frac{1}{x + y}$
B. $\frac{3}{2xy^2}$
C. $\frac{x - y}{x + y}$
D. $\frac{x - y}{x^2 + y^2}$
E. $\frac{1}{x^2y}$
F. $\frac{1}{3x^2 + y}$

Tell which property of multiplication of polynomials justifies each sentence.

SAMPLE: $(2kn^2)(5k) = (5k)(2kn^2)$

What you say: The commutative property

7. $(4 + 3x)x^2 = x^2(4 + 3x)$
8. $(3a)[(5a)(a^2)] = [(3a)(5a)](a^2)$
9. $x^2[(ax)(cx)] = [(ax)(cx)]x^2$
10. $(x + y)(x - y) = (x^2 - y^2)$
11. $(3a + b)[5a^2(a + b)] = [(3a + b)5a^2](a + b)$
12. $(5m^2 + 2mn - n^2)(m + 1) = (m + 1)(5m^2 + 2mn - n^2)$

WRITTEN EXERCISES

Name the multiplicative inverse (reciprocal) for each expression. Assume that no numerator or denominator equals zero.

1. $\dfrac{2n}{m^2}$
2. $\dfrac{4w}{z}$
3. $\dfrac{2a}{3a + b}$
4. $\dfrac{m + n}{m - n}$
5. $\dfrac{3a^2 + b}{3}$
6. x^2y^2
7. $\tfrac{1}{2}m^2$
8. $\dfrac{3xy^2}{-2}$

Show that each sentence is correct.

SAMPLE: $3x[(x + y)(y^2)] = [(3x)(x + y)]y^2$

Solution: $3x[(x + y)(y^2)] = [(3x)(x + y)]y^2$

$\begin{array}{c|c} 3x[xy^2 + y^3] & [3x^2 + 3xy]y^2 \\ 3x^2y^2 + 3xy^3 & 3x^2y^2 + 3xy^3 \end{array}$

9. $3x^2 \cdot 5xy = (3 \cdot 5)(x^2x)y$
10. $[(4ab)(6b)]a^2 = 4ab[(6b)(a^2)]$
11. $4y(3x + x) = (4y)(4x)$
12. $4a(3a + b) = (3a + b)4a$

B 13. $(xy)(x^2y)(xy^2) = (x \cdot x^2 \cdot x)(y \cdot y \cdot y^2)$
14. $[(2.5m)(4m^2)]t = 2.5m[(4m^2)(t)]$
15. $(3x^2 + 8xy + y^2)(x + y) = (x + y)(3x^2 + 8xy + y^2)$
16. $[(a + b)(a - b)](2a + b) = (a + b)[(a - b)(2a + b)]$

Find each product, and tell which property justifies each step.

SAMPLE: $(5a^2)(4a)$ Solution: $(5a^2)(4a)$
$= 5(a^2 \cdot 4)a$ Associative property
$= 5(4 \cdot a^2)a$ Commutative property
$= (5 \cdot 4)(a^2 \cdot a)$ Associative property
$= 20a^3$ Substitution principle

17. $(x^2)(4x^3)$
18. $(ab)(a^2b)$
19. $(9t^2)(6t)$
20. $(6m^2n)(2mn)$
21. $(xyz)(x^2yz^3)$
22. $(3ab^2)(5ab)$

Division

13–8 Division of Monomials

Dividing monomials requires that we understand and use the meaning of factors, division, and exponential notation. For example, 5^7 can be written in any one of several ways.

$$5^7 = 5^6 \cdot 5^1 \qquad 5^7 = 5^4 \cdot 5^3 \qquad 5^7 = 5^2 \cdot 5^3 \cdot 5^2$$
$$5^7 = 5^2 \cdot 5^5 \qquad 5^7 = 5 \cdot 5 \cdot 5 \cdot 5 \cdot 5 \cdot 5 \cdot 5$$

To complete the division $3^5 \div 3^2$ we can arrange our work like this:

EXAMPLE 1.

$$\frac{3^5}{3^2} = \frac{3 \cdot 3 \cdot 3 \cdot 3 \cdot 3}{3 \cdot 3} = \frac{3}{3} \cdot \frac{3}{3} \cdot 3 \cdot 3 \cdot 3 = 1 \cdot 1 \cdot 3 \cdot 3 \cdot 3 = 3^3$$

Do you see that we can also complete the same division in this way?

EXAMPLE 2.

$$\frac{3^5}{3^2} = \frac{3^2 \cdot 3^3}{3^2} = \frac{3^2}{3^2} \cdot 3^3 = 1 \cdot 3^2 = 3^3$$

As you have probably guessed we can complete such a division as $x^5 \div x^3$ by similar methods.

EXAMPLE 3.

$$\frac{x^5}{x^3} = \frac{x \cdot x \cdot x \cdot x \cdot x}{x \cdot x \cdot x} = \frac{x}{x} \cdot \frac{x}{x} \cdot \frac{x}{x} \cdot x \cdot x = 1 \cdot 1 \cdot 1 \cdot x \cdot x = x^2$$

EXAMPLE 4.

$$\frac{x^5}{x^3} = \frac{x^3 \cdot x^2}{x^3} = \frac{x^3}{x^3} \cdot x^2 = 1 \cdot x^2 = x^2$$

You should notice a pattern in each of the examples just presented. This pattern helps us verify a very important rule of exponents for division. The rule is as follows:

For any directed number x except 0, and for all positive integers p and q such that $p > q$,

$$\frac{x^p}{x^q} = x^{p-q}.$$

Multiplication and Division of Polynomials **401**

Now let us consider the divisions $2^2 \div 2^5$ and $x^2 \div x^6$. Look for a pattern that suggests another rule of exponents for division.

EXAMPLE 5. $\dfrac{2^2}{2^5} = \dfrac{2^2}{2^3 \cdot 2^2} = \dfrac{1}{2^3} \cdot \dfrac{2^2}{2^2} = \dfrac{1}{2^3} \cdot 1 = \dfrac{1}{2^3}$

EXAMPLE 6. $\dfrac{x^2}{x^6} = \dfrac{x^2}{x^4 \cdot x^2} = \dfrac{1}{x^4} \cdot \dfrac{x^2}{x^2} = \dfrac{1}{x^4} \cdot 1 = \dfrac{1}{x^4}$

The rule can be stated as follows:

For any directed number *x* except 0, and for all positive integers *p* and *q* such that $p < q$,

$$\dfrac{x^p}{x^q} = \dfrac{1}{x^{q-p}}.$$

To complete the picture, note that we have already used the following fact:

For any directed number *x* except 0, and for all positive integers *p* and *q* such that $p = q$,

$$\dfrac{x^p}{x^q} = 1.$$

We apply the rules of exponents for division to simplify the division. $-12x^3y^4 \div 3xy^2$ as follows:

$$\dfrac{-12x^3y^4}{3xy^2} = \dfrac{-12}{3} \cdot \dfrac{x^3}{x} \cdot \dfrac{y^4}{y^2}$$
$$= -4 \cdot x^{3-1} \cdot y^{4-2}$$
$$= -4x^2y^2$$

ORAL EXERCISES

Name the missing factor to replace the question mark in each sentence.

SAMPLE 1: $x^6 = x^4 \cdot ?$ *What you say:* x^2

SAMPLE 2: $\dfrac{a^5}{a^7} = \dfrac{a^2}{a^4} \cdot ?$ *What you say:* $\dfrac{a^3}{a^3}$, or 1

1. $4^5 = 4^3 \cdot ?$ **2.** $\dfrac{k^6}{k^2} = \dfrac{k^2}{k^2} \cdot ?$ **3.** $1 = \dfrac{b^3}{?}$

4. $a^3 = a^2 \cdot ?$

5. $\dfrac{7^6}{7^4} = \dfrac{7^3}{7^3} \cdot ?$

6. $\dfrac{2^4}{2^7} = \dfrac{2^2}{2^5} \cdot ?$

7. $\dfrac{n^3}{n^4} \cdot ? = \dfrac{n^8}{n^9}$

8. $1 = \dfrac{?}{t^2}$

9. $\dfrac{m^5}{?} = 1$

Give the result of each indicated division.

10. $\dfrac{y \cdot y \cdot y}{y \cdot y}$

11. $\dfrac{a \cdot a \cdot a \cdot a \cdot a}{a \cdot a \cdot a}$

12. $\dfrac{3 \cdot 3 \cdot 3 \cdot 3}{3}$

13. $\dfrac{b \cdot b \cdot b}{b \cdot b \cdot b \cdot b \cdot b}$

14. $\dfrac{8 \cdot 8}{8 \cdot 8 \cdot 8 \cdot 8}$

15. $\dfrac{n}{n \cdot n \cdot n \cdot n}$

16. $\dfrac{x \cdot x \cdot y \cdot y}{x \cdot y}$

17. $\dfrac{x \cdot x \cdot y}{x \cdot x \cdot x \cdot y \cdot y \cdot y}$

18. $\dfrac{a \cdot a \cdot b \cdot c \cdot c \cdot c}{a \cdot b \cdot c}$

Tell what numbers should replace the variables p and q in each of the following.

SAMPLE: $\dfrac{b^p}{b^q} = b^{5-2}$ *What you say:* $p = 5$ and $q = 2$ since $\dfrac{b^5}{b^2} = b^{5-2}$.

19. $\dfrac{a^p}{a^q} = a^{10-3}$

20. $\dfrac{x^p}{x^q} = x^{5-2}$

21. $\dfrac{b^p}{b^q} = \dfrac{1}{b^{6-2}}$

22. $\dfrac{k^p}{k^q} = \dfrac{1}{k^{7-3}}$

23. $\dfrac{t^6}{t^4} = t^{p-q}$

24. $\dfrac{y^8}{y^3} = y^{p-q}$

25. $\dfrac{r^3}{r^7} = \dfrac{1}{r^{q-p}}$

26. $\dfrac{w^7}{w^{10}} = \dfrac{1}{w^{q-p}}$

27. $\dfrac{t^2}{t^5} = \dfrac{1}{t^{p-q}}$

WRITTEN EXERCISES

Complete each indicated division, assuming no denominator has the value 0. Show your work.

SAMPLE: $\dfrac{18x^3 y}{-2x}$ *Solution:* $\dfrac{18x^3 y}{-2x} = \dfrac{18}{-2} \cdot \dfrac{x^3}{x} \cdot y = -9x^2 y$

1. $\dfrac{a^{10}}{a^4}$

2. $\dfrac{y^8}{y^2}$

3. $\dfrac{m^7}{m^{14}}$

4. $\dfrac{8x^2}{2x}$

5. $\dfrac{3n^7}{-n^2}$

6. $\dfrac{x^6}{-x^2}$

7. $\dfrac{q^6}{-q^4}$

8. $\dfrac{-6b^3}{2b}$

9. $\dfrac{30x^2}{2x}$

10. $\dfrac{20r^2}{5r}$

11. $\dfrac{14x^3 y}{-2x}$

12. $\dfrac{21ab^2}{-3b}$

13. $\dfrac{-144bc^2}{-11c}$

14. $\dfrac{-45x^2}{5x}$

15. $\dfrac{-170z^2}{5z}$

16. $\dfrac{-64x^3 y^8}{16xy^5}$

B 17. $\dfrac{-5a^3 b}{-15a^2 b^4}$

18. $\dfrac{(-6abc)^2}{-6abc}$

19. $\dfrac{-5m^2 n}{10m^2 n^2}$

20. $\dfrac{-11ab^2}{-0.11a^2 b}$

21. $\dfrac{-7r^3s}{-21rs^5}$ 25. $\dfrac{36a^2b}{-12a^4b^2}$ 29. $\dfrac{-0.3r^2s}{-3rs^2}$ 33. $\dfrac{-8a^2b^7}{-56a^{12}b^3}$

22. $\dfrac{(-1.5xy)^2}{-1.5xy}$ 26. $\dfrac{-50a^2b}{25(ab)^2}$ 30. $\dfrac{-6xyz}{-9x^2}$ 34. $\dfrac{(-4)^2x^3y^5}{8x^{12}y^{10}}$

23. $\dfrac{-13x^2y^3}{-39x^3y}$ 27. $\dfrac{-4m^2n}{(2mn)^3}$ 31. $\dfrac{2.7p^5r^2}{(3pr)^3}$ 35. $\dfrac{(-3)^3a^{11}b^2}{27a^8b^4}$

24. $\dfrac{6rs^7}{8r^2s^4}$ 28. $\dfrac{-9abc}{-6b^2}$ 32. $\dfrac{-2st^9}{2.5s^2t^7}$ 36. $\dfrac{(-2)^4cd^8}{1.6c^2d^7}$

C 37. $\dfrac{-3a^x}{2a},\ x>1$ 39. $\dfrac{32(ab)^m}{8a^mb^m}$

38. $\dfrac{-2x^ay^b}{x^by^a},\ a>b$ 40. $\dfrac{-57(xy)^r}{-19x^r}$

13–9 Division of a Polynomial by a Monomial

Suppose we want to complete the indicated division $480 \div 40$. We can apply the distributive property and set up the problem as follows:

$$\dfrac{480}{40} = \dfrac{1}{40}(480) = \dfrac{1}{40}(400+80) = \dfrac{400}{40} + \dfrac{80}{40} = 10 + 2 = 12$$

The very same strategy can be used to complete such indicated divisions as $(5x+15y) \div 5$ and $(18a^2 - 6a) \div a$.

EXAMPLE 1.
$$\dfrac{5x+15y}{5} = \dfrac{1}{5}(5x+15y)$$
$$= \dfrac{5x}{5} + \dfrac{15y}{5}$$
$$= \left(\dfrac{5}{5} \cdot x\right) + \left(\dfrac{15}{5} \cdot y\right)$$
$$= (1 \cdot x) + (3 \cdot y)$$
$$= x + 3y$$

EXAMPLE 2.
$$\dfrac{18a^2-6a}{3a} = \dfrac{1}{3a}(18a^2-6a)$$
$$= \dfrac{18a^2}{3a} - \dfrac{6a}{3a}$$
$$= \left(\dfrac{18}{3} \cdot \dfrac{a^2}{a}\right) - \left(\dfrac{6}{3} \cdot \dfrac{a}{a}\right)$$
$$= (6 \cdot a) - (2 \cdot 1)$$
$$= 6a - 2$$

ORAL EXERCISES

Match each indicated division in Column 1 with an equivalent expression in Column 2.

COLUMN 1

1. $\frac{56}{8}$
2. $\frac{10b}{2} + \frac{8}{2}$
3. $\frac{1}{7}(14 + 35x)$
4. $\frac{1}{9}(18 - 27y)$
5. $\frac{42m}{6} + \frac{54}{6}$
6. $12 \overline{)744}$

COLUMN 2

A. $\frac{10b + 8}{2}$
B. $\frac{18}{9} - \frac{27y}{9}$
C. $7m + 9$
D. $\frac{40 + 16}{8}$
E. $12 \overline{)720 + 24}$
F. $2 + 5x$

Give the result of each indicated division.

7. $6\overline{)60 + 18}$
8. $12\overline{)120 + 24}$
9. $\frac{70 + 21}{7}$
10. $\frac{150 + 30}{15}$
11. $\frac{1200 + 240}{120}$
12. $\frac{6a + 10}{2}$
13. $\frac{3m + 27k}{3}$
14. $\frac{3m^2 + 2m}{m}$
15. $\frac{36k^2 + 8k}{4}$
16. $\frac{x^5 + 4x^3}{x^2}$
17. $\frac{24t + 32t^2}{8t}$
18. $\frac{36b^3 + 15b^2}{3b}$

WRITTEN EXERCISES

Simplify each expression, assuming no denominator has the value 0.

SAMPLE: $\frac{1}{3}(9k + 15)$ Solution: $\frac{1}{3}(9k + 15) = \frac{9k}{3} + \frac{15}{3}$
$= 3k + 5$

A
1. $\frac{1}{4}(8m - 16)$
2. $\frac{1}{2}(18y + 10)$
3. $\frac{1}{8}(32y + 8)$
4. $\frac{1}{3}(18a^2 + 12b)$
5. $\frac{1}{9}(-27c^2 - 54)$
6. $\frac{1}{n}(dn + n^2)$
7. $-\frac{1}{3}(18a^2 + 6a - 21)$
8. $-\frac{1}{7}(42x^2 + 35x + 7)$
9. $-\frac{1}{k}(4k^2 - 3k)$

Find the result of each indicated division, assuming no divisor is 0.

SAMPLE: $\frac{7s^3 + 2s^2}{s}$ Solution: $\frac{7s^3 + 2s^2}{s} = \frac{1}{s}(7s^3 + 2s^2)$
$= \frac{7s^3}{s} + \frac{2s^2}{s} = 7s^2 + 2s$

10. $\dfrac{12x^2 + 15y^3}{3}$

11. $\dfrac{2x^4 + 7x^3}{x}$

12. $\dfrac{6y + 36y}{6}$

13. $\dfrac{8n + 3}{n}$

14. $\dfrac{4n^2 - n}{n}$

15. $\dfrac{r^2 - 7r}{r}$

16. $\dfrac{8a^3 - 6a}{2a}$

17. $\dfrac{20 + 5m^2}{5}$

18. $\dfrac{ax^2 + bx}{x}$

B 19. $\dfrac{18x^2 + 12x}{6x}$

20. $\dfrac{35y^2 - 15y}{5y}$

21. $\dfrac{4a^2 + 10a + 8}{2}$

22. $\dfrac{24y^3 - 12y^2 + 15y}{3y}$

23. $\dfrac{50r^3 + 10r^2 - 35r}{5r}$

24. $\dfrac{8x^3 - 4x^2 - 2x}{-x}$

25. $\dfrac{2n^4 - 3n^3 - 4n^2}{-n}$

26. $\dfrac{36x^3y^3 - 6x^2y^2 + 42xy}{-3xy}$

C 27. $\dfrac{-35a^4 - 28a^3 + 56a^2 - 14a}{7a}$

28. $\dfrac{2.4a^2b^2 + 0.6ab^2 + 30a^2b}{-3ab}$

29. $\dfrac{24m^4 - 12m^3n - 6m^2n^2}{-36m^2}$

30. $\dfrac{30x^3y^3 - 45x^2y^2 + 15xy}{30xy^2}$

13–10 Division of a Polynomial by a Polynomial

You may recall that earlier in this book (page 172) we completed divisions such as $682 \div 31$, as shown here. The five numbered steps illustrate the process.

$682 \div 31$ may be written as $30 + 1 \overline{)600 + 80 + 2}$

$$
\begin{array}{r}
\boxed{30 \cdot \,? = 600} \\
20 \\
30 + 1 \overline{)600 + 80 + 2} \\
600 + 20 \quad \text{②} \quad (30 + 1)20 \\
\hline
60 + 2 \quad \text{③} \quad \text{Subtract}
\end{array}
$$

$$
\begin{array}{r}
\boxed{30 \cdot \,? = 60} \\
20 + 2 \\
30 + 1 \overline{)600 + 80 + 2} \\
600 + 20 \\
\hline
60 + 2 \\
60 + 2 \quad \text{⑤} \quad (30 + 1)2 \\
\hline
0 \quad \text{⑥} \quad \text{Subtract}
\end{array}
$$

Chapter 13

Now let us apply a similar five-step process with polynomials that contain variables. For example, consider $(x^2 + 5x + 6) \div (x + 3)$.

$$\boxed{x \cdot ? = x^2}$$

①
$$\begin{array}{r} x \phantom{+ 3\overline{)x^2 + 5x + 6}} \\ x + 3{\overline{\smash{\big)}\,x^2 + 5x + 6}} \\ \underline{x^2 + 3x} \quad ② \quad (x+3)x \\ 2x + 6 \quad ③ \quad \text{Subtract} \end{array}$$

$$\boxed{x \cdot ? = 2x}$$

④
$$\begin{array}{r} x + 2 \\ x + 3{\overline{\smash{\big)}\,x^2 + 5x + 6}} \\ \underline{x^2 + 3x} \\ 2x + 6 \\ \underline{2x + 6} \quad ⑤ \quad (x+3)2 \\ 0 \quad ⑥ \quad \text{Subtract} \end{array}$$

We can check our work by multiplication.

Check: $(x + 3)(x + 2) = x^2 + 5x + 6$

The two examples that follow will help you become more familiar with this division process. Be certain that you understand each step.

EXAMPLE 1.
$$\begin{array}{r} 2x + 1 \\ 3x + 2{\overline{\smash{\big)}\,6x^2 + 7x + 2}} \\ \underline{6x^2 + 4x} \\ 3x + 2 \\ \underline{3x + 2} \\ 0 \end{array}$$

Check: $(3x + 2)(2x + 1) = 6x^2 + 7x + 2$

EXAMPLE 2.
$$\begin{array}{r} x - 8 \\ x - 7{\overline{\smash{\big)}\,x^2 - 15x + 56}} \\ \underline{x^2 - 7x} \\ -8x + 56 \\ \underline{-8x + 56} \\ 0 \end{array}$$

Check: $(x - 7)(x - 8) = x^2 - 15x + 56$

WRITTEN EXERCISES

Complete each indicated division. Check each answer.

A
1. $\dfrac{x^2 + 5x + 6}{x + 2}$
2. $\dfrac{x^2 + 3x + 2}{x + 1}$
3. $\dfrac{x^2 + 7x + 12}{x + 3}$
4. $\dfrac{x^2 + 11x + 28}{x + 7}$
5. $\dfrac{10x^2 - 19x - 15}{5x + 3}$
6. $\dfrac{y^2 - 13y + 42}{y - 6}$
7. $\dfrac{r^2 - 6r - 7}{r + 1}$
8. $\dfrac{n^2 - 8n - 9}{n + 1}$
9. $\dfrac{m^2 - 7m + 12}{m - 4}$
10. $\dfrac{y^2 - 12y + 35}{y - 7}$
11. $\dfrac{6t^2 + 19t + 10}{3t + 2}$
12. $\dfrac{14x^2 + 22x - 12}{7x - 3}$
13. $\dfrac{8b^2 - 22b + 15}{4b - 5}$
14. $\dfrac{x^2 - 16}{x - 4}$
15. $\dfrac{6t^2 + 13t + 5}{3t + 5}$

Complete each indicated division. Rearrange terms where necessary.

SAMPLE: $\dfrac{-30 + x + x^2}{x - 5}$ Solution:
$$\begin{array}{r} x + 6 \\ x - 5 \overline{)x^2 + x - 30} \\ x^2 - 5x \\ \hline 6x - 30 \\ 6x - 30 \\ \hline 0 \end{array}$$

B
16. $\dfrac{4 - 8n + 3n^2}{3n - 2}$
17. $\dfrac{64 - 16z + z^2}{z - 8}$
18. $\dfrac{4x^2 + 12xy + 9y^2}{2x + 3y}$
19. $\dfrac{49 - 16x^2}{4x - 7}$
20. $\dfrac{y^2 - 6yz - 27z^2}{y - 9z}$
21. $\dfrac{w^2 - 11wx - 102x^2}{w - 17x}$
22. $\dfrac{3x^3 + 23x^2 - 14x + 2}{3x - 1}$
23. $\dfrac{9m^2 - 48mn + 15n^2}{3n - 9m}$

CHAPTER SUMMARY

Inventory of Structure and Concepts

1. The general rules of exponents for multiplication state that, for any directed numbers a and b, and for positive integers p and q:

   a. $a^p \cdot a^q = a^{p+q}$ b. $(ab)^p = a^p b^p$ c. $(a^p)^q = a^{pq}$

2. The product of two polynomials is found by multiplying each term of one polynomial by every term of the other, and then adding the like terms.

3. The square of a binomial is equal to the sum of the square of the first term, twice the product of the two terms of the binomial, and the square of the second term. $[(a + b)^2 = a^2 + 2ab + b^2]$

4. The product of two factors, of which the first is the sum of two terms and the second is the difference of the same two terms, is equal to the difference of the squares of the terms. $[(a + b)(a - b) = a^2 - b^2]$

5. For the set of polynomials we found that:
   a. The set is **closed** under multiplication.
   b. Multiplication is **commutative**.
   c. Multiplication is **associative**.
   d. The **identity element** for multiplication is **1**.
   e. Multiplication is **distributive** over addition.
   f. The **reciprocal** of a polynomial is usually not a polynomial.

6. The general rules of exponents for division state that, for any directed number x except 0, and for positive integers p and q:

   a. $\dfrac{x^p}{x^q} = x^{p-q}$, if $p > q$;

   c. $\dfrac{x^p}{x^q} = 1$, if $p = q$.

   b. $\dfrac{x^p}{x^q} = \dfrac{1}{x^{q-p}}$, if $p < q$;

Vocabulary and Spelling

exponents (p. 379)
monomials (p. 382)
polynomials (p. 386)
vertical form (p. 386)
horizontal form (p. 388)
square of a binomial (p. 394)

expand an expression (p. 394)
difference of two squares (p. 395)
multiplicative identity (p. 397)
multiplicative inverse (p. 398)
reciprocal (p. 398)

Chapter Test

Give a simpler name for each of the following, assuming that no denominator equals zero.

1. $x^4 \cdot x^3$

2. $\dfrac{21st^2}{-7st}$

3. $(-2ab^2)^2$

4. $\dfrac{3x^3}{5x^4}$

5. $-2t^3(-2t)^3$

6. $\dfrac{9k^2 - 6k}{3k}$

7. $\dfrac{21r^3 + 24r^4}{3r^2}$ 8. $(7mn^3)(-m^2n)$

Write each product as a polynomial in standard form.

9. $3t^2(t - 5)$
10. $(x - 4)^2$
11. $(3p - t)(3p + t)$
12. $(-2x)(x^2 - 3x + 5)$
13. $(3y - 2)(y + 4)$
14. $(s - 2)(3s^2 - s - 1)$

Find the value of the variable which will make each statement true.

15. $2x + 3(x - 5) = 10$
16. $5(p - 2) - 2(p + 1) = (p + 5)$

Simplify each expression by adding like terms.

17. $(3t)^2 - (2t)(-t)$
18. $4(t - 3) - 5t^2 + t(2 + 6t)$

19. What is the multiplicative inverse of $\dfrac{3}{x + 2}$, if $x \neq -2$?

Complete each indicated division.

20. $\dfrac{6x^2 - x - 2}{2x + 1}$
22. $\dfrac{16 - k^2}{4 - k}$

21. $\dfrac{4p^2 + 4p + 1}{2p + 1}$

23. Give an expression for the volume of a cube that is $(3t)$ inches on each edge.
24. Write a polynomial to represent the area of the shaded portion between the two given rectangles.

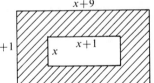

Chapter Review

13-1 Polynomials and Exponents

Write a simpler name for each of the following, using the rule of exponents for multiplication.

1. $k \cdot k \cdot k \cdot k \cdot k$
2. $(ab)(ab)(ab)$
3. $r^2 \cdot r^5$
4. $3 \cdot x \cdot x \cdot x \cdot y \cdot y$
5. $t \cdot t^3$
6. $-2 \cdot x^3 \cdot x^5$

Find replacements for x that will make the following statements true.

7. $2^7 = 2 \cdot 2^x$
8. $(n-2)^3(n-2)^2 = (n-2)^x$
9. $k^2 \cdot k^4 \cdot k = k^x$
10. $3^x \cdot 3^5 = 3^9$

13–2 Multiplication of Monomials

Simplify each indicated product.

11. $(3xy)(5xy^2)$
12. $(-k)(-k)(5k)^2$
13. $(5a)(6b)(2a^2b^3)$
14. $(-2k)(-k)(5k^2)$
15. $(-2y^2)(-2y)^2$
16. $2m^3 \cdot 5m^2 \cdot 3m$

13–3 Power of a Product

Simplify each expression.

17. $(a^2b)^3$
18. $(-2x^4)^3$
19. $(5xy)^2$
20. $(a^2b^3)^2$
21. $(2x)^3 + (2x)(3x^2)$
22. $k^2t^4 + 2(kt^2)^2$
23. $3k^2 + (3k)^2 - k^2$
24. $k^2t^4 + (2kt^2)^2$

13–4 Multiplying a Polynomial by a Binomial

Find each product.

25. $x^2(x^3 - 2x^2 - x + 5)$
26. $-2a(a + ab + b)$
27. $3xy^2(x^2y - 2x^2y)$
28. $(-3x)^2(x - 2x^2 + 1)$

Simplify each of the following expressions.

29. $3r(r - s) - 2s(s - r) + 5rs$
30. $4x(x - 5) - x(5 - x) - x^2$

Find the value of the variable which will make each statement true.

31. $3x + 2(x - 3) = 10$
32. $5(k - 2) + 3(k + 2) = 2(k + 2)$

33. Write a polynomial in standard form for the area of a rectangle if its width is $3x$ and its length is $2x - 5$.

13–5 Multiplying a Polynomial by a Polynomial

Write each product as a polynomial in standard form.

34. $(x + 2)(x + 3)$
35. $(k - 2)(k + 2)$

36. $(2t - 1)(t + 1)$
37. $(y - 5)(5 - y)$
38. $(x^2 - 2x + 1)(x + 1)$
39. $(2m - 3n)(5m + n)$
40. $(r - 5)(r - 5)$
41. $(3x + 4)(2x - 1)$
42. $(2a - 3b)(3b - 2a)$
43. $(xy + 2)(1 - xy)$

44. Write a polynomial in standard form for the area of a rectangle if its width is $2x - 3$ and its length is $3x + 1$.

13–6 Special Products of Polynomials

Expand each of the following.

45. $(x - 5)^2$
46. $(3y - 5)^2$
47. $(2t - \frac{1}{2})^2$
48. $(2k + 1)^2$
49. $(xy + 2)^2$
50. $(xy^2 - 1)^2$
51. $(a - 3)(a + 3)$
52. $(c^2 - 4)(c^2 + 4)$

53. Write a polynomial for the total surface area of a box whose faces are rectangles if the length of the box is $x + 3$, the width is $x - 1$, and the height is $2x - 5$.

13–7 Multiplication Properties for Polynomials

Name the multiplicative inverse for each polynomial, assuming that no denominator equals zero.

54. $\dfrac{2x}{3y}$
55. $\dfrac{t - k}{k}$
56. $3x^2y^3$

Find each product. Tell which property justifies each step of your work.

57. $(x^3)(4x)$
58. $(7kt^3)(3k^2t)$
59. $(ab^2c^3)(a^3b^2c)$

13–8 Division of Monomials

Simplify each indicated division, assuming that no divisor equals zero.

60. $\dfrac{r^6}{r^4}$
61. $\dfrac{t^2}{t^5}$
62. $\dfrac{21ab^2}{-3a^2b}$
63. $\dfrac{z^a}{z}, \quad a > 1$
64. $\dfrac{s^5}{s^4}$
65. $\dfrac{-x^4}{-x^5}$
66. $\dfrac{64x^3y^2}{-16y^3}$
67. $\dfrac{a}{a^x}, \quad x > 1$
68. $\dfrac{-b^3}{b}$
69. $\dfrac{3y}{12y^3}$
70. $\dfrac{5c^2}{15c^2}$
71. $\dfrac{(-3)^2x^3y^2}{-x^2y^4}$

13-9 Division of a Polynomial by a Monomial

Simplify each expression, assuming that no divisor equals zero.

72. $\frac{1}{2}(12x^2 - 6y)$

73. $\dfrac{18s^2 + 12s^4}{12s^2}$

74. $\dfrac{10x - 5y}{5x}$

75. $\dfrac{a^2 - a^3}{a}$

76. $\dfrac{ax^3 - bx}{x}$

77. $\dfrac{3st^2 + 5t}{st}$

78. $\dfrac{t^2 - 3t}{t}$

79. $\dfrac{15ab^2 - 10ab + 5a^2b}{5ab}$

80. $\dfrac{8k - 4t}{-2}$

13-10 Division of a Polynomial by a Polynomial

Complete each indicated division, assuming that no divisor equals zero.

81. $\dfrac{x^2 + 3x + 2}{x + 1}$

82. $\dfrac{2x^2 + 3x - 2}{x + 2}$

83. $\dfrac{n^2 - 3n + 2}{n - 2}$

84. $\dfrac{4 - 4t + t^2}{2 - t}$

85. $\dfrac{x^2 - 9}{x + 3}$

86. $\dfrac{6x^2 - x - 2}{2x + 1}$

Review of Skills

Match the expressions in Column 1 with the appropriate figures from Column 2.

COLUMN 1

1. Line
2. Line segment
3. Right triangle
4. Parallel lines
5. Perpendicular lines
6. 90° angle
7. Ray
8. Equilateral triangle
9. Intersecting lines
10. Obtuse triangle

COLUMN 2

A.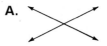

B. ●————●

C. ●

D.

E.

F. △

G.

H.

I.

J.

K.

Complete the following sentences.

11. The measure in degrees one makes in turning completely around is __?__.
12. $3^2 + 4^2 = (\,?\,)^2$
13. $8^2 + (\,?\,)^2 = 10^2$
14. The intersection of two sets that have no element in common is called the __?__ set.
15. A protractor is used to measure __?__.
16. Area is measured in __?__ units.
17. In a right triangle the hypotenuse is the __?__ side of the triangle.
18. The measure in degrees of a right angle is __?__.
19. There are __?__ square inches in a square foot.
20. List the elements contained in the union of the set $\{1, 2, 3, 4, 5\}$ and the set $\{3, 4, 5, 6, 7\}$.

■ ■

CHECK POINT
FOR EXPERTS

Zero as an Exponent

The rules of exponents that we developed for division state that where m and n are positive integers:

$$\frac{x^m}{x^n} = x^{m-n}, \text{ if } m > n, \text{ and } x \neq 0;$$

$$\frac{x^m}{x^n} = \frac{1}{x^{n-m}}, \text{ if } m < n, \text{ and } x \neq 0.$$

Chapter 13

But have you wondered what we do when $m = n$? Consider these examples.

EXAMPLE 1. $5^3 \div 5^3 = \dfrac{5^3}{5^3}$

$= \dfrac{5 \cdot 5 \cdot 5}{5 \cdot 5 \cdot 5}$

$= \dfrac{5}{5} \cdot \dfrac{5}{5} \cdot \dfrac{5}{5}$

$= 1 \cdot 1 \cdot 1 = 1$

EXAMPLE 2. If $a \neq 0$, $a^4 \div a^4 = \dfrac{a^4}{a^4}$

$= \dfrac{a \cdot a \cdot a \cdot a}{a \cdot a \cdot a \cdot a}$

$= \dfrac{a}{a} \cdot \dfrac{a}{a} \cdot \dfrac{a}{a} \cdot \dfrac{a}{a}$

$= 1 \cdot 1 \cdot 1 \cdot 1 = 1$

The two examples help us verify this idea:

If $m = n$ and $x \neq 0$, then $\dfrac{x^m}{x^n} = x^{m-n} = x^0 = 1.$

The symbol 0^0 is undefined.

Questions

Find the value of each expression if $a = 2$, $b = 3$, $x = 4$, and $y = 5$.

1. $x^3 \div x^3$
2. $\dfrac{n^b}{n^b}$, $n \neq 0$
3. $\dfrac{5^a}{5^a}$
4. $(4 \cdot 2)^y \div (4 \cdot 2)^y$
5. $4^{ab} \div 4^{ab}$
6. $(xy)^0$
7. $(12)^{b-b}$
8. $\dfrac{b^0}{y^0}$

What is the simplest form of each of the following? Assume that no variable has 0 as its replacement.

9. $(1000)^0$
10. $(35)^0$
11. $(-10)^2 \div (-10)^2$
12. $(0.372)^0$
13. $(\tfrac{5}{8})^0$
14. $a^5 \div a^0$
15. $c^6 \div c^0$
16. $\dfrac{k^3}{k^0}$
17. $b^0 \div 5$
18. $12 \div w^0$
19. $\dfrac{x^2 \cdot x^3}{x}$
20. $t^5 \cdot t^0$
21. $7b^0$
22. $6x^0$
23. $(-3k)^0$
24. $-5a^0$
25. $-8y^0$
26. $(-7)^0(0)$
27. $(-15)^0 \cdot (0)^3$
28. $b^0 \cdot b^m$

THE HUMAN ELEMENT

Galileo Galilei

Among the names of renowned mathematicians and scientists, that of Galileo Galilei (1564–1642) stands out. The son of a noble of Florence, Italy, Galileo was well educated in several fields. He first studied medicine at the University of Pisa. Later he turned his attention to mathematics and physics. It was in the cathedral of Pisa, when he was seventeen, that the sight of a swinging lamp — which he measured by his pulse beats — led him to formulate the law of the pendulum.

After leaving the university, he devoted time to research in mathematics. At the age of twenty-five he returned to the University of Pisa as a professor of mathematics. It was during this period that Galileo formulated his law of falling bodies. He believed that the earth's gravity pulls heavy objects and light objects to the earth at the same speed. To check this theory, according to legend, Galileo went to the top of the leaning Tower of Pisa and dropped a ten-pound lead ball and a one-pound lead ball before a crowd of witnesses. The ten-pound object and the one-pound object struck the ground at the same moment! According to the beliefs of the time, following the teaching of the Greek philosopher, Aristotle, the ten-pound weight should have fallen at a speed ten times as great as the speed of the one-pound weight.

For eighteen years (1592–1620) Galileo was professor of mathematics at the University of Padua. His interest in astronomy led him to make the first practical use of telescopes for observing the heavens. He ignored the many fables concerning the moon and the Milky Way and instead based his conclusions on scientific observations. It was he who first observed that the moon's surface is rough and mountainous rather than smooth and shining, as claimed by Aristotle. In 1610 he discovered four bright moons of Jupiter. He also noted the peculiar shape of Saturn, resulting from its rings, which were to be discovered years later.

Galileo's contributions to science resulted from his careful observations of nature and his insistence that theories agree with the facts. He was one of the first people to use the modern scientific method. Observations led Galileo to agree with Copernicus that the earth is not the center of the universe. This view brought him into conflict with Church authority, and he spent the last years of his life under house arrest in a villa near Florence.

Navigating a schooner...

Navigating a spaceship to the moon...

Geometric Figures in the Plane

The navigation of an old-time sailing vessel made use of a sextant to measure the angular distance between any two points, and thus enable the captain to determine the position of his ship. Such a use of the sextant is shown in the reproduction of an etching by Winslow Homer (1836–1910), an American artist famous for his sea pictures. Sextants are still of use in navigation, but the development of computers has made possible such complex navigational systems as are used in the Manned Spacecraft Center at Houston, Texas. Seen on the television monitor in the Mission Operations Control Room is a picture of the earth which was telecast from the Apollo 8 spacecraft at a distance of 176,000 miles.

Lines and Curves

14–1 Basic Geometric Figures

You are probably familiar with the idea that points are the simplest of all geometric figures and that they are usually labeled with capital letters. Figure 14–1 shows point A and point B. If we use a straightedge and pencil to join these points, the result is the set of points called **line segment** AB shown in Figure 14–2. In symbols, "line segment AB" is written $\overline{AB}$. By agreement we name a line segment by naming its two endpoints. Of course, $\overline{AB}$ names the same segment as $\overline{BA}$.

Figure 14–1   Figure 14–2

Suppose you were able to extend $\overline{AB}$ on and on in both directions without ever ending. The result would be the set of points called **line** AB. In symbols, "line AB" is written $\overleftrightarrow{AB}$. We can picture $\overleftrightarrow{AB}$ like this:

418 Chapter 14

As you no doubt discovered long ago, if you have two distinct points, there is exactly one line that contains both points. We describe this by saying that two points determine exactly one line. On the line in Figure 14–3, several points are labeled with letters. We can use any two labeled points to name the line. Some examples are $\overleftrightarrow{AH}$, $\overleftrightarrow{HT}$, $\overleftrightarrow{AR}$, and $\overleftrightarrow{TM}$.

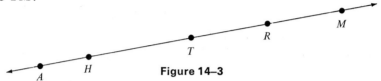

Figure 14–3

We can also name several different line segments shown in this same figure: $\overline{AH}$, $\overline{AT}$, $\overline{HM}$, and so on.

The figures below show intersecting line segments and intersecting lines. Do you agree that in each case the intersection is exactly one point? We can use set notation to describe each intersection.

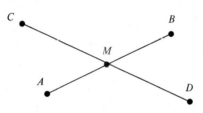 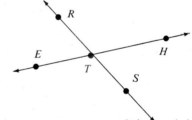

The intersection of $\overline{AB}$ and $\overline{CD}$ is point M.
In symbols: $\overline{AB} \cap \overline{CD} = M$

The intersection of $\overleftrightarrow{EH}$ and $\overleftrightarrow{RS}$ is point T.
In symbols: $\overleftrightarrow{EH} \cap \overleftrightarrow{RS} = T$

You will recall that parallel lines do not intersect. Thus their intersection is the empty set. Similarly, the intersection of parallel line segments is the empty set.

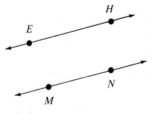 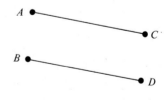

$\overleftrightarrow{EH}$ is parallel to $\overleftrightarrow{MN}$

$\overline{AC}$ and $\overline{BD}$ are parallel

The intersection of $\overleftrightarrow{EH}$ and $\overleftrightarrow{MN}$ is the empty set.
In symbols: $\overleftrightarrow{EH} \cap \overleftrightarrow{MN} = \emptyset$

The intersection of $\overline{AC}$ and $\overline{BD}$ is the empty set.
In symbols: $\overline{AC} \cap \overline{BD} = \emptyset$

Geometric Figures in the Plane 419

ORAL EXERCISES

Give the name of each figure or the meaning of each symbol.

SAMPLE 1: [C————D] *What you say:* line CD

SAMPLE 2: $\overline{JK}$ *What you say:* line segment JK (or segment JK)

1. [X to Y segment]
2. $\overline{XY}$
3. [M————N]
4. T
5. [G to H segment]

6. [A————B with arrow]
7. $\overleftrightarrow{RT}$
8. [M to N line]
9. [A B C collinear]
10. $\overline{QR}$
11. $\overleftrightarrow{AD}$

WRITTEN EXERCISES

Describe in words the intersection of each pair of lines or line segments. (Assume that two lines or segments that look parallel are parallel.)

SAMPLE 1:    Solution: The intersection of $\overleftrightarrow{QR}$ and $\overleftrightarrow{ST}$ is a point.

SAMPLE 2: X●————●Y   Solution: $\overline{XY}$ and $\overline{HK}$ do not intersect.

 1. 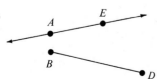   2.

3.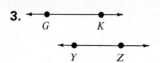

5.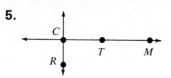

4.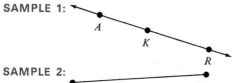

6.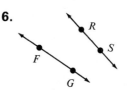

Name each line in three different ways and each line segment in two different ways.

SAMPLE 1:  Solution: $\overleftrightarrow{AK}$, $\overleftrightarrow{KR}$, or $\overleftrightarrow{AR}$

SAMPLE 2: 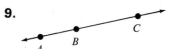 Solution: $\overline{RS}$ or $\overline{SR}$

7.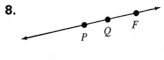

9.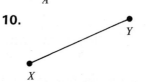

8.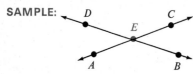

10.

Use set notation to describe the intersection shown in each figure.

SAMPLE: 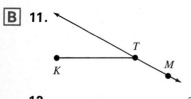 Solution: $\overleftrightarrow{AC} \cap \overleftrightarrow{BD} = E$

B **11.**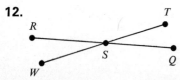

13.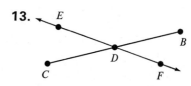

12.

14.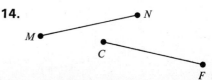

Geometric Figures in the Plane   421

15.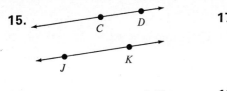

17.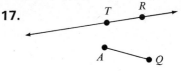

16.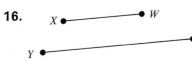

18.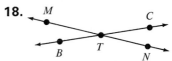

Use the given figure to complete each of the following. Assume that $\overleftrightarrow{HR}$ and $\overleftrightarrow{AT}$ are parallel lines.

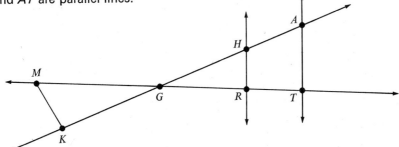

C 19. $\overrightarrow{MT} \cap \overleftrightarrow{KA} = ?$
20. $\overleftrightarrow{HR} \cap \overrightarrow{MT} = ?$
21. $\overleftrightarrow{HR} \cap \overline{MK} = ?$
22. $\overleftrightarrow{AT} \cap \overline{MG} = ?$
23. $\overleftrightarrow{AT} \cap \overleftrightarrow{HR} = ?$

24. $\overline{AH} \cap \overline{HR} = ?$
25. $\overline{MR} \cap \overline{KH} = ?$
26. $\overline{GT} \cap \overleftrightarrow{HR} = ?$
27. $\overline{HR} \cap \overline{AT} = ?$
28. $\overline{MG} \cap \overline{AH} = ?$

14–2 More about Lines and Line Segments

How many different sets of points can you name that are line segments in the figure below? Do you agree that $\overline{AC}$, $\overline{AB}$, and $\overline{BC}$ are segments that can be named?

Now let us consider another figure. You can see that it is possible to name several different segments: $\overline{AK}$, $\overline{KL}$, $\overline{AF}$, $\overline{KF}$ and so on.

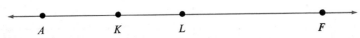

As you probably discovered, some of these line segments have points in common. In other words, they intersect. For example, the intersection of segment AL and segment KF is segment KL.

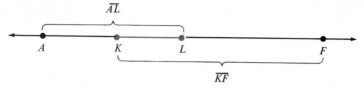

In symbols we write: $\overline{AL} \cap \overline{KF} = \overline{KL}$.

Using this same figure, we can think of joining segment AL and segment KF to form segment AF. You will recall that the joining of sets is called the *union* operation. In symbols we write: $\overline{AL} \cup \overline{KF} = \overline{AF}$.

Similarly, the union of segment AL and segment LF forms segment AF. The intersection of segment AL and segment LF is simply the point L.

In symbols we write: $\overline{AL} \cup \overline{LF} = \overline{AF}$ and $\overline{AL} \cap \overline{LF} = L$.

Do you agree that the union of segments AK, KL, and LF is segment AF?

In symbols we write: $\overline{AK} \cup \overline{KL} \cup \overline{LF} = \overline{AF}$.

ORAL EXERCISES

Tell which word, *union* or *intersection,* should replace the question mark according to the given figure.

SAMPLE:
```
A   B        E    D
```
$\overline{BE}$ is the ? of $\overline{AE}$ and $\overline{BD}$.

What you say: $\overline{BE}$ is the **intersection** of $\overline{AE}$ and $\overline{BD}$.

1.
```
R  S   T   Q
```
$\overline{RT}$ is the ? of $\overline{RS}$ and $\overline{RT}$.

3.
```
X   T  Y    S
```
$\overline{TY}$ is the ? of $\overline{XY}$ and $\overline{ST}$.

2.
```
A    B      C
```
$\overline{AC}$ is the ? of $\overline{AB}$ and $\overline{BC}$.

4.
```
A        D  C
```
D is the ? of $\overline{AD}$ and $\overline{CD}$.

Geometric Figures in the Plane **423**

5.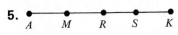

$\overline{MS}$ is the ? of $\overline{AS}$ and $\overline{KM}$.

8.

$\overline{XE}$ is the ? of $\overline{XE}$ and $\overline{KE}$.

6.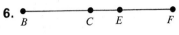

$\overline{CE}$ is the ? of $\overline{CE}$ and $\overline{CF}$.

9.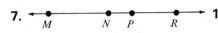

$\overline{AD}$ is the ? of $\overline{AB}$, $\overline{BC}$, and $\overline{CD}$.

7.

10.

$\overleftrightarrow{NP}$ is the ? of $\overleftrightarrow{NP}$ and $\overleftrightarrow{MR}$.

$\overleftrightarrow{XZ}$ is the ? of $\overleftrightarrow{XY}$ and $\overleftrightarrow{YZ}$.

WRITTEN EXERCISES

Name five different line segments in each figure.

SAMPLE:

Solution: $\overline{RS}$, $\overline{ST}$, $\overline{TV}$, $\overline{RT}$, and $\overline{SV}$.

A

1.

4.

2.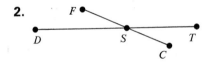

5.

3.

6.

Use the figure below to make a sketch that illustrates each statement.

SAMPLE: The intersection of $\overline{CF}$ and $\overline{DE}$ is $\overline{CE}$.

Solution:

7. The intersection of $\overline{AD}$ and $\overline{BC}$ is $\overline{BD}$.
8. The union of $\overline{DC}$ and $\overline{CE}$ is $\overline{DE}$.

424 Chapter 14

9. The intersection of $\overline{AF}$ and $\overline{BH}$ is $\overline{BF}$.
10. The intersection of $\overline{AC}$ and $\overline{CF}$ is C.
11. The intersection of $\overline{AE}$ and $\overline{CE}$ is $\overline{CE}$.
12. The union of $\overline{CE}$ and $\overline{CF}$ is $\overline{CF}$.
13. The intersection of $\overline{AB}$ and $\overline{FH}$ is $\emptyset$.

B
14. $\overline{AB} \cup \overline{BE} = \overline{AE}$
15. $\overline{BD} \cup \overline{CE} \cup \overline{CD} = \overline{BE}$
16. $\overline{AB} \cup \overline{BD} \cup \overline{DF} = \overline{AF}$
17. $\overline{AF} \cup \overline{DC} = \overline{AF}$
18. $\overline{AE} \cap \overline{DH} = \overline{DE}$
19. $\overline{AC} \cap \overline{CF} = C$
20. $\overline{AC} \cap \overline{EH} = \emptyset$
21. $\overline{AC} \cap \overline{BF} \cap \overline{BH} = \overline{BC}$

Use the figure below to complete each statement. Assume that $\overleftrightarrow{CE}$ and $\overleftrightarrow{DF}$ are parallel lines.

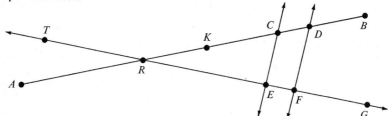

SAMPLE 1: $\overline{TR} \cup \overline{RE} = ?$ Solution: $\overline{TR} \cup \overline{RE} = \overline{TE}$
SAMPLE 2: $\overline{AK} \cap \overline{RC} \cap \overline{RB} = ?$ Solution: $\overline{AK} \cap \overline{RC} \cap \overline{RB} = \overline{RK}$

22. $\overline{EF} \cup \overline{FG} = ?$
23. $\overline{AK} \cup \overline{RD} = ?$
24. $\overleftrightarrow{TG} \cap \overline{AB} = ?$
25. $\overleftrightarrow{CE} \cap \overleftrightarrow{TG} = ?$

26. $\overline{AR} \cup \overline{RK} \cup \overline{KC} = ?$
27. $\overline{TR} \cup \overline{RE} \cup \overline{EF} = ?$
28. $\overline{TF} \cup \overline{RE} \cup \overline{EG} = ?$
29. $\overline{AK} \cap \overline{CE} = ?$

C
30. $\overline{TR} \cap \overleftrightarrow{EF} = ?$
31. $\overline{TR} \cap \overline{RE} \cap \overline{FG} = ?$
32. $\overline{RE} \cap \overleftrightarrow{CE} \cap \overline{EF} = ?$
33. $\overline{TE} \cup \overline{RF} \cup \overline{EF} = ?$
34. $\overline{AR} \cup \overline{RK} \cup ? = \overline{AD}$
35. $\overline{CE} \cap ? = \emptyset$

14–3 Curves: Sets of Points

Suppose you were asked "What is a curve?" Could you give a good definition? In mathematics the word curve has meanings other than those that might usually come to mind.

Let us assume that you were to use a piece of chalk and trace a path across the chalkboard. The chalkboard represents a plane (flat surface) that is a set of points. If you trace the path without lifting the chalk from the chalkboard's flat surface the result is a **plane curve**. We think of a plane curve as a continuous set of points in a plane.

Would you agree that each of the following is a curve according to our "definition"?

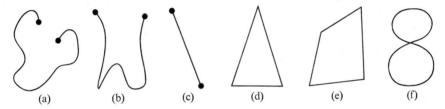

Notice that in (a)–(e) the curve does not cross itself. Figures of this kind are called **simple** curves. In (a)–(c) the curve shown has two distinct endpoints, while for each curve in (d)–(f) any point on the curve can be thought of as both the beginning and the ending. Such figures as (d), (e), and (f) are called **closed** curves. Do you see, then, why the curves in (d) and (e) are called **simple closed curves**? The word **simple** means that the curves do not cross themselves, and the word **closed** indicates that the set has no particular beginning and end — that is, any point on the curve could be considered to be both the beginning and the ending.

A simple closed curve in a plane completely encloses a region called the **interior**. All points not in the interior and not a part of the simple closed curve make up the **exterior**.

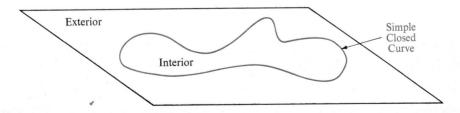

426 Chapter 14

In other words, a simple closed curve separates the plane into three distinct subsets:

        (1) The simple closed curve
        (2) The interior of the simple closed curve
        (3) The exterior of the simple closed curve

ORAL EXERCISES

For each of the following figures, if it is a curve, tell which of the following describes it most completely: curve; simple curve; closed curve; simple closed curve.

SAMPLE 1:   *What you say:* Simple closed curve

SAMPLE 2:   *What you say:* Simple curve

1.   4.   7.

2.   5.   8.

3. 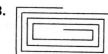   6.    9.

WRITTEN EXERCISES

Classify each figure as a curve, a simple curve, a closed curve, or a simple closed curve.

SAMPLE:

Solution:
The figure is a simple closed curve.

Geometric Figures in the Plane   427

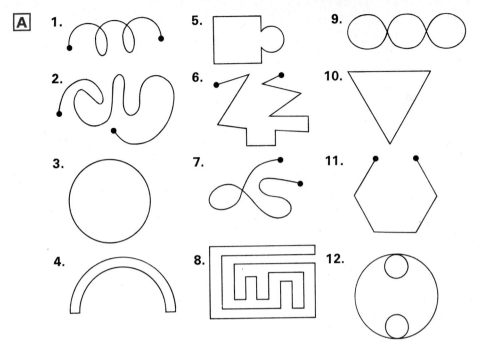

Tell whether each of the following figures is a simple curve or a simple closed curve.

SAMPLE: The letter **M**   Solution: Simple curve

13. The letter **D**
14. A line segment
15. The letter **C**
16. A circle
17. A square
18. The numeral **3**
19. The letter **O**
20. The numeral **7**
21. The letter **S**

The figure below is a simple closed curve. Tell whether each labeled point is on the curve, in the exterior of the curve, or in the interior of the curve.

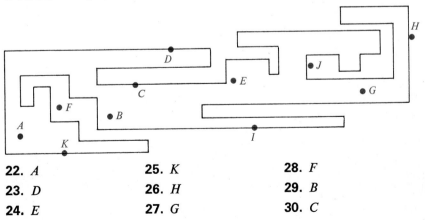

22. A
23. D
24. E
25. K
26. H
27. G
28. F
29. B
30. C

Angles

14-4 Rays and Angles in the Plane

You are familiar with many ideas about angles and how they are drawn. Suppose that you place two dots, *A* and *B*, on your paper. Then, using a straightedge and pencil, draw the set of points that begins at *A*, goes through *B*, and continues on indefinitely. The result is **ray** *AB*. In symbols we write $\overrightarrow{AB}$. Figure 14-4 shows two rays. The letter assigned to the endpoint of the ray and a letter assigned to one other point on the ray are necessary to name it. The endpoint letter is written first.

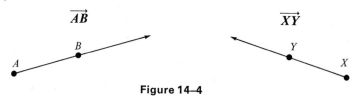

Figure 14-4

If two rays have the same endpoint they form an **angle**. You should recall that the letter assigned to the common endpoint, called the **vertex**, is always the middle letter of the name of the angle. An angle is sometimes referred to by number. The little curved line in the figure is used to emphasize the angle under consideration.

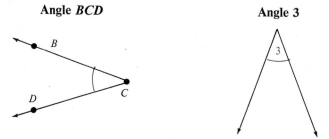

The two rays that form an angle are called the **sides** of the angle. We define an **angle** as the **union of two rays** having a common endpoint. We often use the symbol ∠ in place of the word "angle" when naming an angle.

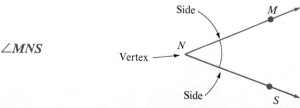

∠*MNS*

Geometric Figures in the Plane   429

We think of an angle as lying on a flat surface or plane. Therefore, angles are called **plane figures**. In the figure below, points R and S lie in the **interior** of $\angle XYZ$; points X, Y and Z lie **on** the angle; points A and B lie in the **exterior** of the angle.

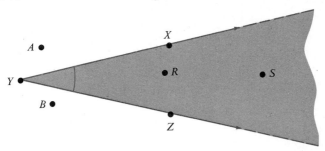

Do you agree that an angle separates a plane into three distinct subsets? These subsets are: the interior of the angle, the angle itself, and the exterior of the angle.

ORAL EXERCISES

Name each angle.

SAMPLE:

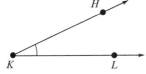

What you say: Angle *HKL* (or Angle *LKH*)

1.

4.
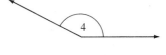

2.
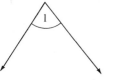

5.
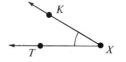

3.
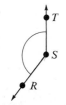

6.
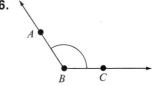

430 Chapter 14

WRITTEN EXERCISES

Use the figure below to complete Exercises 1–4.

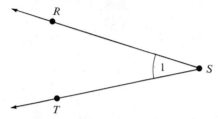

A 1. Name the rays that are the sides of the angle.
2. Name the vertex of the angle.
3. Name the angle in three different ways.
4. Name three points on the angle.

Write the name of each figure in symbols.

5.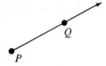

8.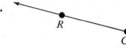

6.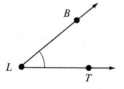

9.

7.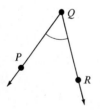

10.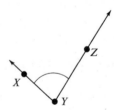

Tell whether each labeled point is in the exterior of the angle, on the angle, or in the interior of the angle.

11.

12.

Geometric Figures in the Plane   431

13. 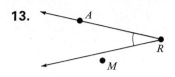   14.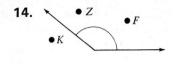

Name three different angles from each given figure.

B 15. 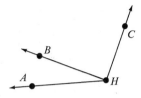   17.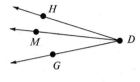

16. 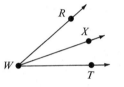   18.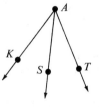

Name at least three different rays for each given figure.

19. 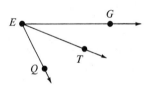   21.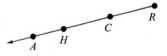

20.   22.

14–5 Angle Measurement and the Protractor

One of the most common and useful angles is the **right angle**. Think of a line $\overleftrightarrow{AB}$ drawn on a piece of paper, and imagine that the paper is folded at point P on the line, in such a way that $\overrightarrow{PA}$ lies exactly on top of $\overrightarrow{PB}$, as shown here.

∠APC is called a **right angle**, and so is ∠BPC. The symbol ⌐ inside an angle is often used to indicate that it is a right angle.

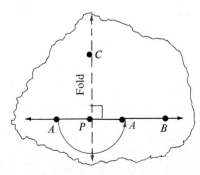

432   Chapter 14

A common unit for measuring angles is the degree, which is $\frac{1}{90}$ of a right angle. Thus the measure of a **right angle** is **90 degrees**, which is usually written **90°**. Instead of writing in words "The measure of angle *APC* is 90 degrees," we can write

$$m\angle APC = 90°$$

Figure 14–5 shows a protractor, which is a useful instrument for finding the measure of an angle.

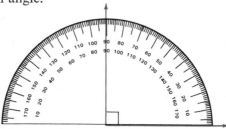

Figure 14–5

Notice that the protractor has two scales. When measuring an angle other than a right angle, you must decide which of two numbers represents the measure of the angle. To use the protractor effectively, then, you first decide whether the measure of the angle is greater than or less than 90 degrees. An angle whose measure is greater than 90 degrees and less than 180 degrees is called an **obtuse angle**; an angle whose measure is less than 90 degrees is called an **acute angle**.

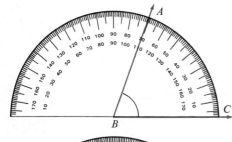

This angle is an acute angle. Its measure is less than 90°. Therefore, the correct measure is 70°. In symbols we write:

$$m\angle ABC = 70°.$$

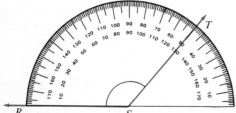

This angle is an obtuse angle. Its measure is between 90° and 180°. Therefore, the correct measure is 130°. In symbols:

$$m\angle RST = 130°.$$

You may already know that the sum of the measures of the angles of a triangle is **180°**. You can demonstrate this by tearing the "angles" off

Geometric Figures in the Plane 433

a triangular-shaped piece of paper and placing them on a protractor as shown in Figure 14–6.

Figure 14–6

The sum of the measures of two or more angles may be greater than 180°. Figure 14–7 shows the use of a circular protractor, which is divided into **360** of the degree units, to find the sum of the measures of two angles.

$m\angle ABC + m\angle CBD$
$= 140° + 75° = 215°$

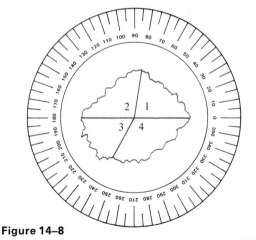

Figure 14–7

In Figure 14–8 is shown a four-sided piece of paper whose four "angles" have been torn off and placed on the circular protractor.

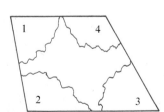

Figure 14–8

Do you think that the sum of the measures of the four angles of any four-sided figure is 360 degrees?

ORAL EXERCISES

Tell whether each angle is acute or obtuse.

1. 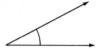   4. 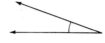   7.

2.   5.    8.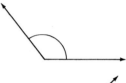

3.   6.    9.

Tell whether each angle is acute or obtuse and give its measure in degrees.

SAMPLE: *What you say:* obtuse; 150°

10.   11.

WRITTEN EXERCISES

Write an appropriate number sentence and calculate each angle measure that is designated by a variable.

SAMPLE: 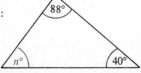   *What you say:* $88 + 40 + n = 180$
$n = 52$
Therefore, the measure of the angle is **52°**.

Geometric Figures in the Plane   **435**

A **1.** 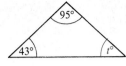   **3.**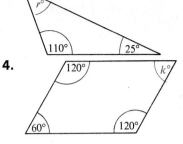

2. 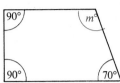   **4.**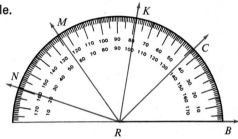

Give the measure of each angle.

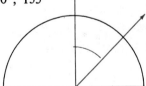

5. $m\angle BRC = ?$ **8.** $m\angle BRN = ?$ **11.** $m\angle CRN = ?$
6. $m\angle BRK = ?$ **9.** $m\angle CRK = ?$ **12.** $m\angle KRM = ?$
7. $m\angle BRM = ?$ **10.** $m\angle CRM = ?$ **13.** $m\angle KRN = ?$

Three measures are suggested for each angle. Tell which is reasonable.

14. 15°; 95°; 70°   **17.** 45°; 90°; 135°

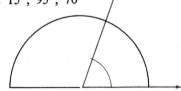   

15. 150°; 30°; 50°   **18.** 120°; 200°; 80°

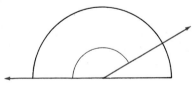

16. 80°; 25°; 3°

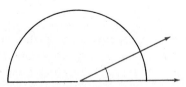

436 Chapter 14

Find the value of each variable that represents the measure of an angle.

SAMPLE: 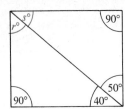   Solution: $90 + 40 + r = 180$
$r = 50$ (degrees)
$90 + 50 + s = 180$
$s = 40$ (degrees)

B 19. 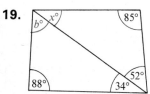   21.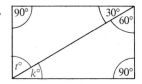

20.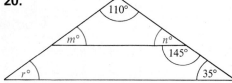

Draw a figure for each of the following.

C 22. $\overline{AB}$
23. Acute angle ABC
24. Obtuse angle RST
25. $\angle MNR$ (acute)

26. Ray MN
27. Right angle XYZ
28. $\angle GHF$ (obtuse)
29. $\angle TQF$ (right)

30. An acute angle which is the union of $\overrightarrow{PR}$ and $\overrightarrow{PQ}$.
31. An obtuse angle which is the union of $\overrightarrow{AR}$ and $\overrightarrow{AK}$.

Triangles

14–6 Kinds of Triangles

You will recall that a triangle is a simple closed curve. A triangle can be defined as the **union** of three line segments which are determined by three points not on the same line.

We often classify triangles by comparing the measures of the sides, or of the angles. For example, a triangle for which **no two sides** are of equal length is called a **scalene triangle**. If the lengths of **at least two** sides of a triangle are equal, it is called an **isosceles triangle**. A triangle with **all three sides** of equal length is called an **equilateral triangle**.

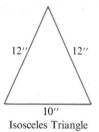
Isosceles Triangle

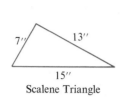
Scalene Triangle

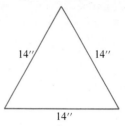
Equilateral Triangle

A convenient way of indicating that two or more parts of a geometric figure have equal measures is to mark such parts with short lines. In Figure 14–9, the short lines tell us that the first triangle is isosceles, with $\overline{AC}$ and $\overline{AB}$ the same length, and that the second triangle is equilateral.

You will recall that a triangle is named by naming its vertices. Thus, in Figure 14–9, we have "triangle ABC" and "triangle PQR." Replacing the word "triangle" by the symbol $\triangle$, we write $\triangle ABC$ and $\triangle PQR$.

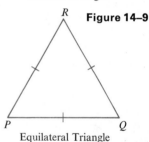

Figure 14–9

Another method of classifying triangles is based on the measures of their angles. If all of the angles of a triangle are acute, it is called an **acute triangle**. If one angle is obtuse, it is called an **obtuse triangle**. Why could a triangle not have more than one obtuse angle?

A **right triangle** has a right angle. The longest side of a right triangle, the side opposite the right angle, is called the **hypotenuse**. The other two sides are called **legs**.

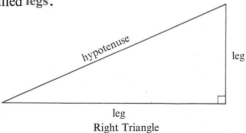
Right Triangle

In the case of an isosceles triangle, it can be shown that the two angles which lie opposite the sides of equal length have equal measures, and that the reverse is also true. That is, if the measures of two **angles** of a triangle are **equal**, then the **sides** opposite those angles will have **equal lengths**.

From this, it follows that if a triangle is **equilateral**, **all** of its angles will have the same measure. Such a triangle is called an **equiangular triangle**. It also follows that if a triangle is **scalene**, then **no two** angles will have the same measure. In each case, the reverse statement is also true: an **equiangular** triangle is **equilateral**, and a triangle in which **no two angles** have the same measure is **scalene**.

ORAL EXERCISES

Classify each triangle as *scalene, isosceles,* or *equilateral.*

1.   4. 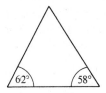   7.

2. 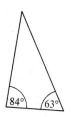   5. 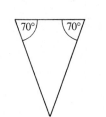   8.

3.   6. 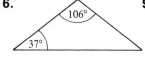   9.

Classify each triangle as *acute, obtuse,* or *right.*

10. 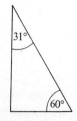   11. 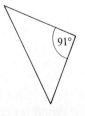   12.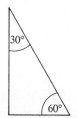

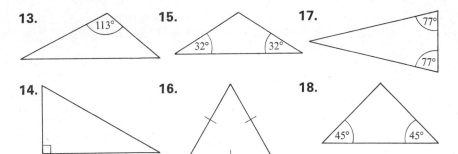

WRITTEN EXERCISES

Name the three line segments that form each triangle. Then name each numbered angle by using letters.

SAMPLE:

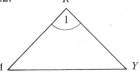

Solution: $\overline{AR}$; $\overline{RY}$; $\overline{AY}$
$\angle 1 = \angle ARY$

A 1.

3.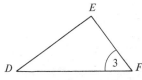

2.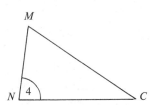

4.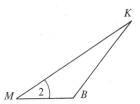

Use the information given to find the unknown measurements.

SAMPLE: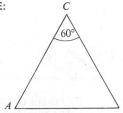

$\triangle ACK$ is an equilateral triangle.
$m\angle CAK = ?$
$m\angle AKC = ?$

Solution: $m\angle CAK = 60°$
$m\angle AKC = 60°$

440 Chapter 14

B **5.**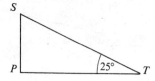
△PST is a right triangle.
m∠SPT = ?   m∠PST = ?

6.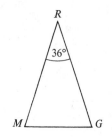
△MRG is an isosceles triangle.
MR = RG and m∠MRG = 36°
m∠RMG = ?   m∠RGM = ?

7.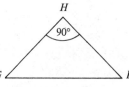
△HGF is an isosceles right triangle.
m∠HGF = ?   m∠GFH = ?
HF = ?

8.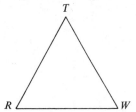
△RTW is an equilateral triangle.
RT = 18 inches
RW = ?   TW = ?
m∠TRW = ?   m∠RTW = ?
m∠RWT = ?

14–7 Right Triangles: The Pythagorean Theorem

If you took a length of rope and tied knots dividing it into twelve equal segments, you could use it to lay out a good approximation to a right angle, as shown in Figure 14–10. The lengths of the sides are **3, 4**, and **5** units, respectively. In Figure 14–11 are two other triangles with lengths of sides as indicated. Notice that the **6–8–10** triangle is really a **3–4–5** triangle with the length of each side doubled. Do you agree that the **9–12–15** triangle is also really a **3–4–5** triangle? The fact that all of these appear to be right triangles is based upon the reverse of the **Pythagorean Theorem**.

Figure 14–10

Geometric Figures in the Plane   441

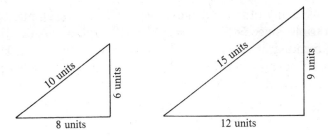

Figure 14-11

Over 2000 years ago the Greek mathematician Pythagoras demonstrated that if squares are constructed on each of the legs and on the hypotenuse of a right triangle the sum of the areas of the two smaller squares equals the area of the larger square. This idea, called the **Pythagorean Theorem**, is illustrated for a 3-4-5 triangle in Figure 14-12.

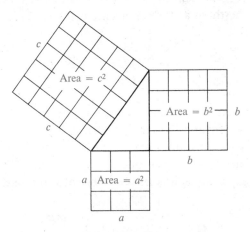

Figure 14-12

For any right triangle, if we represent the legs by a and b and the hypotenuse by c, the Pythagorean Theorem states that

$$a^2 + b^2 = c^2.$$

Any collection of three numbers that can replace the variables in that formula and result in a true statement is called a **Pythagorean triple**. The example below shows that **3, 4, 5** is a Pythagorean triple.

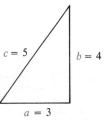

$$a^2 + b^2 = c^2$$
$$3^2 + 4^2 = 5^2$$
$$9 + 16 = 25$$

There are many other Pythagorean triples. For example, **5, 12, 13** is such a triple, since $5^2 + 12^2 = 13^2$. Although others can be found by experimenting, it takes a lot of time and work to do so. Here is a set of variable expressions that will give a Pythagorean triple each time that *n* is replaced by a whole number:

$$\{2n,\ n^2 - 1,\ n^2 + 1\}$$

Although this set of formulas does not yield all possible Pythagorean triples, the use of the formulas will always result in such a triple. Thus, if we choose **6** as the replacement for *n*, we have

$$2n = 2 \cdot 6 = 12;\quad n^2 - 1 = 36 - 1 = 35;\quad n^2 + 1 = 36 + 1 = 37.$$

To check that **12, 35, 37** is a Pythagorean triple, we use the numbers as replacements for the variables in the formula $a^2 + b^2 = c^2$, using the largest of the three as the replacement for *c*.

$$12^2 + 35^2 \stackrel{?}{=} 37^2$$
$$144 + 1225 \stackrel{?}{=} 1369$$
$$1369 = 1369$$

It can be proved that, if the numbers representing the lengths of the sides of a triangle form a Pythagorean triple, then the triangle is a right triangle. Do you see that this is the reverse of the statement of the Pythagorean Theorem? When one of two statements is the reverse of the other, we say that each is the **converse** of the other.

ORAL EXERCISES

Name the hypotenuse, the right angle, and the legs for each right triangle.

SAMPLE:

What you say: $\overline{MT}$ is the hypotenuse.
$\angle MNT$ is the right angle.
$\overline{MN}$ and $\overline{NT}$ are the legs.

1.  **2.**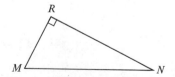

Geometric Figures in the Plane   **443**

3.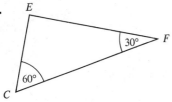

4.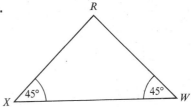

WRITTEN EXERCISES

Show whether each statement is true or false. Then tell whether or not the given number triple is a Pythagorean triple.

SAMPLE: $3^2 + 2^2 \stackrel{?}{=} 4^2$
Triple: 3, 2, 4

Solution: $3^2 + 2^2 \stackrel{?}{=} 4^2$

$9 + 4$	16
13	16

False; 3, 2, 4 is not a Pythagorean triple.

A
1. $5^2 + 12^2 \stackrel{?}{=} 13^2$
   Triple: 5, 12, 13
2. $3^2 + 5^2 \stackrel{?}{=} 6^2$
   Triple: 3, 5, 6
3. $6^2 + 8^2 \stackrel{?}{=} 10^2$
   Triple: 6, 8, 10
4. $8^2 + 15^2 \stackrel{?}{=} 17^2$
   Triple: 8, 15, 17

5. $9^2 + 12^2 \stackrel{?}{=} 15^2$
   Triple: 9, 12, 15
6. $5^2 + 4^2 \stackrel{?}{=} 6^2$
   Triple: 5, 4, 6
7. $4^2 + 6^2 \stackrel{?}{=} 7^2$
   Triple: 4, 6, 7
8. $30^2 + 40^2 \stackrel{?}{=} 50^2$
   Triple: 30, 40, 50

For each figure, write a statement in the form $a^2 + b^2 = c^2$. Then tell whether or not the triangle is a right triangle.

SAMPLE: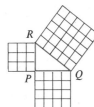

Solution: $a^2 + b^2 = c^2$

$3^2 + 4^2$	5^2
$9 + 16$	25

Since $9 + 16 = 25$, $\triangle PQR$ is a right triangle.

B   9.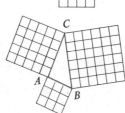

10.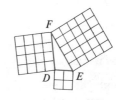

11.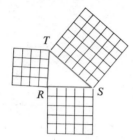

12.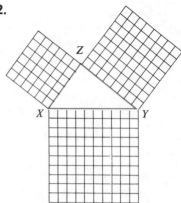

Use the expressions $2n$, $n^2 - 1$, and $n^2 + 1$ to write a Pythagorean triple for each suggested replacement for n. Check your answer by using the relationship $a^2 + b^2 = c^2$.

SAMPLE: $n = 4$ Solution: $2n = 8$; $n^2 - 1 = 15$; $n^2 + 1 = 17$
Pythagorean triple: 8, 15, 17

Check:

$a^2 + b^2$	$= c^2$
$8^2 + 15^2$	17^2
$64 + 225$	289
289	289

[C] **13.** $n = 5$ **16.** $n = 9$ **19.** $n = 17$
14. $n = 12$ **17.** $n = 15$ **20.** $n = 20$
15. $n = 10$ **18.** $n = 18$ **21.** $n = 14$

CHAPTER SUMMARY

Inventory of Structure and Concepts

1. A **line segment** is a set of points. Point A and point B are the endpoints of the segment named $\overline{AB}$.

2. If a line segment is extended indefinitely in both directions the result is a **line**. If point A and point B are any two distinct points on a line it may be named by $\overleftrightarrow{AB}$.

3. The union and intersection operations on sets apply to the sets of points called **geometric figures**.

Geometric Figures in the Plane **445**

4. The intersection of two **non-parallel lines** in a plane is a **point**. **Parallel lines** are lines whose intersection is the empty set (∅).

5. A **curve** is defined (intuitively) as a continuous path traced out by a piece of chalk moving across a flat (plane) surface such as a chalkboard. A curve is a set of points. A **simple curve** is a curve that has two distinct endpoints and does not intersect itself. A **simple closed curve** is a curve that begins and ends at the same point and does not cross itself. A simple closed curve separates a plane into three distinct sets of points: the interior of the curve; the curve itself; the exterior of the curve.

6. If a line segment is extended indefinitely in only one direction, the result is a **ray**. A ray has one endpoint. The ray with point A as its endpoint and point B some other point on the ray is named $\overrightarrow{AB}$.

7. An **angle** is the union of two rays that have the same endpoint called the **vertex**. An angle separates a plane into three distinct sets of points: the interior of the angle; the angle itself; the exterior of the angle.

8. A **triangle** is the union of three line segments joining three points not in the same line.

Vocabulary and Spelling

line segment (*p. 417*)

line (*p. 417*)

intersection (*p. 418*)

intersecting lines (*p. 418*)

parallel lines (*p. 418*)

empty set (*p. 418*)

union (*p. 422*)

curve (*p. 424*)

simple curve (*p. 425*)

simple closed curve (*p. 425*)

interior (*p. 425*)

exterior (*p. 425*)

ray (*p. 428*)

angle (*p. 428*)

vertex (*p. 428*)

side (*p. 428*)

right angle (*p. 431*)

protractor (*p. 432*)

obtuse angle (*p. 432*)

acute angle (*p. 432*)

triangle (*p. 436*)

scalene triangle (*p. 436*)

isosceles triangle (*p. 437*)

equilateral triangle (*p. 437*)

acute triangle (*p. 437*)

obtuse triangle (*p. 437*)

equiangular triangle (*p. 438*)

hypotenuse (*p. 440*)

Pythagorean Theorem (*p. 440*)

Pythagorean triple (*p. 440*)

Chapter Test

To each item in Column 1, match the correct name from Column 2.

COLUMN 1

1. $c^2 = a^2 + b^2$

2. A————B

3. A triangle with all sides of equal length
4. A triangle with no two sides of equal length
5. The union of two rays
6. A triangle all of whose angles have measures less than 90°

7. A•———•B→

8. An instrument for measuring angles
9. A triangle that has one angle with a measure greater than 90° and less than 180°
10. The common endpoint of two rays

COLUMN 2

A. Scalene triangle
B. Obtuse triangle
C. Point
D. Equilateral triangle
E. Isosceles triangle
F. Acute triangle
G. Ray
H. Line segment
I. Line
J. Angle
K. Vertex
L. Protractor
M. Pythagorean Theorem

Determine the value of x for each figure.

11.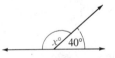
13.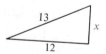
15.
12.
14.

Using the following figure replace the ? in each of Questions 16–18 with an appropriate symbol.

16. $\overline{AC} \cap \overline{BD} = ?$ 17. $\overline{BC} \cap ? = B$ 18. $\overline{AC}$? $\overline{CD} = \overline{AD}$

19. Give an example of a simple closed curve.
20. For each of the following, show whether it is or is not a Pythagorean triple.
   a. 8, 15, 17
   b. 9, 15, 18
   c. 7, 24, 25

Chapter Review

14–1, 2 Lines and Line Segments

Use appropriate symbols to describe each of the following.

1.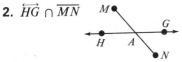

3.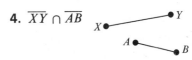

2. $\overleftrightarrow{HG} \cap \overline{MN}$

4. $\overline{XY} \cap \overline{AB}$

Using the following figure, replace the ? in each problem with an appropriate symbol.

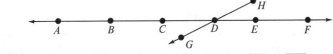

5. $\overline{AC} \cap \overline{BD} = ?$
6. $\overline{AD} \cap \overline{BC} = ?$
7. $\overleftrightarrow{AF} \cap \overleftrightarrow{GH} = ?$

8. $\overline{AC} \; ? \; \overline{BD} = \overline{AD}$
9. $\overline{CF} \; ? \; \overline{EF} = \overline{CF}$
10. $\overline{DH} \; ? \; \overleftrightarrow{BE} = D$

14–3 Curves

Match each figure in Column 1 with one of the terms in Column 2.

COLUMN 1   COLUMN 2

11. •—————•   14. 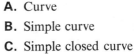   A. Curve
                                              B. Simple curve
12.    15.   C. Simple closed curve
                                              D. None of these
13.    16. ○

The figure shown is a simple closed curve. Tell whether each labeled point is on the curve, in the exterior of the curve, or in the interior of the curve.

17. A
18. B
19. C
20. D
21. E
22. F

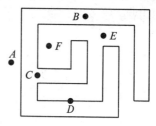

14–4 Rays and Angles in the Plane

23. The result of extending a line segment indefinitely in one direction is called a __?__.
24. The union of $\overrightarrow{AB}$ and $\overrightarrow{AC}$ is called an __?__.
25. The common endpoint of two rays that form an angle is called the __?__ of the angle.

Name each angle.

26.
27.
28.

14–5 Angle Measurement and the Protractor

29. An angle whose measure is greater than 90° and less than 180° is an __?__ angle.
30. The sum of the measures of the three angles of a triangle is __?__ degrees.

Use a protractor to find the measure to the nearest degree of angles A and B in the given triangle.

31. ∠A
32. ∠B

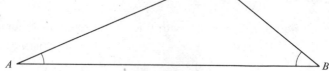

Without using a protractor, compute the value of the variable in each figure.

33. ∠x

34. ∠m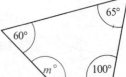

14-6 Kinds of Triangles

Classify each triangle as acute, obtuse, or right.

35.

37.

36.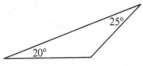

38.

Classify each triangle as scalene, isosceles, or equilateral.

39.

40.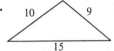

41.

42. In a right triangle the side opposite the right angle is called the __?__.

14-7 The Pythagorean Theorem

Determine the length of the unknown side of each right triangle by using the Pythagorean Theorem.

43.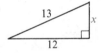

44.

■ ■

CHECK POINT
FOR EXPERTS

Similar Triangles

When two triangles have the same shape, we say they are **similar**. In this illustration $\triangle ABC$ is similar to $\triangle PQR$. We use the symbol $\sim$ to mean "is similar to" and write $\triangle ABC \sim \triangle PQR$.

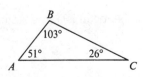
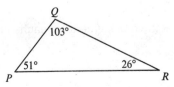

When we work with similar triangles, we are concerned with the relationships between **corresponding parts**. In $\triangle ABC$ and $\triangle PQR$ above:

$\overline{AB}$ corresponds to $\overline{PQ}$ 　　$\angle A$ corresponds to $\angle P$
$\overline{AC}$ corresponds to $\overline{PR}$ 　　$\angle B$ corresponds to $\angle Q$
$\overline{BC}$ corresponds to $\overline{QR}$ 　　$\angle C$ corresponds to $\angle R$

When two triangles are similar, the measures of their **corresponding angles** are **equal**. Study the following pair of similar triangles. Do you agree that the measures of the corresponding angles are equal?

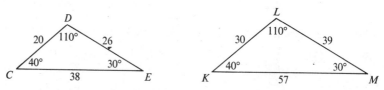

Is there any relationship between the measures of **corresponding sides** of a pair of similar triangles? In $\triangle CDE$ and $\triangle KLM$

$$\frac{CD}{KL} = \frac{20}{30} = \frac{2}{3}; \qquad \frac{CE}{KM} = \frac{38}{57} = \frac{2}{3}; \qquad \frac{DE}{LM} = \frac{26}{39} = \frac{2}{3}.$$

For similar triangles, the ratios between the measures of pairs of corresponding sides are equal.

We can use what we know about similar triangles to solve problems like the one in the following example.

EXAMPLE. If $\triangle KLR \sim \triangle BCD$, what is the length of KL?

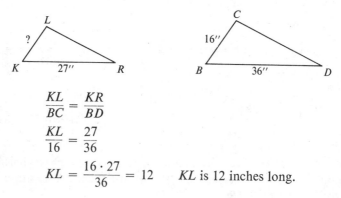

$$\frac{KL}{BC} = \frac{KR}{BD}$$

$$\frac{KL}{16} = \frac{27}{36}$$

$$KL = \frac{16 \cdot 27}{36} = 12 \qquad KL \text{ is 12 inches long.}$$

Questions

What is the replacement for each variable so the result is a true statement?

1. $\dfrac{1}{3} = \dfrac{n}{15}$ 　　2. $\dfrac{3}{15} = \dfrac{5}{t}$ 　　3. $\dfrac{10}{18} = \dfrac{5}{b}$

Geometric Figures in the Plane   451

4. $\dfrac{k}{5} = \dfrac{8}{10}$ 6. $\dfrac{m}{6} = \dfrac{4}{8}$ 8. $\dfrac{y}{3} = \dfrac{8}{2}$

5. $\dfrac{x}{3} = \dfrac{20}{12}$ 7. $\dfrac{3}{a} = \dfrac{10}{20}$ 9. $\dfrac{7}{c} = \dfrac{3}{6}$

Which sides are corresponding in each pair of similar triangles?

10.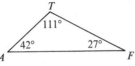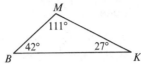

11.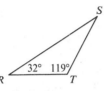

12. 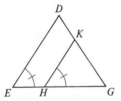   13.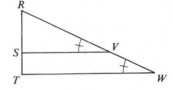

14. △XYZ and △BTR are similar. What are the measures of $\overline{YZ}$ and $\overline{BR}$?

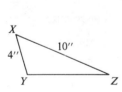 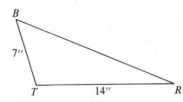

15. △BGH and △BKT are similar. What is the measure of ∠KTH? ∠BHG? ∠BKT? ∠GKT? ∠BGH?

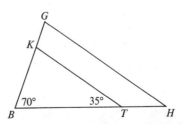

Cumulative Test

Part I. (Chapters 1–7)

Use each of the digits 1, 2, 5, 7 exactly once to write numerals for these whole numbers.

1. The largest number possible.
2. The smallest number possible.
3. The largest even number possible.
4. The largest odd number possible.
5. The smallest odd number possible.
6. The largest number less than 3500 that uses the four digits.
7. The smallest number greater than 2400 that uses the four digits.
8. The largest number between 1300 and 1600 that uses the four digits.

Write the number 20 as:

9. The sum of four consecutive even numbers.
10. The sum of four different odd numbers.
11. The sum of two nonzero even numbers having the greatest possible difference.
12. The sum of two even numbers having the least possible nonzero difference.

Use one word, either *odd* or *even*, to make a true statement.

13. The sum of two even numbers is always an __?__ number.
14. The sum of two odd numbers is always an __?__ number.
15. The sum of an even number and an odd number is always an __?__ number.
16. The sum of any collection of even numbers is always an __?__ number.

If the marks on the following number line are equally spaced, find the coordinate of each point described.

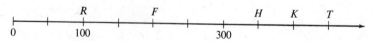

17. The point *H*.
18. The point halfway between *R* and *F*.
19. The point halfway between the origin and *F*.

20. The point one-third of the distance from the origin to T.
21. The point five-eighths of the distance from the origin to K.

Write two equivalent sentences (subtraction or division) for each of the following.

22. $14 = 3\frac{1}{2} \cdot 4$
23. $2\frac{1}{8} + 3\frac{1}{4} = 5\frac{3}{8}$
24. $(15)(6) = 90$
25. $0.26 + 0.03 = 0.29$
26. $s + t = k$
27. $m = bc$

Draw a number line graph for each set.

28. {the whole numbers between 3 and 10}
29. {the numbers between 2 and 7}
30. {the numbers between $\frac{1}{2}$ and $3\frac{1}{2}$, inclusive}
31. {the numbers greater than $\frac{3}{4}$}

Tell whether each set is finite or infinite.

32. $\{0, 2, 4, 6, \ldots\}$
33. $\{0, 2, 4, 6, 8\}$
34. $\{1, 2, 3, \ldots, 100\}$
35. {even numbers between 4 and 6}
36. {whole numbers greater than 1000}
37. $\{\frac{1}{2}, \frac{1}{3}, \frac{1}{4}, \ldots\}$

38. Which one of the sets in Questions 32–37 is the empty set?

Match each word statement in Column 1 with the corresponding number statement from Column 2.

COLUMN 1	COLUMN 2
39. Two increased by six is less than ten.	A. $2 + 6 \neq 10$
40. Six increased by four is equal to ten.	B. $10 < (2)(6)$
41. The sum of two and six is not equal to ten.	C. $2 < 6 < 10$
42. Ten is less than the product of two and six.	D. $10 = 6 + 4$
43. Six is between two and ten.	E. $2 + 6 < 10$
44. Six increased by ten is greater than two.	F. $6 + 10 > 2$

Match each item in Column 1 with the correct set from Column 2. Use $M = \{1, 2, 5, 6\}$, $P = \{2, 3, 4\}$, and $Q = \{3, 4, 7\}$.

COLUMN 1	COLUMN 2
45. $M \cup Q$	A. $\{2\}$
46. $M \cap P$	B. $\emptyset$
47. $P \cup Q$	C. $\{3, 4\}$
48. $Q \cap M$	D. $\{2, 3, 4, 7\}$
49. $M \cup P$	E. $\{1, 2, 3, 4, 5, 6\}$
50. $P \cap Q$	F. $\{1, 2, 3, 4, 5, 6, 7\}$

Complete each table according to the given function rule.

Rule: +3 and ÷ 4

	Input	Output	Number Pair
51.	1	?	?
52.	3	?	?
53.	5	?	?
54.	9	?	?

Rule: $6x - 2$

	x	$f(x)$	$(x, f(x))$
55.	1	?	?
56.	2	?	?
57.	3	?	?
58.	4	?	?

Tell whether or not each set of number pairs is a function.

59. $\{(0, 0), (1, 3), (2, 6), (3, 9)\}$
60. $\{(1, 2), (2, 3), (3, 4), \ldots\}$
61. $\{(1, 3), (2, 3), (4, 3), (5, 3), (6, 3)\}$
62. $\{(3, 1), (3, 2), (3, 3), (3, 4), \ldots\}$

Name the coordinates (number pair) for each labeled lattice point.

63. A
64. B
65. C
66. D
67. E

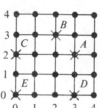

Simplify each expression.

68. $6 + 9(8 - 3)$
69. $\dfrac{15 + (3 \cdot 7)}{9}$
70. $7[10 \div (3 + 2)]$
71. $5[(7 + 4) + 3] + 10$

Write the simplest form for:

72. $\dfrac{x}{7} \cdot 5 \cdot y$
73. $\frac{1}{2}(rs)(6t)$

Find the value of each expression when $a = 5$, $b = 2$, and $c = 10$.

74. $\dfrac{7a + 9}{b}$
75. $(a + b)(c - b)$
76. $(ab)^2$
77. $(ab)(bc) + \dfrac{ac}{b}$
78. $\dfrac{(ab) + (b + c)}{a + b}$
79. $a + b^2 + c^2$

Geometric Figures in the Plane   455

Write each set of numbers in roster form.

80. {the factors of 15}
81. {the multiples of 6}
82. {the prime numbers}
83. {the divisors of 18}

Find the solution set of each equation if the replacement set is {1, 2, 3}.

84. $4m + 6 = 14$
85. $x^2 = 4x - 3$
86. $24 + 2t = 29$
87. $\dfrac{4x}{3} = 1\tfrac{1}{3}$

Find the area and the perimeter of each figure, using one of the following formulas: $P = 4s$; $A = \pi r^2$; $A = lw$; $A = s^2$; $C = \pi d$; $P = 2l + 2w$. (Let $\pi = \tfrac{22}{7}$.)

88.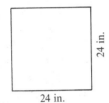
24 in.

89.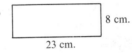
23 cm.   8 cm.

90.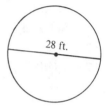
28 ft.

Match each inequality in Column 1 with the graph of its solution set from Column 2. The replacement set is {the numbers of arithmetic}.

COLUMN 1

91. $k > 1\tfrac{1}{2}$
92. $3a + 1 < 7$
93. $2x \geq 5$
94. $n + 2 < 3.5$
95. $2m \leq \tfrac{12}{2} - 1$

COLUMN 2

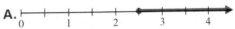

Solve each equation.

96. $6n + 7 = 49 + 3n$
97. $3a - 12 = 60 + 3$
98. $35 = 11 + 8m$
99. $0.7k - 2 = 1.5$
100. $5c - 2 = 1 - 4c$
101. $\dfrac{6t}{9} = 10$
102. $\dfrac{b}{4} = 3.5$
103. $\tfrac{1}{5}w = 2\tfrac{1}{2}$
104. $\dfrac{3m}{2} = 9$
105. $4c + 3c = 45$
106. $5k + 3 + k = 21$
107. $3b - 15 = 0$
108. $\tfrac{2}{3}m + \tfrac{1}{3}m = 4\tfrac{1}{2}$
109. $2(x + 7) + x = 28$
110. $5t + t = 1.2$

456 Chapter 14

Match each item in Column 1 with the related item from Column 2.

COLUMN 1

111. Commutative property of addition
112. Identity element for addition
113. Identity element for multiplication
114. Commutative property of multiplication
115. Transitive property of equality
116. Symmetric property of equality
117. Addition property of equality
118. Multiplication property of equality
119. Distributive property
120. Subtraction property of equality
121. Associative property of addition
122. Division property of equality

COLUMN 2

A. $a \cdot 1 = 1 \cdot a = a$
B. $a \cdot b = b \cdot a$
C. $a(b + c) = ab + ac$
D. If $a = b$, then $\dfrac{a}{c} = \dfrac{b}{c} \ (c \neq 0)$.
E. If $a = b$ and $b = c$, then $a = c$.
F. $a + 0 = 0 + a = a$
G. If $a = b$, then $b = a$.
H. If $a = b$, then $a - c = b - c$.
I. $a + b = b + a$
J. $a + (b + c) = (a + b) + c$
K. If $a = b$, then $a + x = b + x$.
L. If $a = b$, then $a \cdot c = b \cdot c$.

123. The relative lengths of the sides of a rectangle are as shown. If the perimeter of the rectangle is 108 feet, what are its dimensions?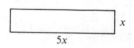

124. The sum of one-fourth a certain number and one-half the same number is 21. Find the number.

125. When three times a certain number is increased by 12, the result is six times the number. Find the number.

126. Five less than four times a certain number is equal to the number increased by four. Find the number.

Part II. (Chapters 8–14)

Complete the following to make true statements. Replace each first blank with one word, either **left** or **right**. Replace each second blank with one symbol, either $>$ or $<$.

1. -3 is to the __?__ of -10; therefore, -3 __?__ -10.
2. $8\frac{1}{2}$ is to the __?__ of 19; therefore, $8\frac{1}{2}$ __?__ 19.
3. -1.8 is to the __?__ of -1.2; therefore, -1.8 __?__ -1.2.

Use one word, either **positive** or **negative**, in each blank to make the statement true.

4. When $k > 0$, then k is __?__, and $-k$ is __?__.
5. When $k < 0$, then k is __?__, and $-k$ is __?__.

Simplify each expression.

6. $3(-5) + 4(-2)$ **9.** $-5 + [-2(3 + 5)]$ **12.** $(3)(-12)(\frac{1}{7})(0)$
7. $(-5)(7)(2)(-1)$ **10.** $-3[-1(4 - 9)]$ **13.** $(-2)(-8)(-5)$
8. $-4(8 - 6)$ **11.** $(-3)^4 + 20$ **14.** $(-\frac{1}{2})(-4)(-\frac{1}{3})$

Match each item in Column 1 with the correct numbers in Column 2 with reference to the number line for directed numbers.

COLUMN 1   COLUMN 2
15. A negative number $3\frac{1}{2}$ units from 0.   **A.** -5 and 3
16. A directed number that is its own additive inverse.   **B.** $3\frac{1}{2}$
17. Two directed numbers that are each 9 units from 0.   **C.** $\frac{2}{3}$ and $3\frac{1}{3}$
18. Two directed numbers that are each $1\frac{1}{3}$ units from 2.   **D.** 0
19. Two directed numbers that are each 4 units from $^-1$.   **E.** $-3\frac{1}{2}$
20. A positive number $3\frac{1}{2}$ units from 0.   **F.** 9 and -9

Match each inequality in Column 1 with the graph of its solution set from Column 2. Each replacement set is {the directed numbers}.

COLUMN 1   COLUMN 2
21. $x > -\frac{1}{2}$ **A.**

22. $m \leq 1\frac{1}{2}$ **B.**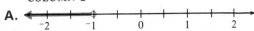

23. $t < ^-1$ **C.**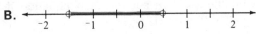

24. $^-1\frac{1}{2} < y < \frac{1}{2}$ **D.**

25. $k \neq 0$ **E.**

26. $r \geq -\frac{1}{2}$ **F.**

Tell which statements are true and which are false.

27. $8 - 12 \neq 12 - 8$ **32.** $3 < -5(2 + 4)$
28. $-4 + 10 \neq 10 - 4$ **33.** $-2 \geq -\frac{15}{3}$
29. $-\frac{10}{7} > 1.25$ **34.** $-6\frac{1}{2} + 6\frac{1}{2} \geq 7 - 7$
30. $-3(4 + 6) > -10 + 5$ **35.** $-(4 + 3) \neq -10 + 3(5 + \frac{2}{3})$
31. $-0.05 \leq 0 < 0.05$ **36.** $0.2 < 0.02 \leq 0.002$

Find the solution set for each equation if the replacement set is {the directed numbers}.

37. $3a = 4 - 19$
38. $\dfrac{x}{5} = -4 + 7$
39. $4.5 - 8.2 = w$
40. $3\tfrac{3}{4}y = -5(3)$
41. $t + (7)(-3) = 31$
42. $3x + 8 + 4x = 15$
43. $9b + 12 = 3b + 6$
44. $-1\tfrac{1}{3}w = 36 + 2\tfrac{2}{3}w$
45. $\dfrac{7r}{4} + \dfrac{r}{4} = -14$

Add these polynomials.

46. $3m - 2n - 3$
    $m + 3n + 6$
    $m - 2$

47. $3t^2 + 2t + 7$
    $t^2 + t$
    $-2t^2 - 4$

48. $-x^2 - 5x + 2$
    $4x^2 + 2x - 9$
    $-2x^2 + 3$

49. $(7a^2 + 2b^2)$ and $(4a^2 + 3ab - 5b^2)$
50. $(2t^4 + t^2 + 3t)$ and $(t^3 - t^2 + 2)$

Subtract each lower polynomial from the upper polynomial.

51. $14x - 2$
    $8x + 9$

52. $20k^2 + 4k - 6$
    $3k^2 - 5k + 1$

53. $8n^2 - 3n + 4$
    $n - 3$

Simplify each expression. (Remove parentheses and combine similar terms.)

54. $(8y + 11) - (3y + 1)$
55. $(a + b) - (a - b)$
56. $(2b^2 - b + 6) - (b^2 + 4)$
57. $(2rs - 4) - [rs + (6 - 3rs)]$

Find the solution set for each equation if the replacement set is {the directed numbers}.

58. $(3m - 1) + (m + 4) = 19$
59. $3a^2 - (a^2 + 2a - 6) = 11 - a^2$

Multiply these polynomials.

60. $4x^2y + 2x$
    $5xy^2$

61. $2x^2 - 2x + 3$
    $x - 1$

62. $a^2 + 3ab - b^2$
    $a + 2b$

63. $(3y + 4)(2y - 1)$
64. $(x - y)(x + y)$
65. $(3x - 7)(x + 1)$

Expand each of the following.

66. $(x + 2)^2$
67. $(3y - x)^2$

Complete each indicated division.

68. $\dfrac{22xy^4}{2y}$
69. $\dfrac{32a^2b}{-4b}$
70. $\dfrac{-64x^2y^3}{16xy^2}$
71. $\dfrac{21r^3s^5 - 7r^2s^3}{-7r^2s^3}$

72. $x - 3 \overline{) x^2 - 5x + 6}$ 73. $x + 5 \overline{) x^2 + 3x - 10}$ 74. $x - 4 \overline{) x^2 - 16}$

Complete each table according to the given function rule.

Rule: $x^2 - 5$

	x	$f(x)$	$(x, f(x))$
75.	-2	?	?
76.	-1	?	?
77.	0	?	?
78.	1	?	?
79.	2	?	?
80.	3	?	?

Rule: $2x^2 + 3x - 4$

	x	$f(x)$	$(x, f(x))$
81.	-6	?	?
82.	-4	?	?
83.	-2	?	?
84.	0	?	?
85.	2	?	?
86.	4	?	?

Use the given function equation and replacement set to write the function as a set of number pairs.

87. $f(x) = x^2 + 3$; $\{-5, -4, -3, 0, 3, 4, 5\}$
88. $f(t) = 4t^2 - 2$; $\{0, 2, 4, 6, 8, \ldots\}$
89. $f(m) = 4 - m^2$; $\{-1, -\frac{1}{2}, 0, \frac{1}{2}, 1\}$
90. A tank of cylindrical shape is 15 feet deep. Find the area of the bottom of the tank if its volume is 1200 cubic feet. (Formula: $V = Bh$)
91. A businessman is charged 6% interest on a loan of $2400. How much interest does he pay if the loan is for two years? (Formula: $I = prt$)
92. The distances between points A, B, C, and D are given in the figure. Find the distance from A to D, through points B and C.

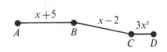

93. The area of the rectangle in the figure is given by the polynomial $2s^2 + t^2$. The area of the triangle is $s^2 - 3st - t^2$. Find the area of the shaded portion.

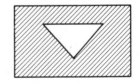

94. The volume of water in a tank is 960 cubic feet. The water is to be pumped into a new tank 9 feet wide, 12 feet long, and 10 feet high. Explain how you know the new tank is large enough to hold all the water.

FORMULAS

Circle	$A = \pi r^2$, $C = 2\pi r$	Cube	$V = s^3$
Parallelogram	$A = bh$	Rectangular Box	$V = lwh$
Right Triangle	$A = \frac{1}{2}bh$, $c^2 = a^2 + b^2$	Cylinder	$V = \pi r^2 h$
Square	$A = s^2$	Pyramid	$V = \frac{1}{3}Bh$
Trapezoid	$A = \frac{1}{2}h(b + b')$	Cone	$V = \frac{1}{3}\pi r^2 h$
Triangle	$A = \frac{1}{2}bh$	Sphere	$V = \frac{4}{3}\pi r^3$
Sphere	$A = 4\pi r^2$		

TABLE OF WEIGHTS AND MEASURES

AMERICAN SYSTEM OF WEIGHTS AND MEASURES

LENGTH
- 12 inches = 1 foot
- 3 feet = 1 yard
- $5\frac{1}{2}$ yards = 1 rod
- 5280 feet = 1 land mile
- 6076 feet = 1 nautical mile

AREA
- 144 square inches = 1 square foot
- 9 square feet = 1 square yard
- 160 square rods = 1 acre
- 640 acres = 1 square mile

VOLUME
- 1728 cubic inches = 1 cubic foot
- 27 cubic feet = 1 cubic yard

WEIGHT
- 16 ounces = 1 pound
- 2000 pounds = 1 ton
- 2240 pounds = 1 long ton

CAPACITY

Dry Measure
- 2 pints = 1 quart
- 8 quarts = 1 peck
- 4 pecks = 1 bushel

Liquid Measure
- 16 fluid ounces = 1 pint
- 2 pints = 1 quart
- 4 quarts = 1 gallon
- 231 cubic inches = 1 gallon

METRIC SYSTEM OF WEIGHTS AND MEASURES

LENGTH	10 millimeters (mm)	= 1 centimeter (cm)	$\doteq$ 0.3937	inch
	100 centimeters	= 1 meter (m)	$\doteq$ 39.37	inches
	1000 meters	= 1 kilometer (km)	$\doteq$ 0.6	mile
CAPACITY	1000 milliliters (ml)	= 1 liter (l)	$\doteq$ 1.1	quart
	1000 liters (l)	= 1 kiloliter (kl)	$\doteq$ 264.2	gallons
WEIGHT	1000 milligrams (mg)	= 1 gram (g)	$\doteq$ 0.035	ounce
	1000 grams	= 1 kilogram (kg)	$\doteq$ 2.2	pounds

Glossary

Absolute value. The absolute value of a nonzero number is either the number or its opposite, whichever is positive. The absolute value of 0 is 0. (p. 375)

Acute angle. An angle whose measure is less than 90°. (p. 432)

Acute triangle. A triangle all of whose angles are acute. (p. 437)

Addend. In addition, the numbers to be added. (p. 11)

Addition property of equality. For all numbers r, s, and t, if $r = s$, then $r + t = s + t$. (p. 194)

Addition property of inequality. For any directed numbers r, s, and t, if $r < s$, then $r + t < s + t$. (p. 326)

Additive identity element. Zero, because for any number n, $n + 0 = n$. (p. 174)

Additive inverse. See **Opposite of a directed number.**

Additive property of opposites. The sum of any directed number r, and its opposite, $-r$, is zero; $r + (-r) = 0$. (p. 250)

Additive property of zero. For every number r, $r + 0 = 0 + r = r$. (p. 174)

Algebraic expression. Any variable expression or numerical expression. (p. 96)

Angle. The union of two different rays which have the same endpoint, called the vertex of the angle; the rays are called the sides of the angle. (p. 428)

Associative properties: For all numbers r, s, and t,
$$r + (s + t) = (r + s) + t; \quad \text{(p. 165)}$$
$$r(st) = (rs)t. \quad \text{(p. 166)}$$

Axiom. A statement accepted as true without proof; also called postulate. (p. 157)

Base. In the expression 5^4, 5 is the base. (p. 108)

Binomial. A polynomial consisting of two terms. (p. 345)

Cardinal number. A whole number used to indicate quantity, but not order. (p. 154)

Closed curve. A curve which has no endpoints. (p. 425)

Closure property. A given set of numbers is closed under an operation if each result is a unique element of the given set. (p. 161)

Coefficient. Any factor of a product is the coefficient of the product of the remaining factors. In general, the entire numerical factor of an algebraic product is called its coefficient. (p. 102)

Commutative properties. For all numbers r and s,
$$r + s = s + r; \quad \text{(p. 164)}$$
$$rs = sr. \quad \text{(p. 165)}$$

Congruent. Having the same size and shape. (p. 214)

Consecutive order. The natural order used for counting. (p. 1)

Converse. The converse of the statement "If a, then b," is "If b, then a." (p. 442)

Coordinate of a point. The number matched with the point on the number line. (p. 4)

Curve. A set of points which may be described as a continuous path traced by a point moving on a plane. (p. 425)

Degree. One-ninetieth of a right angle. (p. 432)

Degree of a polynomial in one variable. The number indicated by the largest-valued exponent of any of its terms. (p. 346)

Digits. The symbols used in writing numerals in a place-value system of numeration. In the decimal system, the ten digits are 0, 1, 2, 3, ... , 9. (p. 7)

Disjoint sets. Sets having no members in common. (p. 50)

Distributive property. For all numbers r, s, and t, $r(s + t) = rs + rt$ and $(s + t)r = sr + tr$. (p. 169)

Division. Dividing by a number is the same as multiplying by its reciprocal. (p. 299)

Division property of equality. For all numbers r and s, and for all t except zero, if $r = s$, then $\frac{r}{t} = \frac{s}{t}$. (p. 198)

Division by zero. Division of any number by zero is said to be undefined. (p. 175)

Edge of a polyhedron. The intersection of two faces of the polyhedron. (p. 214)

Element of a set. Any object of the set; also called a member of the set. If a is an element of set S, we write $a \in S$. (p. 35)

Empty set. A set whose number property is zero; the usual symbol is ∅. (p. 40)

Endpoint of a line segment. See **Line segment.**

Equation. A number sentence which uses the symbol "=" to state that two expressions name the same number. (p. 8)

Equiangular triangle. A triangle having all of its angles equal in measure. (p. 438)

Equilateral triangle. A triangle having all of its sides equal in length. (p. 437)

Equivalent inequality. An inequality which has the same solution set as the given inequality. (p. 326)

Even number. A whole number which has 2 as a factor. (p. 48)

Existence property. The property that there is at least one object of a specified type. (p. 161)

Exponent. In the expression 5^4, the 4 is the exponent of the base 5. The exponent 4 tells the number of times 5 is used as a factor. (p. 108)

Exponential notation. Any product expressed with exponent(s). (p. 108)

Exterior of an angle. The set of all points neither on the angle nor in its interior. (p. 429)

Exterior of a curve. The set of all points which are neither inside nor on the curve. (p. 425)

Face of a polyhedron. Any of the regions which bound the polyhedron. (p. 214)

Factor. A number which is multiplied by another number to form a product. (p. 14)

Finite set. A set whose number property can be named by a whole number. (p. 40)

First coordinate of a point. The first number of the ordered number pair of coordinates of the point. (p. 68)

Formula. A general rule, usually stated in equation form, which is used to solve a particular kind of problem. (p. 138)

Function. A set of ordered pairs, in which no two pairs have the same first element. (p. 65)

Function rule. A rule used to describe a function. (p. 177)

$f(x)$-scale. The vertical scale on a lattice or number plane. (p. 71)

Graph of a number. The point on the number line that is paired with the number. (p. 18)

Graph of an ordered number pair. The point on a lattice or number plane that is paired with the ordered number pair. (p. 68)

Greater than. A directed number a is greater than a directed number b ($a > b$) if the point paired with a is to the right of the point paired with b on the number line. (p. 225)

Hypotenuse. The longest side of a right triangle; the side opposite the right angle. (p. 347)

Identity element for addition. See **Additive identity element.**

Identity element for multiplication. See **Multiplicative identity element.**

Improper subset. If a subset A of set S contains every element of S, then A is an improper subset of S. (p. 45)

Inequality. A statement that two numerical expressions do not name the same number. (p. 24)

Infinite set. A set whose number property cannot be named by a whole number. (p. 40)

Integer. A member of the set $\{\ldots, {}^-3, {}^-2, {}^-1, 0, 1, 2, 3, \ldots\}$. (p. 227)

Interior of an angle. The region between the two rays that are the sides of the angle. (p. 429)

Interior of a curve. The region of the plane enclosed by the curve. (p. 425)

Intersecting lines. Two lines which have one point in common. (p. 418)

Intersection of sets. The intersection of set A and set B, in symbols, $A \cap B$, is the set of all elements common to both A and B. (p. 51)

Inverse operations. Operations which produce opposite effects; for example, addition and subtraction are inverse operations. (p. 10)

Isosceles triangle. A triangle having at least two sides equal in length. (p. 437)

Lattice. An array of points used to graph ordered pairs of whole numbers. (p. 68)

Legs of a right triangle. The two sides that are segments of the rays which form the right angle; that is, the two sides which are not the hypotenuse. (p. 347)

Less than. A directed number a is less than a directed number b ($a < b$) if the point paired with a is to the left of the point paired with b on the number line. (p. 225)

Like terms. See **Similar terms.**

Line. A line segment which has been extended indefinitely in both directions. (p. 417)

Line segment. For two points, A and B, line segment AB, in symbols $\overline{AB}$, is the set of points including endpoints A and B, and all points between A and B; sometimes called segment AB. (p. 117)

Line of symmetry. A line which may be drawn on a figure in such a way that the two resulting parts are mirror images of each other. The figure is said to be symmetric about the line. (p. 240)

Magnitude of a number. The distance on the number line between the origin and the point named by the number. (p. 218)

Member of a set. Any object of the set; see also **Element of a set.** (p. 35)

Members of an equation (inequality). The expressions joined by the symbols of equality (inequality). (p. 126)

Monomial. A polynomial consisting of only one term. (p. 345)

Multiple. The product of a number and an integer is called a multiple of the number. Since $6 \times 3 = 18$, 18 is a multiple of both 6 and 3. (p. 105)

Multiplication property of equality. For all numbers r and s, and for all t except zero, if $r = s$, then $tr = ts$. (p. 200)

Multiplication property of inequality. For any numbers r, s and t, with $r < s$:

$$\text{if } t > 0, \text{ then } rt < st;$$
$$\text{if } t = 0, \text{ then } rt = st = 0;$$
$$\text{if } t < 0, \text{ then } rt > st. \quad \text{(pp. 329–330)}$$

Multiplicative identity element. 1, because for any number n, $n \cdot 1 = n$. (p. 175)

Multiplicative inverse. See **Reciprocal.**

Multiplicative property of one. For every number r, $r \times 1 = 1 \times r = r$. (p. 175)

Multiplicative property of zero. For every number r, $r \times 0 = 0 \times r = 0$. (p. 174)

Natural numbers. The numbers used for counting. (p. 311)

Negative direction. On the number line, the direction from the origin to a point on the negative side of the line. (p. 244)

Negative numbers. Numbers for points to the left of zero on the number line. (p. 217)

Network. A series of connected curves. (p. 343)

Number pair. See **Ordered number pair.**

Number property of a set. The number that tells how many objects are in the set. (p. 39)

Numbers of arithmetic. Zero and all numbers which name points to the right of zero on the number line. (p. 2)

Numerical expression. Any symbol which names a number. (p. 8)

Obtuse angle. An angle whose measure is greater than 90° and less than 180°. (p. 432)

Obtuse triangle. A triangle having one angle which is obtuse. (p. 437)

Odd number. A whole number which does not have two as a factor. (p. 48)

One-to-one correspondence. The pairing of the members of two sets in which each member of one set is matched with one and only one member of another set. (p. 39)

Open sentence. A number sentence that contains one or more variables. (p. 126)

Opposite of a directed number. The number which is the same distance from zero, but on the opposite side of zero on the number line; also called the additive inverse of the number. Zero is its own opposite. (p. 248)

Opposite of a polynomial. The opposite of a polynomial is a polynomial each of whose terms is the opposite of the corresponding term of the given polynomial. (p. 357)

Order property. For any two numbers r and s, exactly one of the following is true: $r < s$, $r = s$, or $r > s$. (p. 325)

Ordered number pair. A pair of numbers in which order is important, used generally in functions. (p. 64)

Ordinal number. A number used to indicate order, or position in a series; for example, *first*, *second*, *third*, etc. (p. 154)

Origin. On a number line, the point that is paired with zero. (p. 1)

Parallel lines. Lines in a plane which do not intersect. (p. 148)

Plane. A flat surface that extends indefinitely. (p. 424)

Plane curve. A continuous set of points in a plane. (p. 425)

Plane figure. A figure which lies on a plane, or flat surface. (p. 429)

Polyhedron. An enclosed region bounded by planes, called its faces. (p. 214)

Polynomial. An indicated sum of monomials. (p. 346)

Positive direction. On the number line, the direction from the origin to a point on the positive side of the line. (p. 244)

Positive numbers. Numbers for points to the right of zero on the number line. (p. 217)

Postulate. See Axiom.

Prime number. A whole number greater than 1, whose only factors are itself and 1. (p. 48)

Probability of an event. The ratio of the number of results that produce the event to the total number of possible results. The probability of event e is written $P(e)$. (p. 280)

Product. The result of multiplying two or more factors. (p. 14)

Proper subset. If subset A of set S does not contain *every* element of S, then A is a proper subset of S. (p. 45)

Property of the opposite of a sum. The opposite of the sum of two directed numbers is the sum of their opposites; $-(r + s) = (-r) + (-s)$. (p. 253)

Pythagorean Theorem. For any right triangle, if a and b represent the lengths of the legs and c represents the length of the hypotenuse, $a^2 + b^2 = c^2$. (p. 441)

Pythagorean triple. A set of three positive integers which can replace the variables in the formula $a^2 + b^2 = c^2$ and make a true statement. These three numbers may represent the lengths of the sides and hypotenuse of a right triangle. (p. 441)

Quadrants. The four sections into which a number plane of pairs of directed numbers is divided by the x-scale and the $f(x)$ scale. (p. 340)

Rational number. A number which may be written in the form $\frac{a}{b}$ where a and b are integers and $b \neq 0$. (p. 312)

Ray. A line segment extended indefinitely in only one direction. The symbol $\overrightarrow{AB}$ names a ray with point A as its endpoint and point B as some other point on the ray. (p. 428)

Reciprocal. Two numbers whose product is 1 are reciprocals (multiplicative inverses) of each other. For $a \neq 0$, the reciprocal of a is $\frac{1}{a}$; 0 has *no* reciprocal. (p. 298)

Reciprocal property. For every directed number r, except 0, there is a number $\frac{1}{r}$ such that $r \times \frac{1}{r} = \frac{1}{r} \times r = 1$. (p. 298)

Reflexive property of equality. For any number r, $r = r$. (p. 158)

Repeating decimal. A decimal numeral which contains a succession of blocks of one or more digits which repeat indefinitely. (p. 312)

Replacement set. For an open sentence, the set from which replacements for the variable may be chosen. (p. 71)

Right angle. An angle whose measure is 90°. (p. 432)

Right triangle. A triangle one of whose angles is a right angle. (p. 347)

Root. A solution of an open sentence. (p. 128)

Roster. A list of the members of a set. (p. 36)

Scalene triangle. A triangle having no two sides equal in length. (p. 436)

Scientific notation. Notation in which a number is expressed as the product of a positive number less than ten and a power of 10. For example, 9,400,000 in scientific notation is written 9.4×10^6. (p. 122)

Second coordinate of a point. The second number of the ordered pair of coordinates of the point. (p. 68)

Segment. See **Line segment**.

Set. A collection of objects described so that it is always possible to tell whether or not a particular object belongs to the set. (p. 35)

Sides of an angle. The two rays that form the angle. (p. 428)

Similar figures. Figures that have the same shape. (p. 449)

Similar terms. Terms, such as $5ab$ and $8ab$, which contain the same variable(s) as factor(s). (p. 191)

Simple closed curve. A curve which has no endpoints and does not cross itself at any point. (p. 425)

Simple curve. A curve which does not cross itself at any point. (p. 425)

Solution of an open sentence. From the replacement set for the variable, any member which makes the statement true; also called a root. (p. 128)

Solution set. The set of all solutions of an open sentence; also called truth set. (p. 128)

Solve. To find all of the replacements for the variable(s) in an open sentence which make the sentence true. (p. 128)

Standard form of a polynomial. A polynomial *in one variable* takes the standard form $ax^m + bx^n + cx^p + \cdots + d$ where a, b, c, and d represent directed numbers and m, n, and p are different positive integers written in descending order. A polynomial in *more than one variable* is written in standard form by ordering the terms according to the exponents of one of the variables. (p. 346)

Statement. A mathematical sentence which is either true or false. (p. 8)

Subset. Set A is a subset of set S if each member of A is also a member of S. (p. 45)

Substitution principle. For any numbers m and n, if $m = n$, then either one may be used in place of the other. (p. 161)

Subtraction of directed numbers. Subtracting a directed number is equivalent to adding its opposite; $a - b = a + (-b)$. (p. 264)

Subtraction property of equality. For all numbers r, s, and t, if $r = s$, then $r - t = s - t$. (p. 194)

Successor of a number. The next larger number in the set. (p. 2)

Sum. The number that is the result of adding two or more addends. (p. 11)

Symmetric property of equality. For any numbers r and s, if $r = s$, then $s = r$. (p. 158)

Symmetry. See **Line of symmetry**.

Term. A mathematical expression using numeral(s) or variable(s) or both to indicate a product. Since $\frac{k}{4} = \frac{1}{4}k$, $\frac{k}{4}$ is a term of the expression $\frac{k}{4} + 7t$. (p. 102)

Terminating decimal. A decimal numeral which can be expressed with a finite number of digits. (p. 312)

Transform. To change a given sentence into an equivalent sentence. (p. 195)

Transitive property of equality. For any numbers r, s, and t, if $r = s$, and $s = t$, then $r = t$. (p. 158)

Transitive property of inequality. For any directed numbers r, s, and t, if $r < s$, and $s < t$, then $r < t$. (p. 325)

Triangle. A union of three line segments which are determined by 3 points not on the same line. (p. 436)

Trinomial. A polynomial consisting of three terms. (p. 345)

Truth set. See **Solution set.**

Union of two sets. The union of sets A and B, in symbols $A \cup B$, is the set of all elements that are in A or B. (p. 51)

Uniqueness property. The property that there is at most one object of a specified type. (p. 161)

Unlike terms. Terms which have different variable factors. (p. 192)

Variable. A symbol, usually a letter, which is used to represent members of a specified set. (p. 96)

Variable expression. An expression containing at least one variable. (p. 96)

Vertex of an angle. The common endpoint of the two rays which form the angle. (p. 428)

Vertex of a network. An intersection of two or more of the curves in the network. (p. 343)

Vertex of a polyhedron. A point where three or more edges of the polyhedron intersect. (p. 214)

Whole numbers. Zero and the numbers used for counting. (p. 1)

x-scale. The horizontal scale on a lattice or number plane. (p. 71)

PICTURE CREDITS

Page x / *top*, Courtesy of Italian Government Travel Office — *bottom*, Robert W. Young, Design Photographers International; Page 34 / *top*, Culver Pictures, Inc. — *bottom*, Courtesy of French Embassy, Press and Information Division; Page 61 / Courtesy of Columbia University Library, D. E. Smith Collection; Page 62 / *top*, Courtesy of National Park Service, U.S. Dept. of the Interior — *bottom*, Courtesy of Central Mortgage and Housing Corp., Montreal; Page 89 / Courtesy of Columbia University Library, D. E. Smith Collection; Page 90 / *top*, Courtesy IBM, New York, photo by Lee Boltin (c) Time, Inc. 1967 — *bottom*, Courtesy of *Patriot Ledger*, Quincy, Mass., photo by Everett A. Tatreau; Page 123 / Courtesy of Columbia University Library, D. E. Smith Collection; Page 124 / *top*, Courtesy of Esther C. Goddard — *bottom*, Courtesy of National Aeronautics and Space Administration; Page 156 / *top*, Courtesy of Mexican National Tourist Council — *bottom*, The Mount Wilson and Palomar Observatories; Page 189 / Courtesy of Columbia University Library, D. E. Smith Collection; Page 190 / Courtesy of American Telephone and Telegraph Company; Page 216 / *top*, Courtesy of United States Navy — *bottom*, Courtesy of Grumman Aircraft Engineering Corp.; Page 242 / *top*, Courtesy of Radio Corporation of America — *bottom*, Courtesy of United States State Department; Page 282 / *top*, Culver Pictures, Inc. — *bottom*, Laurence Lustig; Page 314 / *top*, Reproduced by permission of Frederick A. Praeger, Inc. and Heineman Educational Books, Ltd., from Sherwood Taylor's *An Illustrated History of Science*, (c) 1955. Drawing by A. E. Thomson — *bottom*, Courtesy of Radio Corporation of America; Page 343 / Courtesy of Columbia University Library, D. E. Smith Collection; Page 344 / *top*, Riccard Grimoldi — *bottom*, Societa Italiana Autostrade; Page 377 / Ewing Galloway; Page 378 / *top*, The Smithsonian Institution — *bottom*, Courtesy of American Textile Manufacturer's Institute, Inc.; Page 415 / Courtesy of Columbia University Library, D. E. Smith Collection; Page 416 / *top*, Library of Congress — *bottom*, Courtesy of National Aeronautics and Space Administration.

Index

Absolute value, 375
Acute angle, 432
  triangle, 437
Addend, 11
Addition, 10
  of directed numbers, 243–244
  identity element for, 174, 251
  of polynomials, 345–353
  property of equality, 193–194
  property of inequality, 326
Additive identity element, 174, 251
  inverse, 250
  property of opposites, 250
  property of zero, 174, 251
Algebraic expression, 96
Angle(s), 428–433
  acute, 432
  corresponding, 450
  exterior of, 428
  interior of, 428
  measurement of, 431–433
  obtuse, 432
  right, 431–432
  side of, 428
  vertex of, 428
Archimedes, 61
Associative properties, 165–166, 257, 287
Axiom, 157

Base, 108
Basic operations, 10
Binomial, 345
  square of, 394

Cardinal numbers, 154–155
Closed curve, 425
  interior of, 425
  exterior of, 425
  simple, 425
Closure property, 161, 257, 287
Coefficient, 102
Combining similar terms, 191–192, 321–322
Commutative properties, 164–165, 257, 287
Comparing directed numbers, 225
Congruent, 214
Consecutive order, 1
Converse, 442
Coordinate of a point on a number line, 4, 218
Coordinates of a point on a lattice, 68
Corresponding angles, 450
  parts, 450
  sides, 450
Cube, 108, 214
Cubic numbers, 155
Curve(s), 424–426

Decimal(s), repeating, 312
  terminating, 312
Degree, 432
  of a polynomial, 346
Difference of two squares, 395
  See also Subtraction
Digit, 7
Directed number(s), 218–307
  adding, 243–244
  comparing, 225
  dividing, 295–297
  graphing, 218, 314
  magnitude of, 218
  multiplying, 283–290
  opposite of, 247–248
  properties of, 257, 287
  representations of, 227
  subtracting, 260–264
Direction on the number line, 221
Disjoint sets, 50
Distributive property
  of multiplication over addition, 169, 287
  of multiplication over subtraction, 171–172
Division, 15
  of directed numbers, 295–297
  of polynomials, 403–406
  property of equality, 198
  by zero, 174, 296–297
Dodecahedron, 214

Edge of a polyhedron, 214
Element of a set, 35
Empty set, 40, 418
Endpoints of a line segment, 417
Equality
  addition property of, 193–194
  division property of, 198
  multiplication property of, 200
  subtraction property of, 194
  reflexive property of, 158
  relation, 23
  symmetric property of, 158
  transitive property of, 158
  See also Equation(s)
Equation(s), 8, 125
  solving, 128–129, 194–195, 198, 202–203, 207, 316, 318, 321–322, 365
Equiangular triangle, 438
Equilateral triangle, 437
Equivalent inequality, 326
  statement, 11, 15
Euler, Leonhard, 343
Euler's formula, 215
Even number(s), 47, 104–105
Existence property, 161

467

Expanding an expression, 394
Exponent(s), 108
　rules for, 380, 384, 400–401
　zero as, 414
Expression(s), algebraic, 96
　factoring, 102
　simplifying, 91–95
　term of, 102
　variable, 96
Exterior of an angle, 429
　of a curve, 425

Face of a polyhedron, 214
Factor, 14, 102
Factoring expressions, 102
Fibonacci, 187
Fibonacci numbers, 189
Fibonacci Sequence, 189
Finite set, 40
First coordinate, 68
Formulas
　area, 100–101, 111–112, 138–139, 334
　circumference, 138–139
　interest, 100, 335
　perimeter, 138–139
　volume, 100, 111–112, 138–139, 333
Function(s), 63–65
　graph of a, 71
　machines, 64, 70–71, 266–267
　using directed numbers, 266–267, 302–303
　using polynomials, 366–367
　using variables, 176–177

Galileo Galilei, 415
Gauss, Carl Frederick, 89
Generalized form of an even number, 104
　of a multiple of a number, 105
　of an odd number, 105
Geometric figures, 417–442
Graph of a directed number, 218, 314
　of an inequality, 143, 230, 232
　of a number, 18
　of a number pair, 68, 341
　of a set, 42
　of a solution set, 128
Grouping symbols, 92–94

Hypotenuse, 437

Icosahedron, 214
Identity element for addition, 174, 251
　for multiplication, 175, 284
Improper subset, 45
Inequality(ies), 23–24, 125
　addition property of, 326
　equivalent, 326
　graphing, 143, 230–232
　multiplication property of, 329–330
　solving, 140, 143–144, 230, 232, 326, 329–330

transitive property of, 325
truth table for, 141, 230
Infinite set, 40
Integer(s), 227
　graph of, 227, 341
Interior of an angle, 429
　of a curve, 425
Intersection
　of lines and segments, 418, 421–422
　of sets, 51
Inverse operations
　addition and subtraction, 10–12
　multiplication and division, 14–15
Isosceles triangle, 437

Königsburg Bridge Problem, 343

Lattice(s) 68, 70–71
　for directed numbers, 341
Leg of a right triangle, 437
Leonardo of Pisa, 189
Like terms, 191
Line(s), intersecting, 418, 421–422
　parallel, 418, 421–422
　segment, 417, 421–422
　of symmetry, 240

Magnitude of a directed number, 218
Measuring angles, 431–433
Member of an equation (inequality), 126
　of a set, 35
Monomial(s), 345
Moves on the number line, 221–222
Multiple of a number, 105–106
Multiplication, 14
　of directed numbers, 283–293
　identity element for, 175, 284
　of monomials, 382
　of polynomials, 386–391
　property of equality, 200
　property of inequality, 329–330
　by a reciprocal, 299
Multiplicative identity element, 175, 284
　property of one, 175, 287
　property of zero, 174, 287

Natural numbers, 311
Negative direction, 221
　numbers, 217
　See also Directed number(s)
Network, 343
Nomographs, addition, 75
　for directed numbers, 269–270
　for exponents, 112–113
　multiplication, 79
Notation, exponential, 108
　scientific, 122
Number(s) of arithmetic, 2
　cardinal, 154–155
　cubic, 155
　even, 47, 104–105

Index 469

natural, 311
negative, 217
odd, 48, 105
ordinal, 154–155
pair, 68
positive, 217
prime, 48
property of a set, 39
rational, 312
square, 154–155
triangular, 59–60
whole, 1, 47
Number line, 1
  directed numbers on, 217
  direction on, 221
Number pair, 48
  graph of a, 68
  ordered, 64–71
Numerical expression, 8

Obtuse angle, 432
  triangle, 437
Octahedron, 214
Odd numbers, 48, 105
One-to-one correspondence, 39
Open sentence, 126
Operation(s), basic, 10
  inverse, 10–15
  on a number line, 20–23
  rule for order of, 92
Opposite(s), additive property of, 250
  of a directed number, 247–248
  of a polynomial, 357
  of a sum, 253
Order of operations, 92
  property, 26
Ordered number pair, 64–71
  graph of, 68, 341
Ordinal numbers, 154–155
Origin, 1, 217
Outcome of an event, 280

Parallel lines, 418
Parentheses, 92
Pascal, Blaise, 123
Pascal's Triangle, 123
Plane, 424
  curve, 425
  figure, 429
Polygons, congruent, 214
  regular, 214
Polyhedron, 214
  regular, 214
Polynomial(s), 346
  adding, 348–353
  degree of, 346
  dividing, 403–406
  multiplying, 391
  opposite of, 259
  properties of addition for, 353
  properties of multiplication for, 397–398

standard form of a, 346
subtracting, 359–360
Positive direction, 221
  numbers, 217
Postulate, 157
Power of a product, 383–384
Prime numbers, 48
Probability, 280
Problems
  area, 100–101, 111–112, 138–139,
    147–148, 334–335, 396
  circumference, 100, 138–139, 147–148
  using inequalities, 147–148
  interest, 100, 335
  method of solving, 134–136, 333, 363
  number, 147–148, 205–206, 208–209,
    302, 396
  perimeter, 100, 138–139, 147–148
  using polynomials, 363–365, 388–389,
    392–393
  volume, 108, 111–112, 138–139, 333–335
Product, 14
  of directed numbers, 283–291
  See also Multiplication
Proper subset, 45
Property(ies)
  additive, of opposites, 250
  additive, of zero, 174, 251
  associative, 165–166, 257, 287
  closure, 161, 257, 287
  commutative, 164–165, 257, 287
  distributive, 169, 171, 287
  of equality, *see under* Equality(ies)
  of inequality, *see under* Inequality(ies)
  multiplicative, of one, 175, 287
  multiplicative, of zero, 174, 287
  number, of a set, 39
  of the opposite of a sum, 253
  order, 26, 325
Protractor, 432
Pythagoras, 441
Pythagorean Theorem, 440
Pythagorean triple, 441

Quadrants, 340

Rational numbers, 312
Ray, 428
Reciprocal(s), 298
  multiplication by, 299
  property of, 298–299
Reflexive property of equality, 158
Regular polygon, 214
Relation symbols, 143–144
Repeating decimal, 312
Replacement set, 71, 96
Representations of directed numbers, 227
Right angle, 431–432
  triangle, 437, 440–442
Root, 128, 299
Roster form, 36

Scale, $f(x)$-, 71
  horizontal, 68
  logarithmic, 80
  vertical, 68
  x-, 71
Scalene triangle, 436
Scientific notation, 122
Second coordinate, 68
Segment, 417
Set(s), 35
  disjoint, 50
  element of, 35
  empty, 40
  finite, 40
  graph of, 42–43
  infinite, 40
  intersection of, 51
  number property of, 39
  replacement, 71, 96
  roster form of, 36
  solution, 128
  subset, 45
  truth, 128
  union of, 51
Side of an angle, 428
Sides of a triangle, 436
Similar terms, 191–192
  combining, 191–192, 321–322
Similar triangles, 449
Simple closed curve, 425
Simple curve, 425
Simplifying expressions, 91–95
Solution
  of equations, *see under* Solving equations
  of inequalities, *see under* Solving inequalities
  set, 128
Solving
  equations, 128–129, 194–195, 198, 202–203, 207, 316, 318, 321–322, 365
  inequalities, 140–141, 143–144, 230, 232, 326, 329–330
  problems, methods of, 134–136, 333, 363
  See also Problems
Square of a binomial, 394–395
Square numbers, 154–155
Standard form of a polynomial, 346
Statement, 8, 125–126
  equivalent, 11, 15
Subset, 45
  improper, 45
  proper, 45
Substitution principle, 161
Subtraction, 10
  of directed numbers, 260–264
  of polynomials, 359–360
  property of equality, 194
Successor of an even number, 105
  of an odd number, 105
  of a whole number, 2
Sum, 11
  of the measures of angles, 432–433
  opposite of a, 253
  See also Addition
Symbols, ix
Symmetric property of equality, 158
Symmetry, 240
  line(s) of, 240
  in operation tables, 187

Terminating decimal, 312
Terms, combining, 191–192, 321–322
  of an expression, 102
  like, 191
  of a polynomial, 345
  similar, 191–192
  unlike, 192
Tetrahedron, 214
Topology, 343
Transformation, 194
Transitive property of equality, 158
  of inequality, 325
Triangle(s), 436
  acute, 437
  corresponding parts of, 450
  equiangular, 438
  equilateral, 437
  isosceles, 437
  obtuse, 437
  right, 437
  scalene, 436
  similar, 449
  sum of measures of the angles, 432
Triangular numbers, 59–60
Trinomial, 345
Truth set, 128
  table, 128, 230

Union of lines and segments, 422
  of sets, 51
Uniqueness property, 161
Unlike terms, 192

Variable, 96
  expression, 96
Venn diagram, 36, 50–52
Vertex of an angle, 428
  of a network, 343
  of a polyhedron, 214

Whole number(s), 1, 47

Zero, additive property of, 251
  division by, 174
  as an exponent, 414
  multiplicative property of, 174, 287

answers
for odd-numbered exercises
Elementary
PART 1 Algebra
Denholm/Dolciani/Cunningham

Houghton Mifflin Company
BOSTON New York Atlanta Geneva, Ill. Dallas Palo Alto

1973 Impression

Copyright © 1970 by Houghton Mifflin Company. All rights reserved. No part of this work may be reproduced or transmitted in any form or by any means, electronic or mechanical, including photocopying and recording, or by any information storage or retrieval system, without permission in writing from the publisher. Printed in the U.S.A.

Answers to Odd-Numbered Exercises

Chapter 1. Basic Concepts about Numbers and Numerals

Pages 3–4 Written Exercises **A 1.** 4, 5, 6 **3.** 9, 10, 11, 12 **5.** 4, 5, 6, 7, 8 **7.** 6 and 7 **B 9.** 0 and 1 **11.** 1, 2, and 3 **13.** 39 **C 15.** 216

17. ⊢—+—+—+—+—+—→
0 $\frac{1}{6}$ $\frac{1}{3}$ $\frac{1}{2}$ $\frac{2}{3}$ $\frac{5}{6}$ 1

19. ⊢—+—+—+—+—+—+—→
0 $\frac{1}{3}$ $\frac{4}{6}$ 1 $\frac{8}{6}$ $\frac{10}{6}$ 2

Pages 6–7 Written Exercises **A 1.** W: 0; F: $\frac{1}{4}$; S: $\frac{3}{4}$; K: 1; B: $1\frac{3}{4}$ **3.** G: $\frac{1}{8}$; H: $\frac{5}{8}$; J: $\frac{7}{8}$; T: $1\frac{1}{8}$ **5.** R: 0.3; S: 0.4; H: 0.7; K: 0.9; V: 1.1 **7.** M **9.** M **B 11.** 3, 4 **13.** 4, 5, 6, 7 **15.** 3, 4, 5, 6, 7 **17.** 1, 2, 3, 4, 5, 6 **19.** 5 **21.** $\frac{5}{2}$ **23.** 1 **25.** 3 **27.** $7\frac{1}{5}$ **29.** 5.4 **C 31.** $\frac{18}{5}$ **33.** $\frac{23}{3}$ **35.** $\frac{31}{8}$

Pages 9–10 Written Exercises **A 1.** true **3.** true **5.** false **7.** true **9.** false **11.** false **13.** false **15.** true **B 17.** 3 **19.** 2 **21.** 8 **23.** 3 **25.** 1 **27.** 7 + 10 + 8 = 5 × 5 **29.** 48 + 62 = 15 + 95 **31.** 63 ÷ 9 = 3 + 4 **33.** $5\frac{1}{2}$ + $3\frac{1}{2}$ = 18 ÷ 2 **C 35.** $3\frac{1}{4}$ − $1\frac{3}{4}$ = $1\frac{1}{4}$ + $\frac{1}{4}$ **37.** 1.25 − 0.22 = 0.64 + 0.39 **39.** false **41.** false **43.** false

Pages 13–14 Written Exercises **A 1.** 23 **3.** $8\frac{7}{8}$ **5.** 64 **7.** 11.2 **9.** $3\frac{3}{5}$ **11.** 6.9 **13.** $8\frac{3}{8}$ **B 15.** 17 **17.** 8 **19.** 6 **21.** 45 **23.** 7.8 **25.** $5\frac{5}{8}$; $5\frac{5}{8}$ − $3\frac{1}{4}$ = $2\frac{3}{8}$; $5\frac{5}{8}$ − $2\frac{3}{8}$ = $3\frac{1}{4}$ **27.** 45; 117 − 72 = 45; 117 − 45 = 72 **29.** 64.2; 64.2 − 36.5 = 27.7; 64.2 − 27.7 = 36.5 **31.** 72.2; 72.2 − 52.5 = 19.7; 72.2 − 19.7 = 52.5 **33.** $11\frac{11}{16}$; $11\frac{11}{16}$ − $8\frac{5}{8}$ = $3\frac{1}{16}$; $11\frac{11}{16}$ − $3\frac{1}{16}$ = $8\frac{5}{8}$ **35.** 751; 751 − 642 = 109; 751 − 109 = 642 **C 37.** 15 + 35 = 50 **39.** $8\frac{1}{3}$ + $6\frac{1}{3}$ = $14\frac{2}{3}$ **41.** 65 + 20 = 85 **43.** $6\frac{3}{8}$ + $2\frac{1}{2}$ = $8\frac{7}{8}$

Pages 16–17 Written Exercises **A 1.** 98 **3.** $6\frac{3}{4}$ **5.** 5 **7.** 32.66 **9.** 12 **11.** 8 **13.** 2 **15.** 3 **B 17.** 0.62 **19.** $\frac{1}{3}$ **21.** 87; 87 ÷ $14\frac{1}{2}$ = 6; 87 ÷ 6 = $14\frac{1}{2}$ **23.** 15; 195 ÷ 13 = 15; 195 ÷ 15 = 13 **25.** 9.35; 9.35 ÷ 5.5 = 1.7; 9.35 ÷ 1.7 = 5.5 **27.** $1\frac{8}{9}$; 17 ÷ 9 = $1\frac{8}{9}$; 17 ÷ $1\frac{8}{9}$ = 9 **C 29.** 12; 108 ÷ 12 = 9; 108 ÷ 9 = 12 **31.** 7 × 45 = 315 **33.** $\frac{7}{3}$ × $\frac{1}{2}$ = $\frac{7}{6}$ **35.** 1.23 × 4.6 = 5.658

Pages 19–20 Written Exercises **A 1.**  **3.** ⊢●+●+●+●+●++→
0 1 2 3

5. ⊢+●+●+●●+→
0 1 2 3

7. ⊢+●+●+●+●++●+→
0 1 2 3

9. ⊢+●+●+●+●+●→
0 1 2 3 4 5

11. ⊢+●+●+●●+●+→
0 1 2 3

13. D **15.** B

B 17. ⊢+●+●+●+●++→
0 1 2 3 4 5 6 7

19. ⊢+●+●+●+●+●+→
0 1 2 3 4 5 6 7 8

21. ⊢+●+○━━━━○+→
0 1 2 3 4 5 6 7 8 9

23. ⊢+●+●━━━●+→
0 1 2 3 4 5 6 7 8

25. ●+→
0 1 2 3

C 27. ⊢+●+●+●+→
0 1 2 3 4 5 6 7

29.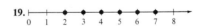

Pages 22–23 Written Exercises

A 1. ⊢—3—→⊢—5—→
0 1 2 3 4 5 6 7 8 9

3. ⊢—$2\frac{1}{2}$—→⊢—3—→
0 1 2 3 4 5 6

1

2 ANSWERS TO ODD-NUMBERED EXERCISES

5.

7.

9. $9 - 6 = 3$ 11. $6 \div 1\frac{1}{2} = 4$

B 13.

15.

17.

19.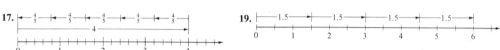

Pages 25–26 Written Exercises A 1. true **3.** false **5.** true **7.** false **9.** true **11.** true **13.** true **15.** true **17.** true **19.** false **21.** = **23.** > **25.** = **B 27.** < **29.** > **31.** = **33.** > **35.** 3 **37.** 9 **C 39.** 20 **41.** 8 (or any number other than 7) **43.** 2 (or any number other than 3) **45.** 7 (or any number between 6 and 8) **47.** 11 (or any number greater than 10)

Page 28 Chapter Test 1. 20 and 21 **3.** 5, 6, 7, 8 **5.** 0 **7.** $\frac{7}{12}$ **9.** 32 **11.** 111 **13.** 20

15.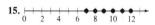

17.

19. =

Pages 29–31 Chapter Review 1. 12, 13, 14 **3.** 24 **5.** $A: \frac{1}{2}$; $B: \frac{5}{8}$; $C: \frac{7}{8}$; $D: 0.8$; $E: 1.0$; $F: 1.4$ **7.** $5\frac{1}{4}$; $2\frac{2}{3}$ **9.** $5\frac{1}{2}$; 15 **11.** $5\frac{1}{2}$ **13.** 31 **15.** $34 + 15 = 49$ **17.** 5 **19.** 21 **21.** 6.2

23.

25.

27. $1\frac{1}{4} + 1\frac{1}{4} = 2\frac{1}{2}$

29.

31. false **33.** true **35.** <

Pages 31–32 Review of Skills 1. 13 **3.** 144 **5.** 6.13 **7.** $\frac{7}{8}$ **9.** 10 **11.** 66 **13.** 2.4 **15.** $1\frac{3}{8}$ **17.** 30 **19.** 345,000 **21.** 19.22 **23.** $\frac{3}{8}$ **25.** 7 **27.** 3 **29.** 11 **31.** $1\frac{1}{3}$ **33.** 5 **35.** addend **37.** 1, 2, 5, 10; 1, 2, 4, 8, 16; 1, 2, 3, 4, 6, 8, 12, 24 **39.** 128 **41.** 15, 9 **43.** 21, 34 **45.** $\frac{5}{4}, \frac{5}{8}$ **47.** 20, 7 **49.** 37

Page 33 Questions 1. 99 **3.** 999 **5.** 11 **7.** 234, 243, 324, 342, 423, 432 **9.** 654, 564, 546, 456

Chapter 2. The Language and Symbols of Mathematics

Pages 37–39 Written Exercises A 1. F **3.** A **5.** G **7.** C **9.** B **11.** {a, b, c, d, e} **13.** {3, 4, 5, 6, 7, 8, 9} **15.** {4} **17.** {0, 1, 2, ..., 999} **19.** {t, o, m, r, w} **21.** false; true **B 23.** true; false; true **25.** true; true; false **27.** false; true **29.** true **31.** {Sunday, Monday, Tuesday, ..., Saturday} **33.** {whole numbers between 2 and 8} **35.** {whole numbers between 7 and 13} **C 37.** {11, 13, 15, ..., 99} **39.** {5, 7, 9} **41.** {10, 11, 12, ..., 99} **43.** {100, 102, 104, ..., 300}

Pages 41–42 Written Exercises A 1. D **3.** A **5.** C **7.** $n(J) = 1$ **9.** $n(C) = 15$ **11.** $n(K) = 0$ **13.** $n(W) = 0$ **15.** $n(P) = 4$ **B 17.** ∅; finite **19.** {1, 3, 5, ..., 89}; finite **21.** ∅; finite **C 23.** {0, 1, 2, 3, ...}; infinite **25.** $\{\frac{1}{2}, \frac{1}{3}, \frac{1}{4}, \frac{1}{5}, ...\}$; infinite **27.** ∅; finite

Pages 43–44 Written Exercises A 1. {Q, Y, B} **3.** {M, B} **5.** ∅ **7.** {M} **9.** {A, F, Y, K, B} **11.** {K, Y} **13.** {M, R, W, Q, A} **15.** {N} **B 17.** {M, R, W, Q} **19.** {R, W, Q}

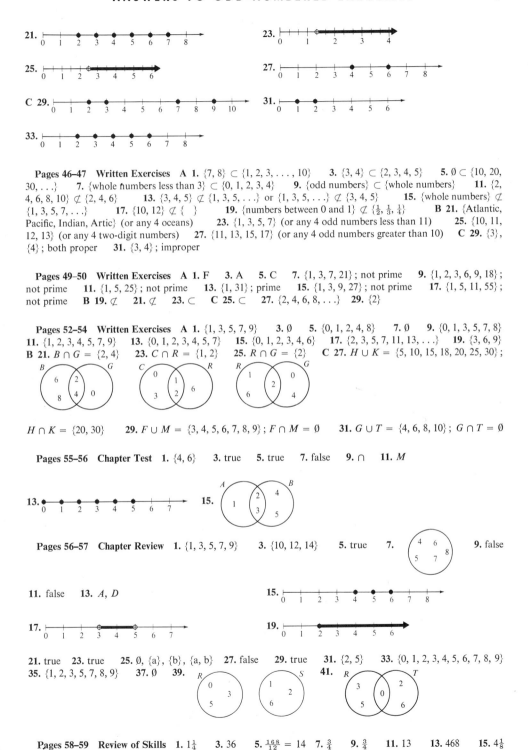

ANSWERS TO ODD-NUMBERED EXERCISES

Pages 59–60 Questions 1. 15 21 28 36

3. Fifth: $15 = 1 + 2 + 3 + 4 + 5$
Sixth: $21 = 1 + 2 + 3 + 4 + 5 + 6$
Seventh: $28 = 1 + 2 + 3 + 4 + 5 + 6 + 7$
Eighth: $36 = 1 + 2 + 3 + 4 + 5 + 6 + 7 + 8$
Ninth: $45 = 1 + 2 + 3 + 4 + 5 + 6 + 7 + 8 + 9$
Tenth: $55 = 1 + 2 + 3 + 4 + 5 + 6 + 7 + 8 + 9 + 10$
5. n; 120; 210; 1326; 5050

Chapter 3. Numbers, Functions, and Number Pairs

Pages 66–67 Written Exercises A 1. 7; 8; 9 **3.** 32; 56; 88 **5.** $\frac{5}{10}$; $\frac{9}{10}$; $\frac{13}{10}$; 1 **7.** 18; 27; 38; 59 **9.** C **11.** A **13.** B **15.** D **B 17.** $(5, 4\frac{3}{10})$ **19.** $7\frac{7}{10}$; $(7\frac{7}{10}, 7)$ **21.** $14\frac{3}{10}$; $(15, 14\frac{3}{10})$ **23.** 5; 4 **25.** 7; (7, 5) **27.** 4.3; (5.6, 4.3) **C 29.** -3 **31.** $\times 0$ **33.** $\times 0.1$ **35.** no **37.** yes **39.** $\times 1$ or $\div 1$ **41.** $\times 2$ or $\div \frac{1}{2}$

Pages 69–70 Written Exercises A 1.   **3.** 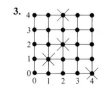 **5.**

7.   **B 9.** {(4, 0), (4, 1), (4, 2), (4, 3), (4, 4)} 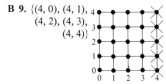   **11.** {(0, 2), (1, 2), (2, 2), (3, 2), (4, 2)}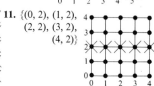

13. {(0, 0), (1, 0), (2, 0), (3, 0), (4, 0)} 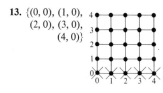   **15.** {(3, 0), (3, 1), (3, 2), (3, 3), (3, 4), (4, 0), (4, 1), (4, 2), (4, 3), (4, 4)} 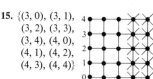   **17.** {(2, 0), (2, 1), (2, 2), (2, 3), (2, 4), (3, 0), (3, 1), (3, 2), (3, 3), (3, 4)}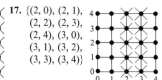

Pages 72–75 Written Exercises A 1. 8, (12, 8); 9, (14, 9); 10, (16, 10); 11, (18, 11); 12, (20, 12)
3. $\frac{1}{3}$, $(2, \frac{1}{3})$; $\frac{2}{3}$, $(4, \frac{2}{3})$; 1, (6, 1); $\frac{4}{3}$, $(8, \frac{4}{3})$; $\frac{5}{3}$, $(10, \frac{5}{3})$ **5.** {(1, 10), (4, 13), (9, 18), (16, 25), (25, 34)} **7.** {(10, 1), (20, 2), (30, 3), (40, 4), (50, 5), (60, 6)} **9.** {(0, 3), (1, 5), (2, 7), (3, 9), (4, 11), (5, 13), (6, 15), (7, 17)}
11. {(10, 10), (8, 9), (6, 8), (4, 7), (2, 6), (0, 5)} **B 13.** 11; 12; 13 **15.** 5; 7; 9; 11 **17.** 0; 6; 12; 18; 24; 30

C 19. {(0, 2), (2, 3), (4, 4), (6, 5)}

ANSWERS TO ODD-NUMBERED EXERCISES   **5**

21. {(1, 3), (3, 4), (5, 5), (7, 6), (9, 7)}

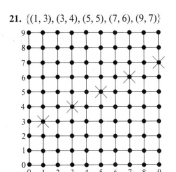

23. {(8, 4), (4, 2), (10, 5), (2, 1), (6, 3)}

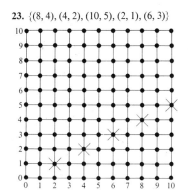

Pages 78–79   Written Exercises   **A**   **1.** 110   **3.** 59   **5.** 100   **7.** $\frac{1}{2}$ **9.** $\frac{1}{2}$ **11.** 2   **13.** 1.0   **15.** 1.2
17. 1.9   **B**   **19.** 0.040   **21.** 0.016   **23.** 0.015   **25.** 0.018   **27.** 0.015

Pages 81–83   Written Exercises   **A**   **1.** $\frac{72}{9} = 8$; $\frac{72}{8} = 9$ **3.** $45 = 9 \times 5$ **5.** $\frac{57}{3} = 19$; $\frac{57}{19} = 3$ **7.** $15 \times 5 = 75$ **9.** $13 = 4 \times 3\frac{1}{4}$ **11.** $0.36 = 0.4 \times 0.9$ **13.** $\frac{196}{28} = 7$; $\frac{196}{7} = 28$ **15.** $2\frac{1}{2} \times 6 = 15$ **17.** $\frac{228}{12} = 19$; $\frac{228}{19} = 12$ **19.** 16   **21.** 30   **23.** 9   **B**   **25.** $\frac{1}{16}$ **27.** $\frac{1}{4}$ **29.** $\frac{1}{2}$ **31.** 42   **33.** 45   **35.** $\frac{2}{9}$ **37.** 18
39. 0.24   **41.** 0.24   **C**   **43.** 0.02   **45.** 0.20   **47.** 0.3   **49.** 0.7

Pages 84–85   Chapter Test   **1.** true   **3.** false   **5.** $\frac{3}{8}$ **7.** 639   **9.** 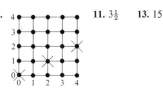   **11.** $3\frac{1}{2}$ **13.** 15
15. 6

Pages 85–87   Chapter Review   **1.** yes   **3.** no   **5.** {(1, 6), 3, 8), (5, 10), (6, 11)}   **7.** {(1, 1), (3, 3), (5, 5), (6, 6)}   **9.** 11; 23; 5   **11.**   **13.**   **15.** {(1, 2), (3, 3), (4, 1), (5, 4)}   **17.** 9
19. B   **21.** D   **23.** $3\frac{1}{2}$ **25.** $1\frac{5}{8}$ **27.** $\frac{24}{4} = 6$; $\frac{24}{6} = 4$ **29.** $\frac{168}{8} = 21$; $\frac{168}{21} = 8$ **31.** 7   **33.** 420   **35.** $\frac{1}{4}$

Page 87   Review of Skills   **1.** 12   **3.** 13   **5.** 28   **7.** 10   **9.** 42   **11.** 52   **13.** 60   **15.** 2   **17.** 5   **19.** 5
21. 18   **23.** 25   **25.** 18   **27.** 2   **29.** 27   **31.** 125   **33.** 81   **35.** 13   **37.** 5   **39.** 64   **41.** 125   **43.** 36
45. 10   **47.** 3   **49.** 6   **51.** 11   **53.** 72   **55.** 24 sq. ft.   **57.** 25 sq. in.   **59.** B   **61.** F   **63.** E

Chapter 4.   Variables, Expressions, and Sentences

Pages 93–94   Written Exercises   **A**   **1.** C   **3.** G   **5.** D   **7.** A   **9.** $(5 \times 2) + 7 = 17$ **11.** $5 + (4 \div 2) = 7$ **13.** $(8 \div 2) + 7 = 11$ **15.** $(\frac{3}{2} + \frac{5}{2}) \div 2 = 2$ **17.** $(20 \div 2) \div 5 = 2$ **B**   **19.** 0   **21.** 3
23. $\frac{11}{9}$ **25.** 27   **27.** $\frac{5}{8}$ **29.** $\frac{3}{2}$ **31.** 3   **33.** $\frac{3}{2}$ **35.** 134

Pages 95–96   Written Exercises   **A.**   **1.** 5   **3.** 75   **5.** 35   **7.** 6   **9.** 72   **11.** 120   **13.** 13   **15.** 4   **17.** 8
19. $45 - \frac{17}{6}$ **21.** $\frac{2 + 6}{10 - 2}$ **23.** $\frac{19 + 3}{(4)2}$ **B**   **25.** 15   **27.** $\frac{3}{4}$ **29.** 13   **31.** 5   **33.** 64   **35.** 1   **37.** 16
C   **39.** $\frac{17}{33}$ **41.** 56

Pages 98–100   Written Exercises   **A**   **1.** $12\frac{1}{3}$ **3.** 62   **5.** 192   **7.** $\frac{25}{7}$ **9.** 4   **11.** 116   **13.** 1225   **15.** 80
B   **17.** 80   **19.** 56   **21.** 49   **23.** $\frac{266}{75}$ **25.** 300   **27.** $\frac{9}{4}$ **29.** $\frac{9}{37}$ **31.** 7, 8, 9, 10   **33.** 9, 18, 27   **35.** $3, 4, \frac{25}{3}$
37. 18, 36, 54, 72   **39.** $(3 + n) - 5$ **41.** $\dfrac{k}{5 + r}$ **43.** $(x + 15) - m$ **C**   **45.** 56   **47.** 57   **49.** $19\frac{1}{3}$ **51.** $8\frac{19}{29}$

ANSWERS TO ODD-NUMBERED EXERCISES

Pages 100–101 Problems A 1. 33.6 ft. **3.** $180 **5.** 3510 sq. in. **7.** 528 in. **9. a.** $66\frac{1}{2}$ sq. in. **b.** 121 sq. in. **c.** 68 sq. yds. **d.** 125.96 sq. in. **B 11.** 104 **13. a.** 1792 sq. in. **b.** 324 sq. cm. **15.** 1934.4 per sq. ft. **C 17.** 17,472 lbs.

Pages 103–104 Written Exercises A 1. $4m, 6, 12$ **3.** $\frac{1}{4}, \frac{1}{2}z, z$ **5.** $a, ab, 1.5b$ **7.** a, b, ab **9.** $r, \frac{1}{4}, \frac{r}{4}, \frac{3}{4}$ **B 11.** $n+2, b(n+2), a(n+2), ab$ **13.** $a+b, \frac{x}{2}, x$ **15.** 2, 8, 4 **17.** $6xy$ **19.** $21mz$ **21.** $1.5rs$ **23.** $\frac{2nt}{3}$ **25.** $\frac{4\pi d}{5}$ **27.** $24yz$ **C 29.** $7(qr); 7q(r); 7r(q)$; etc. **31.** $14(mn); 14m(n); 2(7mn)$; etc. **33.** $3\left(\frac{t}{4}\right)$; $\frac{1}{4}(3t); \frac{3}{4}(t)$; etc. **35.** $\frac{1}{9}(2m); \frac{2}{9}(m); 2\left(\frac{m}{9}\right)$; etc. **37.** $\frac{1}{2}(abc); \frac{ab}{2}(c); \frac{ac}{2}(b)$; etc. **39.** $\frac{2}{3}(mk); 2m\left(\frac{k}{3}\right)$; $\frac{1}{3}(2mk)$; etc. **41.** $\frac{7}{8}(xy); \frac{1}{8}(7xy); 7x\left(\frac{y}{8}\right)$; etc.

Pages 106–107 Written Exercises A 1. true **3.** false **5.** true **7.** false **9.** $5 \cdot 3$; multiple of 5 **11.** $9 \cdot 13$; multiple of 9 **13.** $4 \cdot 9$; multiple of 9 **15.** $7 \cdot 7$; multiple of 7 **17.** $9 \cdot 6$; multiple of 9 **19.** $5 \cdot 3$; multiple of 5 **21.** $2 \cdot 31$; even **23.** $(2 \cdot 9) + 1$; odd **25.** $2 \cdot 63$; even **B 27.** $2 \cdot 15$; even **29.** $(2 \cdot 7) + 1$; odd **31.** $(2 \cdot 2) + 1$; odd **33.** odd **35.** odd **37.** odd **39.** odd **C 41.** even **43.** even **45.** even

Pages 109–110 Written Exercises A 1. $y \cdot y \cdot y \cdot y$ **3.** $rt \cdot rt$ **5.** $xy \cdot xy \cdot xy$ **7.** $13 \cdot k \cdot k \cdot k \cdot k$ **9.** $\frac{a \cdot b \cdot b}{c \cdot c}$ **11.** $7 \cdot r \cdot r \cdot s \cdot s \cdot s$ **13.** $\frac{a \cdot a \cdot a \cdot b}{m \cdot m \cdot m}$ **15.** $2 \cdot 2 \cdot 2 \cdot d \cdot d$ **17.** $(xyz)(xyz)$ **19.** $(ab)^3$ **21.** $(kt)^2$ **23.** m^3x^3 **25.** c^3d^2 **27.** $\frac{3^3g^2}{m^2}$ **29.** $(b-h)^3$ **B 31.** $10r^2$ **33.** $5w^3$ **35.** z^6 **37.** 40 **39.** 70,000 **41.** 3 **43.** 81 **45.** 3375 **C 47.** 161 **49.** 48 **51.** 19 **53.** $2xy(y); 2y(xy); 2x(y^2); 2y^2(x)$; etc. **55.** $x\left(\frac{xy^2}{2}\right); xy\left(\frac{xy}{2}\right); \frac{1}{2}xy(xy); \frac{1}{2}x^2y(y)$; etc. **57.** $mnr\left(\frac{r}{3}\right); \frac{1}{3}(mnr^2); mr\left(\frac{nr}{3}\right); mr^2\left(\frac{n}{3}\right)$; etc. **59.** $6(a+2)(a+2)^2; 6(a+2)(a+2)(a+2); (a+2)(6)(a+2)^2; 6(a+2)^2(a+2)$; etc. **61.** $\frac{3}{2}m(mn^3); 3m^2\left(\frac{n^3}{2}\right); 3mn\left(\frac{mn^2}{2}\right)$; etc. **63.** $2.5\pi r(r); \pi r(2.5r); 2.5\pi(r^2); 2.5r^2(\pi)$; etc. **65.** no **67.** no **69.** no

Pages 111–112 Problems A 1. 343 cu. in. **3.** $156\frac{1}{4}$ sq. in. **B 5.** 1047.2 cu. in. **7.** $523\frac{1}{3}$ cu. cm.

Pages 114–115 Written Exercises A 1. $2^3 = 8$ **3.** $3^3 = 27$ **5.** $6^2 = 36$ **7.** $3^2 + 7^4$ **9.** $b^3 + c^2$ **11.** $2^4 + 2$ **13.** $7a^5$ **15.** $(x+y)^3$ **17.** $\left(\frac{a}{5}\right)^3 - \left(\frac{a}{5}\right)^2$ **19.** 6^7 **21.** n^6 **23.** 2^8 **25.** $(xyz)^6$ **27.** 7^4 **29.** $\left(\frac{ab}{3}\right)^7$ **31.** mrn^2s^2 **B 33.** $16 = 16$ **35.** $512 = 512$ **37.** $32 > 25$ **39.** $10{,}000{,}000 \neq 1200$ **41.** $2000 \neq 1{,}000{,}000$

Pages 117–118 Chapter Test 1. 7 **3.** 55 **5.** 95 **7.** 107 **9.** $24\frac{1}{2}$ **11.** 3 **13.** 26 **15.** $3kb$ **17.** $(p+q) + ab$ **19.** 76 cm. **21.** 31.8 mi. **23.** $\frac{n}{5}(2)$ **25.** $\frac{3xy}{5}$ **27.** $2 \cdot 13$ **29.** $2 \cdot 27$ **31.** $ab \cdot ab \cdot ab \cdot ab$ **33.** $10v^3$ **35.** n^3m^3

Pages 118–119 Chapter Review 1. parentheses; brackets; fraction bars **3.** variables **5.** open **7.** 4, 5, 6, 7 **9.** $(2+7) + 3$ **11.** $\frac{r+s}{5}$ **13.** factors **15.** terms **17.** 1, 3, 5, 7 **19.** $2n + 1$ **21.** $r + 2$ **23.** 3^4 **25.** 7^5 **27.** $(ab)^3$ **29.** 7^3 **31.** 12^2 **33.** 5^5 **35.** k^8

Pages 120–121 Review of Skills 1. 40 **3.** 49 **5.** 56 **7.** 72 **9.** 144 **11.** 810 **13.** 5 **15.** 9 **17.** 2 **19.** 8 **21.** 20 **23.** 18 **25.** 1484 **27.** 179.4 **29.** 745 **31.** $37.94 **33.** $1 \cdot 9; 3 \cdot 3$ **35.** $1 \cdot 4; 2 \cdot 2$ **37.** $1 \cdot 15; 3 \cdot 5$ **39.** $1 \cdot 30; 2 \cdot 15; 3 \cdot 10; 5 \cdot 6$ **41.** $1 \cdot 20, 2 \cdot 10; 4 \cdot 5$ **43.** $1 \cdot 28; 2 \cdot 14; 4 \cdot 7$ **45.** 1 **47.** 2 **49.** $7\frac{3}{4}$ **51.** $1\frac{7}{8}$ **53.** $5\frac{1}{4}$ **55.** 12 **57.** 9 **59.** $3\frac{1}{2}$ **61.** 1 **63.** 20 **65.** 27 **67.** 4

Page 122 Questions 1. a. 10^6 **b.** 10^7 **c.** 10^3 **d.** 10^8 **e.** 10^2 **f.** 10^4 **g.** 10^9 **h.** 10^{10} **i.** 10^6 **3. a.** 4860 **b.** 324.5 **c.** 439,250 **d.** 8,700,000 **e.** 36,500 **f.** 10.07

ANSWERS TO ODD-NUMBERED EXERCISES

Chapter 5. Open Sentences

Pages 127–128 Written Exercises A 1. D **3.** A **5.** E **7.** F **9.** $3x + 5 < 25$ **11.** $x - \frac{3}{4} > \frac{1}{2}$ **13.** $7x^3 > m$ **B 15.** $3(x + 8) = 30$ **17.** $x^4 - 2 = 25$ **19.** 3 more than m is the same as 10. **21.** 31 is less than some number decreased by 25. **23.** Some number that is a repeated factor three times is greater than the sum of 7 and 2. **25.** 4 more than some number squared is the same as 18.

Pages 130–131 Written Exercises A 1. Solution set: {5} **3.** Solution set: {1} **5.** Solution set: $\{\frac{1}{2}\}$ **7.** Solution set: {14} **9.** Solution set: {2} **11.** Solution set: {whole numbers} **13.** {7} **15.** {9} **17.** {1.5} **19.** {1.6} **21.** $\{\frac{1}{4}\}$ **23.** {48} **B 25.** {2} **27.** ∅ **29.** {10} **31.** {0, 2, 4, 6} **C 33.** Solution set: {1, 3, 5, 7}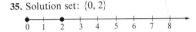

35. Solution set: {0, 2}  **37.** Solution set: ∅

Pages 133–134 Written Exercises A 1. $5t$ **3.** $y - 4$ **5.** $3x + 8$ **7.** $d + 3$ **9.** $n - 2$ **11.** $10 - 3t$ **13.** $2y - 5$ **15.** $5(a + b)$ **17.** $xy + 3a$ **19.** $7r + 5$ **B 21.** $\frac{t}{3}$ **23.** $h + 2$ **25.** $2d + 5$ **27.** $2y + 8$

Pages 136–137 Problems A 1. \$13 **3.** \$625 **5.** 45 **B 7.** 15 ft. wide; 75 ft. long **9.** 11 games won; 8 games lost **11.** 50 yds. long; 40 yds. wide **13.** 23 lbs; 38 lbs. **15.** 75 mi. on Saturday; 150 mi. on Sunday **17.** 10 and 11

Pages 139–140 Written Exercises A 1. $C = 2\pi r$; 88 in. **3.** $P = 2l + 2w$; 76 in. **5.** $A = lw$; 23.56 sq. in. **7.** $P = 4s$; 14 in. **B 9.** $P = 2l + 2w$; 14 in. **11.** $C = \pi d$; 66 in. **C 13. a.** 132 in. **b.** 158 in.

Pages 142–143 Written Exercises A 1. false; true; true **3.** false; true; true **5.** false; false; false **7.** false; true; true; true **9.** true; true; false; false **11.** true; false; false **13.** true; true; false **15.** false; false; true; true
17. {the numbers of arithmetic greater than 4} 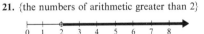 **19.** {the numbers of arithmetic less than 3}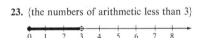

21. {the numbers of arithmetic greater than 2} 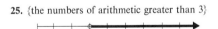 **23.** {the numbers of arithmetic less than 3}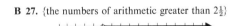

25. {the numbers of arithmetic greater than 3} 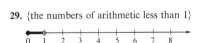 **B 27.** {the numbers of arithmetic greater than $2\frac{1}{2}$}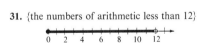

29. {the numbers of arithmetic less than 1}  **31.** {the numbers of arithmetic less than 12}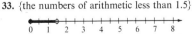

33. {the numbers of arithmetic less than 1.5}

Pages 144–146 Written Exercises A 1. true; true; true **3.** false; false; false **5.** true; false; false **7.** true; true; true **9.** false; false; true; true **11.** false; false; false; true **13.** false; false; true **B 15.** false; false; false **17.** false; false; true; true **19.** false; false; true **21.** true; true; true; true **23.** B **25.** D **27.** C **29.** {the numbers of arithmetic less than or equal to $1\frac{3}{4}$} **31.** {the numbers of arithmetic greater than or equal to $\frac{2}{3}$} **33.** {the numbers of arithmetic, except 1} **C 35.** {the numbers of arithmetic less than or equal to $\frac{1}{2}$} **37.** {the numbers of arithmetic greater than or equal to $1\frac{1}{2}$} **39.** {the numbers of arithmetic

8 ANSWERS TO ODD-NUMBERED EXERCISES

greater than or equal to 3} **41.** {the numbers of arithmetic greater than or equal to 2.2} **43.** {the numbers of arithmetic greater than or equal to 2 and less than or equal to 9} **45.** {the numbers of arithmetic less than 5}

47. 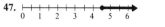 **49.**

51. ├─┼─┼─┼─┼─┼─┼─┼─┼─┼─●─►
    0 1 2 3 4 5 6 7 8 9 10

53. ├─┼─┼─┼─┼─┼─┼─○─┼─►
    0 1 2 3 4 5 6 7 8

Page 148 Problems A 1. {the numbers of arithmetic less than 400} **3.** {the numbers of arithmetic less than or equal to 18} **5.** {the numbers of arithmetic greater than or equal to 0.8} **B 7.** {the numbers of arithmetic less than or equal to 5320} **9.** {the numbers of arithmetic greater than or equal to 10}

Pages 149–150 Chapter Test 1. $5 - x = 3$ **3.** $x + 5 > 15$ **5.** {0, 1, 2, 3} **7.** {6, 7, 8, 9, 10}

9. ●─┼─○─┼─┼─┼─┼─►
    0 1 2 3 4 5 6

11. ├─┼─┼─┼─┼─●─┼─►
    0 1 2 3 4 5 6

13. $15 - 2x$ **15.** $A = 3.14 \cdot 5^2$ **17.** 21 yrs.

Pages 150–152 Chapter Review 1. $x + 5 = 24$ **3.** $x^2 + 2 = 15$ **5.** {2} **7.** {60} **9.** {18} **11.** $x = 8$ **13.** $2n - 5$ **15.** $6y + 15$ **17.** $2y + 10$ **19.** 4 runs **21.** 40 ft. **23.** 47.10 in. **25.** 139.25 sq. in. **27.** {0, 1, 2, 3} **29.** ∅ **31.** ├─┼─┼─┼─●─┼─►
    0 1 2 3 4 5 6
33. ●─┼─┼─┼─┼─┼─►
    0 1 2 3 4 5 6

35. ●─┼─○─┼─┼─┼─►
    0 1 2 3 4 5 6
37. ●─┼─┼─┼─┼─┼─●─┼─►
    0 1 2 3 4 5 6 7
39. $n \geq 13$

Page 153 Review of Skills 1. true **3.** true **5.** true **7.** true **9.** true **11.** true **13.** false **15.** false **17.** true **19.** true **21.** true **23.** false **25.** false **27.** false **29.** true **31.** {5} **33.** {0} **35.** ∅ **37.** {the numbers of arithmetic} **39.** 54 **41.** 55 **43.** 1 **45.** 54 **47.** 0 **49.** 53

Page 155 Questions 1. 1, 4, 9, 16, 25, 36, 49, 64, 81, 100 **3.** 15; 23; 173 **5.** 1; 8; 27; 64; 125; 216; 343; 512; 729; 1000 **7. a.** odd numbers **b.** odd numbers

Chapter 6. Operations, Axioms, and Equations

Pages 159–160 Written Exercises A 1. $18 = 10 + 8$ **3.** $x^2 \cdot x = x^3$ **5.** $3a + 6 = 3(2 + a)$ **7.** $2^2 + k = \dfrac{10 + 6}{4} + k$ **9.** $3 \cdot 8 = 20 + 4$ **11.** $4 \cdot 12 = 3 \cdot 4^2$ **B 13.** $3 + 1 = 4$ **15.** $a = b = rs$ **17.** $8 = 2 \cdot 2^2$ **C 19.** symmetric; transitive **21.** transitive; transitive

Pages 163–164 Written Exercises A 1. not closed for any operation **3.** closed for add. and mult.; not closed for div. and sub. **5.** closed for add., sub., and mult.; not closed for div. **7.** closed for mult., not closed for add., sub., and div. **9.** closed for mult.; not closed for add., sub., and div. **11.** closed for add. and mult.; not closed for sub. and div. **13.** not closed for any operation **15.** $\frac{1}{28}$ **17.** 18.4 **19.** $\frac{7}{4}$

Pages 167–168 Written Exercises A 1. $17.4 = 17.4$; comm. of add. **3.** $720 = 720$; comm. of mult. **5.** $1.8 = 1.8$; assoc. of add. **7.** $5\frac{5}{8} = 5\frac{5}{8}$; comm. of add. **9.** $1\frac{3}{8} = 1\frac{3}{8}$; assoc. of add. **B 11.** $42 = 42$; comm. of mult. **13.** $11.2 = 11.2$; comm. of mult. **15.** $67.2 = 67.2$; assoc. of mult. **17.** $134.4 = 134.4$; comm. of mult. **19.** $35.84 = 35.84$; assoc. of mult. **21.** $41.6 = 41.6$; comm. of mult. **23.** $14.6 = 14.6$; comm. of add. **25.** $16.2 = 16.2$; assoc. of add. **27.** comm. of add.; assoc. of add.; sub. prin.; sub. prin. **29.** comm. of mult.; assoc. of mult.; sub. prin.; sub. prin. **C 31.** ≠ **33.** = **35.** = **37.** ≠ **39.** = **41.** = **43.** =

Pages 170–171 Written Exercises A 1. $56 = 56$ **3.** $75 = 75$ **5.** $28 = 28$ **7.** $140 = 140$ **9.** $24\frac{6}{10} = 24\frac{6}{10}$ **11.** $10 = 10$ **13.** 1890 **15.** 732 **17.** 924 **19.** 918 **21.** 2360 **23.** 16,852 **B 25.** = **27.** ≠ **29.** ≠ **31.** ≠ **33.** ≠ **C 35.** ≠ **37.** = **39.** = **41.** sub. prin.; dist. prop.; sub. prin.; sub. prin. **43.** sub. prin.; dist. prop.; sub. prin.; sub. prin.

ANSWERS TO ODD-NUMBERED EXERCISES

Pages 173–174   Written Exercises   A 1. $56 = 56$ **3.** $51 = 51$ **5.** $107 = 107$ **7.** $31.00 = 31.00$ **9.** $49 = 49$ **B 11.** 76 **13.** 144 **15.** 32 **17.** 432 **19.** 23 **21.** $14\frac{5}{8}$ **23.** $14\frac{5}{8}$ **25.** 49 **C 27.** $10 + 2 = 12$ **29.** $10 + 3 = 13$ **31.** $20 + 3 = 23$

Page 176   Written Exercises   A 1. $\{1\}$ **3.** $\{1\}$ **5.** $\{0\}$ **7.** $\{0\}$ **9.** $\{14\}$ **11.** $\{16\}$ **13.** $\{0\}$ **15.** $\{9\}$ **B 17.** $\{0\}$ **19.** {the whole numbers}   **21.** {the whole numbers}   **23.** $\{225\}$ **25.** $\emptyset$ **27.** $\emptyset$ **C 29.** {the whole numbers, except 0}   **31.** {the whole numbers, except 0}   **33.** $\{5\}$ **35.** $\{1\}$ **37.** $\emptyset$

Pages 178–181   Written Exercises   A 5. $16; (4, 16)$ **7.** $36; (6, 36)$ **11.** $(15, 1)$ **13.** $1; (25, 1)$ **15.** $1; (35, 1)$ **17.** $0; (0, 0)$ **21.** $80; (8, 80)$ **23.** $168; (12, 168)$ **B 25.** $\{(0, 6), (1, 5), (2, 2)\}$ **27.** $\{(0, 2), (1, 3), (2, 6)\}$

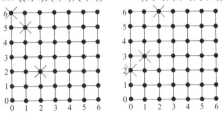

29. $\{(1, 1), (2, 1), (3, 1), (4, 1), (5, 1)\}$ **31.** $\{(0, 0), (1, 0), (2, 2), (3, 6)\}$ **33.** $(4, 0), (5, 1), (6, 2), (7, 3), (8, 4)$

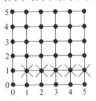 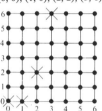

35. $(0, 17), (1, 16), (2, 13), (3, 8), (4, 1)$ **37.** $(0, 0), (1, \frac{2}{3}), (2, \frac{4}{3}), (3, 2), (4, \frac{8}{3}), (5, \frac{10}{3})$ **C 39.** $11; 18; 27$ **41.** $12; 30; 56; 90$ **43.** $5.14; 6.14; 7.14; 8.14$ **45.** $7\frac{1}{7}; 8\frac{1}{8}; 9\frac{1}{9}; 10\frac{1}{10}$

Page 183   Chapter Test   1. I **3.** K **5.** M **7.** I **9.** N **11.** D **13.** F **15.** K **17.** L **19.** M **21.** $(0, 12), (1, 11), (2, 8), (3, 3)$

Pages 184–185   Chapter Review   1. reflexive   **3.** transitive   **5.** symmetric   **7.** unique   **9.** A **11.** B **13.** comm. of add.   **15.** assoc. of mult.   **17.** assoc. of add.   **19.** true   **21.** true   **23.** $(10 \cdot 7) + (18 \cdot 7)$ **25.** $2(x + y)$ **27.** $a \cdot 5 + b \cdot 5$ **29.** No; for example, $4 + (2 \cdot 3) = 10$, but $(4 + 2)(4 + 3) = 42$ **31.** 7 **33.** 15 **35.** 6 **37.** 2 **39.** {the whole numbers}   **41.** $\{0\}$ **43.** $\emptyset$ **45.** $\{0\}$ **47.** See at right.   **49.** $1; 4; 9; 121$

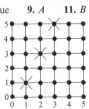

Page 186   Review of Skills   1. 16 **3.** 11 **5.** 3 **7.** no solution   **9.** 20 **11.** 5 **13.** 12 **15.** 0 **17.** 36 **19.** 20 **21.** 4 **23.** 7 **25.** 2 **27.** 13 **29.** comm. of mult.   **31.** dist. prop.   **33.** assoc. of add.   **35.** assoc. of mult.

Pages 187–188   Questions   1. a. yes   **b.** yes   **c.** yes, 1   **3. a.** yes   **b.** yes   **c.** no

Chapter 7. Equations and Problem Solving

Pages 192–193   Written Exercises   A 1. $13k$ **3.** $5y$ **5.** $13n + 5$ **7.** $16xy$ **9.** $16t + 6$ **11.** $10mn + 4b$ **13.** $11y$ **15.** $3m$ **17.** $7w + 6$ **19.** $6h + 8k$ **B 21.** $21z$ **23.** $4t + 3r$ **25.** $20k + 15$ **27.** $w + 5y + 27$ **29.** $12c + 3cd$ **31.** $7m + 8n$ **33.** $11r + 11t$ **35.** $6a + 2$ **37.** $37 + 14m$ **39.** $43c + 5d$ **41.** $8a + 6c + 6$ **C 43.** $24a + 24x + 140$ **45.** $39a + 1$

10 ANSWERS TO ODD-NUMBERED EXERCISES

Pages 196–197 Written Exercises **A** 1. {3} 3. {9$\frac{1}{3}$} 5. {4$\frac{3}{8}$} 7. {7} 9. {0.43} 11. {6} 13. {5}
15. {8} 17. {6} 19. {5} 21. {3} 23. {2.7} **B** 25. {1} 27. {3} 29. {$\frac{8}{3}$} 31. {23} 33. {0}
35. {24} 37. {90} 39. {25} 41. {5.3} 43. {8.7} 45. {9.98} **C** 47. {$\frac{6}{5}$} 49. {$\frac{7}{5}$} 51. {273}
53. {1.32}

Pages 199–200 Written Exercises **A** 1. {12} 3. {2$\frac{1}{2}$} 5. {8} 7. {70} 9. {$\frac{2}{5}$} 11. {0.035}
13. {0.7} 15. {1.75} 17. {$\frac{1}{10}$} 19. {3} 21. {7} 23. {6} 25. {30} **B** 27. {7} 29. {1$\frac{3}{4}$} 31. {0}
33. {2} 35. {20} **C** 37. {1} 39. {0.5} 41. {$\frac{1}{3}$}

Pages 201–202 Written Exercises **A** 1. {180} 3. {6} 5. {255} 7. {37.8} 9. {32.4} 11. {4.5}
13. {84} 15. {6} 17. {18} 19. {24} 21. {4} **B** 23. {50} 25. {0} 27. {25} 29. {7} **C** 31. {10}
33. {100} 35. {0} 37. {$\frac{5}{12}$} 39. {100}

Pages 204–205 Written Exercises **A** 1. {3} 3. {10} 5. {6} 7. {4} 9. {2$\frac{1}{2}$} 11. {$\frac{2}{3}$} 13. {0}
15. {5} 17. {5} 19. {2} **B** 21. {7} 23. {6$\frac{1}{3}$} 25. {4} 27. {$\frac{1}{7}$} 29. {$\frac{1}{2}$} 31. {15} 33. {15}
35. {10} **C** 37. {0.1} 39. {16} 41. {$\frac{20}{7}$} 43. {$\frac{8}{7}$}

Pages 205–206 Problems 1. 3$\frac{1}{2}$ 3. 18 5. 34 ft. wide; 136 ft. long 7. 15, 16 9. 15, 17 11. 30 gal. in each small drum; 60 gal. in large drum 13. 21, 26, 24

Page 208 Written Exercises **A** 1. {4} 3. {13} 5. {1} 7. {9} 9. {$\frac{3}{4}$} 11. {$\frac{5}{4}$} 13. {4} 15. {5$\frac{2}{5}$}
17. {2} 19. {2} 21. {2} 23. {1} 25. {5} **B** 27. {$\frac{1}{2}$} 29. {$\frac{5}{4}$} 31. {$\frac{7}{13}$} **C** 33. {4} 35. {3}
37. {8} 39. {4$\frac{1}{2}$} 41. {$\frac{14}{13}$}

Pages 208–209 Problems 1. 5 3. 4 5. 3 7. 3 9. 10 ft. × 25 ft.

Pages 210–211 Chapter Test 1. {119} 3. {50} 5. {11} 7. {7$\frac{1}{2}$} 9. {45} 11. {54} 13. {55}
15. {7} 17. Let n represent the number; $n + 15 = 4n$ 19. Let x represent the number; $15 + \frac{1}{2}x = 19$

Pages 211–212 Chapter Review 1. 7x 3. 7x + 2y 5. 4y 7. 13x + 34 9. {20} 11. {1} 13. {5}
15. {$\frac{7}{3}$} 17. {12} 19. {102} 21. {9} 23. {117} 25. {9} 27. {8} 29. 42 yrs. old 31. 13 33. {$\frac{1}{5}$}
35. {5} 37. {6} 39. 7

Page 213 Review of Skills 1. H 3. J 5. I 7. $^+$8° 9. $^+$140 11. $^+$7° 13. fell; 3°

Chapter 8. Negative Numbers

Pages 220–221 Written Exercises **A** 1. $^-$1 3. $^+\frac{2}{3}$ 5. $^+$1$\frac{2}{3}$ 7. $^+$2 9. 0 11. $^-$8 13. $^-$300
15. $^-$18.2 17. $^+\frac{3}{4}$ 19. $^-$4$\frac{1}{3}$ 21. $^+$20 23. $^+$15 25. $^-$10.1 **B** 27. $^-$9 29. $^+$4$\frac{5}{8}$ 31. $^+$0.025 33. $^+$17;
$^-$17 35. $^+$7.2; $^-$7.2 **C** 37. $^+$4; $^+$24 39. $^-$7; $^+$3

Pages 223–225 Written Exercises **A** 1. $^+$1

3. $^-$2  5. 0

7. $^+$2 9. $^-$4 11. $^-$1

13. $^-$3 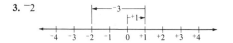 15. $^+$5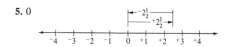

17. $^-$4  **B** 19. $^+$5

ANSWERS TO ODD-NUMBERED EXERCISES

21. positive   **23.** positive   **25.** negative   **C 27.** ⁻1   **29.** 0   **31.** 0

Pages 226–227   **Written Exercises**   **A 1.** left; <   **3.** right; >   **5.** right; >   **7.** >   **9.** <   **11.** >
13. >   **15.** <   **17.** <   **19.** >   **21.** >   **B 23.** true   **25.** true   **27.** true   **29.** true   **31.** false   **33.** true
35. true   **C 37.** true   **39.** false   **41.** true   **43.** true   **45.** false   **47.** false

Pages 228–229   **Written Exercises**   **A 1.** false   **3.** true   **5.** true   **7.** true   **9.** true   **11.** false   **13.** true
15. true   **17.** true   **19.** true   **21.** 9   **23.** 4   **25.** 5   **27.** $\frac{2}{5}$ **29.** 3.00   **B 31.** ⁺135 = ⁺135   **33.** ⁻12 = ⁻12   **35.** ⁺$\frac{1}{2}$ = ⁺$\frac{1}{2}$ **37.** ⁻16 = ⁻16   **39.** ⁺64 = ⁺64   **C 41.** ⁻32 = ⁻32   **43.** ⁺99   **45.** ⁻10   **47.** ⁻999

Pages 231–232   **Written Exercises**   **A 1.** false, false, true; {⁺5}   **3.** false, false, false; ∅   **5.** true, true, true, true; {⁺10, ⁺30, ⁻10, ⁻30}   **7.** true, true, true, false; {⁻1, ⁻$\frac{1}{2}$, 0}   **9.** false, false, true; {⁺15}   **11.** false, false, false; ∅   **13.** {⁻3, ⁻2, ⁻1, 0}

15. {⁻3, ⁻2, ⁻1, 0, ⁺1, ⁺2}   **17.** {⁻3, ⁻2, ⁻1, 0, ⁺1}

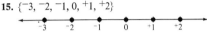   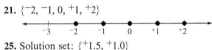

B 19. {⁻3, ⁻2, ⁻1}   **21.** {⁻2, ⁻1, 0, ⁺1, ⁺2}

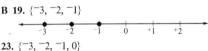

23. {⁻3, ⁻2, ⁻1, 0}   **25.** Solution set: {⁺1.5, ⁺1.0}

27. Solution set: {⁻0.5, 0, ⁺1.5}   **C 29.** Solution set: {⁻1.9, ⁺2.2}   **31.** Solution set: {⁻$\frac{3}{5}$, ⁺$\frac{1}{2}$, ⁺$\frac{2}{3}$, 0}

Pages 233–235   **Written Exercises**   **A 1.**

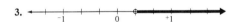

3.    **5.**

7. {the directed numbers less than 0}   **9.** {the directed numbers greater than ⁺$\frac{1}{3}$}   **11.** {⁻8, and the directed numbers greater than ⁻8}   **13.** {the directed numbers between ⁺$\frac{1}{2}$ and ⁺2$\frac{1}{2}$}   **B 15.** {the directed numbers between ⁻1$\frac{1}{4}$ and ⁺$\frac{3}{4}$}   **17.** {the directed numbers between ⁻$\frac{3}{5}$ and ⁺$\frac{4}{5}$}   **19.** {⁻$\frac{1}{3}$, and the directed numbers between ⁻$\frac{1}{3}$ and ⁺1$\frac{1}{3}$}   **21.** {the directed numbers less than ⁺0.7}   **C 23.** {⁺2 and the directed numbers less than ⁺2}   **25.** {the directed numbers between ⁻2$\frac{1}{2}$ and ⁺5}   **27.** {the directed numbers greater than ⁻12}   **29.** {the directed numbers between 0 and ⁺3$\frac{1}{3}$}   **31.** {⁻3.75 and the directed numbers between ⁻3.75 and ⁺1.04}   **33.** {⁻36 and the directed numbers less than ⁻36}

Page 236   **Chapter Test**   **1.** ⁻2   **3.** ⁺1   **5.** ⁺2   **7.** >   **9.** ⁻10   **11.** {⁻3, ⁻2, ⁻1, 0, ⁺1, ⁺2, ⁺3, ⁺4}

13. {⁻9, ⁻8}   **15.**

Pages 237–238   **Chapter Review**   **1.** ⁻1$\frac{3}{4}$ **3.** ⁺$\frac{1}{4}$ **5.** ⁺1   **7.** ⁻1   **9.** ⁺1$\frac{1}{2}$ **11.** ⁻3$\frac{3}{4}$ **13.** ⁻16, ⁺4
15. ⁻13   **17.** positive   **19.** >   **21.** <   **23.** true   **25.** true   **27.** true   **29.** ⁻99   **31.** {⁻4, ⁻3, ⁻2, ⁻1, 0, ⁺1}   **33.** {⁻4, ⁻3}

35. 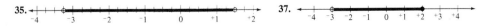   **37.**

39. {the directed numbers greater than ⁻1$\frac{2}{3}$}

Page 239   **Review of Skills**   **1.** ⁻3   **3.** ⁻1   **5.** ⁻12   **7.** ⁺5   **9.** 0   **11.** ⁻4   **13.** 5   **15.** 20 + 15 = 35
17. $x + 5$ **19.** 15   **21.** 23   **23.** 21   **25.** $n = 12$ **27.** $n = 7$ **29.** $n = 5$ **31.** 10; 15; 7; 15

12   ANSWERS TO ODD-NUMBERED EXERCISES

Chapter 9.   **Addition and Subtraction of Directed Numbers**

Pages 245–246   Written Exercises   A   **1.** ⁻6; ⁻6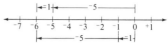

3. ⁺4; ⁺4 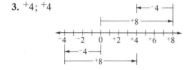   **5.** ⁻1; ⁻1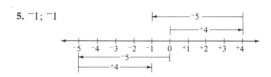

7. ⁻9; ⁻9    **9.** ⁺5½; ⁺5½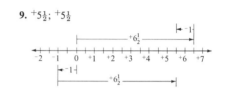

11. positive, ⁺15   **13.** positive, ⁺22   **15.** positive, ⁺58   **17.** negative, ⁻12   **19.** positive, ⁺8   **21.** positive, ⁺9.6   **23.** positive, ⁺5½   **25.** negative, ⁻1   **27.** positive, ⁺3¼

29.

+	⁺1	⁻2	⁺3	⁻4
⁺1	⁺2	⁻1	⁺4	⁻3
⁻2	⁻1	⁻4	⁺1	⁻6
⁺3	⁺4	⁺1	⁺6	⁻1
⁻4	⁻3	⁻6	⁻1	⁻8

B **31.** ⁺161   **33.** ⁺13   **35.** ⁻496   **37.** ⁻81   C **39.** ⁺14   **41.** ⁺18   **43.** ⁻21   **45.** ⁻6   **47.** ⁺14   **49.** ⁻105

Page 247   Problems   **1.** $19.00   **3.** 2700 ft.   **5.** negative $5, or overdrawn by $5   **7.** 10   **9.** $5 profit

Page 249   Written Exercises   A   **1.** −(⁺3); ⁻3   **3.** −(⁻12); ⁺12   **5.** −(⁻25); ⁺25   **7.** −(⁻2⅛); ⁺2⅛   **9.** −(0); 0   **11.** −(⁻8); ⁺8   **13.** ⁻17, ⁺17   **15.** ⁻12.5, ⁺12.5   **17.** ⁻6⅛, ⁺6⅛   B **19.** the opposite of ⁻21; ⁺21   **21.** the opposite of ⁺4; ⁻4   **23.** the opposite of ⁻3⅗; ⁺3⅗   **25.** the opposite of 0; 0   **27.** the opposite of ⁺9.2; ⁻9.2   C **29.** ⁻5   **31.** ⁺2   **33.** ⁺14

Pages 251–253   Written Exercises   A   **1.** ⁻2 + ⁺2 = 0   **3.** ⁺1½ + ⁻1½ = 0   **5.** ⁻10 + ⁺10 = 0   **7.** ⁻1¼ + ⁺1¼ = 0   **9.** yes   **11.** no   **13.** no   **15.** no   **17.** yes   **19.** {⁺19}   **21.** {0}   **23.** {⁻3}   B **25.** {⁺1⅛}   **27.** {⁻⅑}   **29.** {⁺27}   **31.** {⁻4.6}   C **33.** true   **35.** true   **37.** false   **39.** true   **41.** true

Pages 254–255   Written Exercises   A   **1.** ⁻11   **3.** ⁻32   **5.** ⁻⁷⁄₉   **7.** ⁻28   **9.** ⁻63   **11.** ⁻4.6   **13.** ⁻3   **15.** 9   **17.** ⁻23 = ⁻23   **19.** ⁻61 = ⁻61   **21.** ⁻7 = ⁻7   B **23.** ⁻2¾ = ⁻2¾   **25.** 7 = 7   **27.** 0.12 = 0.12   **29.** {⁻3}   **31.** {5}   **33.** {⁻7}   **35.** {⁻2}   **37.** {14}   **39.** {⁻5}   **41.** {⁻11}   **43.** {⁻8}   **45.** {2}   **47.** {⁻2}   **49.** {1}   **51.** {⁻24}

Pages 257–259   Written Exercises   A   **1.** Comm. prop.   **3.** Add. prop. of opp.   **5.** Add. prop. of opp.   **7.** Assoc. prop.   **9.** Add. prop. of zero   **11.** 35 + ⁻35 = 0   **13.** 19 + ⁻19 = 0   **15.** 45 + 0 = 45   **17.** (⁻4 + 4) + (5 + ⁻5) = 0   **19.** (⁻7 + 7) + ⁻16 = ⁻16   **21.** ⁻18 = (⁻1 + 1) + (⁻18)   **23.** 8 + (⁻3 + 3) = 8   **25.** 18   **27.** ⁻14   **29.** 17   **31.** 14   **33.** 18   **35.** ⁻14   **37.** 7½   **39.** 37   **41.** 8   **43.** 2.8   B **45.** 5 = 5   **47.** ⁻3⅓ = ⁻3⅓   **49.** 11 = 11   **51.** 3   **53.** ⁻9   C **55.** 14   **57.** ⁻23   **59.** 15.1   **61.** 54   **63.** 53

Pages 262–263   Written Exercises   A   **1.** 6 + a = 15; {9}   **3.** 8 + y = 20; {12}   **5.** 10 + r = 3; {⁻7}   **7.** 9 + b = 14; {5}   **9.** 20 + n = 30; {10}   **11.** 10 + c = 14; {4}   **13.** 12 + k = 3; {⁻9}   **15.** 18 + z = 15; {⁻3}

ANSWERS TO ODD-NUMBERED EXERCISES

17. ⁻5   **19.** ⁻3   **23.** $1\frac{1}{3}$

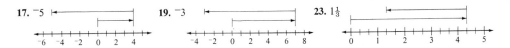

21. $3\frac{1}{2}$ **25.** 0

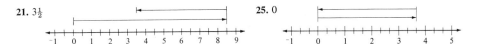

27. 5.1

B 29. $24 + {}^-10 = b$; {14}   **31.** $12 + {}^-17 = x$; {⁻5}   **33.** $w = 9 + {}^-9$; {0}   **35.** $3\frac{1}{3} + ({}^-7\frac{2}{3}) = s$; {⁻4$\frac{1}{3}$}   **37.** $4.8 + {}^-2.6 = c$; {2.2}   **39.** $18.2 + {}^-3.7 = s$; {14.5}   **C 41.** =   **43.** ≠   **45.** ≠   **47.** =   **49.** ⁻2   **51.** 8

Pages 264–265 Written Exercises A 1. $^-18 + m = 3$; {21}   **3.** $^-6 + c = 15$; {21}   **5.** $^-11 + a = 5$; {16}   **7.** $^-5 + k = {}^-2$; {3}   **9.** $3 + w = {}^-11$; {⁻14}   **11.** $1 + t = {}^-3\frac{1}{2}$; {⁻4$\frac{1}{2}$}   **13.** 12   **15.** ⁻14   **17.** 13   **19.** 47   **21.** ⁻31   **23.** 7   **B 25.** ⁻7   **27.** 19   **29.** 8   **31.** ⁻34   **33.** 10   **35.** ⁻0.40   **C 37.** 17   **39.** ⁻13   **41.** 15   **43.** ⁻6   **45.** ⁻2   **47.** 4

Pages 268–269 Written Exercises A 1. 0; (3, 0)   **3.** 0; (1, 0)   **5.** 0; (⁻1, 0)   **7.** ⁻2; 0; 2; 4; 6   **9.** ⁻1; 1; 3   **11.** ⁻6; ⁻13; ⁻8; ⁻15   **13.** 6; 9; ⁻3, ⁻5   **15.** no   **17.** yes   **B 19.** B   **21.** A   **23.** C   **C 25.** {(3, ⁻6), (6, ⁻3), (9, 0), (12, 3), (15, 6), (18, 9)}   **27.** {(⁻15, ⁻5), (⁻10, 0), (⁻5, 5), (0, 10), (5, 15), (10, 20), (15, 25)}

Pages 271–273 Written Exercises A 1. {⁻3}   **3.** {1}   **5.** {⁻3}   **7.** {⁻3}   **9.** {0}   **11.** {⁻1}   **13.** {⁻5}   **15.** {1}   **17.** {0}   **19.** {⁻1}   **21.** {2}   **23.** {5}   **25.** {0}   **27.** {5}   **29.** $^-2 + 2 = 0$ **31.** $^-3 + 3 = 0$ **B 33.** {⁻2$\frac{1}{2}$}   **35.** {$\frac{3}{4}$}   **37.** {$\frac{1}{2}$}   **39.** {1}   **41.** {1$\frac{1}{2}$}   **43.** {⁻1$\frac{1}{2}$}   **45.** {⁻$\frac{5}{4}$}   **C 47.** {⁻$\frac{3}{4}$}   **49.** {$\frac{5}{4}$}   **51.** {⁻$\frac{1}{4}$}   **53.** {$\frac{1}{4}$}   **55.** {1$\frac{1}{4}$}

Pages 274–275 Chapter Test 1. {⁻9}   **3.** {⁻2}   **5.** {94}   **7.** {⁻27}   **9.** {⁻23}   **11.** 6   **13.** ⁻18   **15.** 14   **17.** true   **19.** true   **21.** ⁻3; ⁻7   **23.** ⁻13; 0   **25.** 20

Pages 275–278 Chapter Review 1. ⁺3

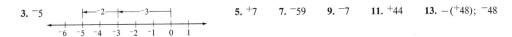

3. ⁻5   **5.** ⁺7   **7.** ⁻59   **9.** ⁻7   **11.** ⁺44   **13.** −(⁺48); ⁻48

15. ⁻13   **17.** ⁻2   **19.** ⁺14   **21.** ⁻3   **23.** {⁻54}   **25.** true   **27.** true   **29.** true   **31.** {⁻54}   **33.** {⁻52}   **35.** {⁻12}   **37.** Assoc. prop.   **39.** Add. prop. of zero   **41.** ⁻5   **43.** ⁻28   **45.** $5.9 + n = 4.8$; {⁻1.1}   **47.** $2\frac{1}{4} + k = 1\frac{1}{8}$; {⁻1$\frac{1}{8}$}   **49.** ⁻5   **51.** ⁻17   **53.** $^-5 + x = 3$; {8}   **55.** $^-25 + k = 17$; {42}   **57.** $9 + y = {}^-8$; {⁻17}   **59.** ⁻5   **61.** ⁻8   **63.** ⁻15   **65.** ⁻12   **67.** 10   **69.** ⁻5; 8; ⁻14   **71.** {(15, 2), (7, ⁻6), (⁻3, ⁻16), (⁻17, ⁻30), (0, ⁻13)}   **73.** {(7, 3), (2, ⁻2), (⁻3, ⁻7), (4, 0), (⁻9, ⁻13)}   **75.** {⁻7}   **77.** {3}   **79.** {2}   **81.** {5}

Page 278 Review of Skills 1. 0   **3.** ⁻14   **5.** 5; 0; ⁻5; ⁻10   **7.** ⁻15   **9.** $9\frac{3}{5}$ **11.** $1\frac{3}{7}$ **13.** $3\frac{3}{32}$ **15.** 5.0985   **17.** A   **19.** B   **21.** C   **23.** 16   **25.** 9   **27.** 1   **29.** 243   **31.** $\frac{5}{4}n$ **33.** $6x + 4y$ **35.** 1, 5   **37.** 5   **39.** 13   **41.** 12   **43.** 8   **45.** 90   **47.** 24   **49.** <   **51.** =   **53.** <

Page 281 Questions 1. $\frac{1}{3}$; $\frac{1}{3}$; $\frac{2}{3}$ **3.** $\frac{1}{3}$; $\frac{2}{3}$ **5.** $\frac{1}{13}$; $\frac{1}{13}$; $\frac{1}{52}$; $\frac{1}{4}$; $\frac{1}{4}$; $\frac{1}{4}$

Chapter 10. Multiplication and Division of Directed Numbers

Pages 285–286 Written Exercises A 1. ⁻15   **3.** 168   **5.** ⁻30   **7.** $-\dfrac{2s}{3}$ **9.** ⁻6k **11.** $-\dfrac{t}{2}$ **13.** ⁻56t

15. ⁻0.13   17. ⁻0.448   19. $\frac{4}{5}$ 21. 0   23.

×	0	⁻1	⁻2	⁻3
0	0	0	0	0
1	0	⁻1	⁻2	⁻3
2	0	⁻2	⁻4	⁻6
3	0	⁻3	⁻6	⁻9

25. ⁻180   27. ⁻735   29. 20
31. ⁻36   33. ⁻5   35. ⁻19   37. 0   **B** 39. {⁻60}   41. {⁻90}   43. {27}   45. {12}   47. {13}   49. {⁻6}
51. {⁻1}   **C** 53. {⁻6}   55. {⁻12}   57. {8}

Pages 288–289   Written Exercises   A 1. 12   3. ⁻60   5. ⁻120   7. ⁻150   9. ⁻5   11. ⁻5   13. ⁻105 = ⁻105   15. 16 = 16   17. ⁻140 = ⁻140   19. 40 = 40   21. ⁻42 = ⁻42   23. ⁻20 = ⁻20   **B** 25. ⁻216 = ⁻216   27. 12 = 12   29. ⁻78 = ⁻78   31. 0 = 0   33. ⁻30 = ⁻30   **C** 35. ⁻868   37. ⁻147   39. ⁻6309   41. ⁻2169   43. ⁻53,560

Pages 292–293   Written Exercises   A 1. negative; ⁻80   3. zero   5. negative; ⁻27   7. positive; 84   9. positive; 90   11. zero   13. negative; ⁻27   15. zero   17. positive; 300   19. ⁻14   21. 40   23. ⁻12   **B** 25. 1   27. 10   29. 6   31. ⁻24   33. ⁻2   35. ⁻120   37. 256   39. 200   41. 20   43. 30   45. 25   47. 400   **C** 49. ⁻8   51. 10   53. 1   55. 1   57. ⁻8   59. 100   61. 10,000   63. negative   65. positive   67. closed   69. not closed; (⁻1)(⁻2) = 2   71. closed

Pages 294–295   Written Exercises   A 1. $9x + 14$ 3. $7k + 8t$ 5. $15a - 3b$ 7. $3rs - 8$ 9. $7xyz + 2k$ 11. $16a^2 - 2b$ **B** 13. $5a + 5b$ 15. $⁻6y + 2y^2$ 17. $2s + 3$ 19. $3ab + 10a$ **C** 21. $10x + 20$ 23. $31y - 6$ 25. ⁻174   27. ⁻60   29. 16   31. 108   33. ⁻588   35. 117

Pages 297–298   Written Exercises   A 1. ⁻2   3. ⁻2   5. ⁻7   7. 6   9. ⁻49   11. ⁻8   13. 8   15. ⁻25   17. 3   19. ⁻0.26   21. 2   23. ⁻0.75   25. ⁻0.625   27. ⁻0.3   29. ⁻1.25   31. ⁻0.8   33. ⁻0.25   35. 0.16   37. ⁻0.49   **B** 39. true   41. false   43. false   45. true   47. ⁻25   49. 6.2

Pages 300–302   Written Exercises   A 1. $⁻\frac{9}{10}$ 3. $⁻\frac{10}{7}$ 5. $\frac{3}{5}$ 7. 1   9. 8   11. $⁻\frac{3}{20}$ 13. $⁻\frac{4}{15}$ 15. ⁻24   17. 1   19. $⁻\frac{8}{3}$ 21. $\frac{1}{6}$ 23. ⁻1   25. {⁻2}   27. {⁻8}   29. {5}   31. {$2\frac{1}{2}$}   33. {3}   35. {0}   **B** 37. {$\frac{1}{5}$}   39. {$⁻\frac{1}{9}$}   41. {$⁻\frac{1}{4}$}   43. $\frac{c}{12}$ 45. ⁻$2b$ 47. $\frac{5k}{⁻8}$ 49. $\frac{-x}{7}$ 51. $\frac{2}{3}h$ **C** 53. closed   55. not closed; $\frac{1}{0}$ has no meaning   57. not closed; $\frac{-2}{-1} = 2$

Page 302   Problems 1. $\frac{10}{-3}$ 3. 5, ⁻5   5. ⁻9   7. $⁻\frac{3}{2}$

Pages 304–306   Written Exercises   A 1. $⁻\frac{2}{5}$; (⁻2, $⁻\frac{2}{5}$)   3. $\frac{1}{5}$; (1, $\frac{1}{5}$)   5. 1; ($⁻\frac{1}{3}$, 1)   7. 1; ($⁻\frac{1}{2}$, 1)   9. 1; ($\frac{1}{2}$, 1)   11. 1; (⁻2, 1)   13. 1; (2, 1)   15. 2; 0; ⁻2   17. $\frac{6}{5}$; $\frac{16}{15}$; 1; $1\frac{4}{15}$ 19. $\frac{3}{2}$; $\frac{1}{2}$; 0; $⁻\frac{1}{2}$ **B** 21. {(⁻6, 36), (⁻4, 16), (⁻2, 4), (0, 0), (2, 4), (4, 16), (6, 36)}   23. {(⁻3, ⁻27), (⁻2, ⁻8), (⁻1, ⁻1), (0, 0), (1, 1), (2, 8), (3, 27)}   25. {(0, 0), (⁻1, ⁻1), (⁻2, ⁻4), (⁻3, ⁻9), (⁻4, ⁻16), (⁻5, ⁻25), (⁻6, ⁻36), . . .}   27. {(⁻5, 6.5), (⁻3, 3.9), (0, 0), (1, ⁻1.3), (2, ⁻2.6), (7, ⁻9.1), (12, ⁻15.6)}   29. {($⁻\frac{1}{5}$, ⁻20), ($⁻\frac{1}{4}$, ⁻16), ($⁻\frac{1}{3}$, ⁻12), ($⁻\frac{1}{2}$, ⁻8), ($\frac{1}{2}$, 8), ($\frac{1}{3}$, 12), ($\frac{1}{4}$, 16), ($\frac{1}{5}$, 20)}

Pages 307–308   Chapter Test 1. ⁻35   3. 24   5. 17   7. 16   9. ⁻6   11. {⁻5}   13. {$\frac{1}{5}$}   15. {⁻12.5}   17. $⁻\frac{4}{21}$ 19. ⁻8; ⁻7; ⁻6

Pages 308–310   Chapter Review 1. ⁻28   3. ⁻60   5. ⁻24   7. ⁻39   9. {7}   11. {50}   13. 9   15. 8   17. ⁻9; ⁻6; ⁻3; 0; 3; 6   19. 4; ⁻8; 16   21. 21   23. 72   25. ⁻1t 27. ⁻4xy 29. $6x + 6y$ 31. 1   33. ⁻11   35. $⁻\frac{3}{4}$ 37. ⁻0.4   39. 0.12   41. 4   43. $\frac{b}{-a}$ 45. $\frac{8}{-9}$ 47. $⁻\frac{1}{8}$ 49. $⁻\frac{10}{3}$ 51. {⁻4}   53. {$\frac{1}{8}$}   55. ⁻4; 4; ⁻5

ANSWERS TO ODD-NUMBERED EXERCISES 15

Pages 310–311 Review of Skills 1. 0 3. 10 5. 1 7. 0 9. ⁻12 11. ⁺12 13. $-\frac{3}{4}$ 15. $\frac{3}{4}$ 17. ⁻2
19. 6 21. ⁻2 23. 16 25. ⁻8 27. 5n 29. $\frac{5}{4}x$ 31. 3 33. < 35. > 37. < 39. > 41. >
43. false 45. true 47. true 49. false

Page 313 Questions 1. 0.875 3. $0.\overline{2}$ 5. ⁻0.6 7. $0.\overline{428571}$ 9. $\frac{47}{200}$ 11. $\frac{52}{5}$ 13. $-\frac{1}{4}$ 15. $\frac{1}{10,000}$
17. $\frac{1}{100}$ 19. $\frac{413}{1000}$ 21. $\frac{37}{99}$ 23. $\frac{32}{99}$ 25. $\frac{91}{33}$ 27. $\frac{3815}{99}$

Chapter 11. Solving Equations and Inequalities

Page 317 Written Exercises A 1. 0; 13; {13} 3. ⁻4½; 0; 2½; {2½} 5. {21} 7. {11} 9. {2¼}
11. {19} 13. {3} 15. {⁻19} 17. {5¼} 19. {5.9} **B** 21. {⁻7½} 23. {⁻9⁄8} 25. {1⁄16} 27. {⁻7}
29. $x = m - t$ 31. $x = b - c$ 33. $x = -s - a$ **C** 35. $s = r + a$ 37. $w = h + s$ 39. $p = w - q$

Pages 319–321 Written Exercises A 1. {7} 3. {5⁄2} 5. {⁻4} 7. {4} 9. {⁻5} 11. {3} 13. {⁻80}
15. {⁻3⁄10} 17. $m = \frac{a}{w}$ 19. $h = \frac{p}{d}$ 21. $a = \frac{v}{t}$ 23. $r = \frac{d}{t}$ **B** 25. {⁻5⁄3} 27. {1⁄10} 29. {5⁄2} 31. {5⁄2}
33. {⁻36} 35. {⁻8} 37. $l = \frac{A}{W}$ 39. $r = \frac{c}{2\pi}$ **C** 41. $y = \frac{b}{a}$; ⁻3 43. $n = \frac{b}{a}$; ⁻3 45. $p = \frac{-b}{k}$; $-\frac{1}{4}$
47. $r = \frac{a}{-t}$; ⁻2 49. $m = \frac{bk}{a}$; 24 51. $g = \frac{-k}{-t}$; ⁻24

Pages 323–325 Written Exercises A 1. {14⁄9} 3. {36} 5. {⁻10} 7. $m = \frac{s}{t + k}$ 9. $k = \frac{w}{3 + n}$
11. $m = \frac{10}{d - 9}$ 13. $k = \frac{-t}{m + b}$ 15. {5⁄3} 17. {⁻3⁄7} 19. {2} 21. {⁻4⁄9} 23. {⁻3} 25. {2} 27. {⁻3⁄8}
29. {2} 31. {3} 33. {8⁄3} 35. {8} **B** 37. {19} 39. {⁻3⁄5} 41. {6} 43. {2} **C** 45. {2} 47. {3}
49. {⁻4} 51. {5}

Pages 327–328 Written Exercises A 1. E 3. A 5. D 7. $y > ^-7$

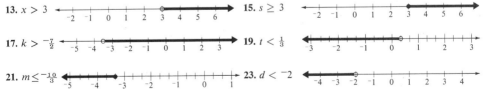

11. $b \geq ^-4$

13. {the directed numbers > ⁻4} 15. {the directed numbers > 1} 17. {the directed numbers > ⁻5}
19. {the directed numbers < 12½} 21. {the directed numbers > ⁻12} 23. {the directed numbers ≥ ⁻7}
B 25. {the directed numbers ≤ 33} 27. {the directed numbers > 1} 29. {the directed numbers ≥ 0}
31. {the directed numbers ≥ ⁻6} 33. {the directed numbers > 1} **C** 35. {the directed numbers ≥ ⁻21}
37. {the directed numbers ≥ 8} 39. {the directed numbers ≤ ⁻3⁄2}

Pages 331–332 Written Exercises A 1. $12 \cdot 4 < 25 \cdot 4$ 3. $11 \cdot 5 > ^-12 \cdot 5$ 5. $\frac{1}{2} \cdot 2 > ^-3 \cdot 2$ 7. $4 \cdot ^-2 <$
$^-2 \cdot ^-2$ 9. $^-9 \cdot ^-2 > 10 \cdot ^-2$ 11. $^-m \cdot ^-1 > 4 \cdot ^-1$

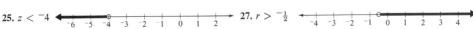

29. {the directed numbers > 5⁄2} 31. {the directed numbers ≥ 10⁄3} 33. {the directed numbers ≤ 20⁄3}
35. {the directed numbers > ⁻15} 37. {the directed numbers ≤ 5} 39. {the directed numbers > 5}
B 41. {the directed numbers ≤ ⁻25} 43. {the directed numbers > ⁻2} 45. {the directed numbers > ⁻12}
47. {the directed numbers ≥ ⁻28} 49. {the directed numbers < 5} 51. {the directed numbers > ⁻2}
53. {the directed numbers < ⁻30} **C** 55. {the directed numbers < 2} 57. {the directed numbers ≤ 13½}

ANSWERS TO ODD-NUMBERED EXERCISES

Pages 334–335 Problems A 1. $8\frac{2}{17}$ ft. **3.** $l > 6\frac{3}{4}$ (ft.) **5.** 4 in. **7.** yes; $(20)(18)(8) \leq 3200$; $2800 \leq 3200$ **9.** $1875 **11.** $1500

Pages 336–337 Chapter Test 1. {5} **3.** $\{\frac{1}{2}\}$ **5.** $\{\frac{-7}{2}\}$ **7.** {20} **9.** {$^-$12} **11.** $x = a + b$ **13.** $x = \dfrac{b+c}{a}$

15. $x > {}^-3$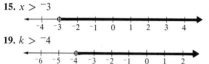
17. $x > {}^-5$
19. $k > {}^-4$
21. $x \geq {}^-2$
23. yes **25.** 18 ft. or 6 yd.

Pages 337–339 Chapter Review 1. {15} **3.** $\{{}^-1\frac{1}{2}\}$ **5.** {$^-$12} **7.** $x = c - a$ **9.** $\{\frac{-10}{3}\}$ **11.** $\{{}^-1\frac{1}{2}\}$ **13.** {$^-$80} **15.** $x = \dfrac{c-b}{a}$ **17.** {$^-$5} **19.** {180} **21.** {$^-$16} **23.** {5} **25.** $x = \dfrac{c}{a-b}$ **27.** Addition property **29.** $y < {}^-7$
31. $k < 0$
33. $p \leq {}^-3$
35. $^-15 < 10$
37. $x \geq {}^-2$
39. $k > 2$
41. $x \leq {}^-2$
43. $n < {}^-4$
45. 20 in.

Pages 339–340 Review of Skills 1. 10^2 **3.** 10^1 **5.** $5 \cdot 10^2 + 4 \cdot 10 + 3$ **7.** $1 \cdot 10^3 + 3 \cdot 10^2 + 7 \cdot 10 + 9$ **9.** $5x$ **11.** $18k$ **13.** 30 **15.** 230 **17.** 13 **19.** 0 **21.** 58 **23.** 83 **25.** 25 **27.** 120 **29.** 0 **31.** 45 **33.** $^-\frac{3}{4}$ **35.** $\frac{1}{2}$ **37.** 5 **39.** {1} **41.** {7} **43.** {$^-$7} **45.** 5; $^-$5 **47.** 4 **49.** 17

Pages 341–342 Questions 1. a. first quadrant; $\{A(4, 1), B(4, 5), C(1, 4)\}$ **b.** second quadrant; $\{D({}^-4, 5), E({}^-2, 3), F({}^-4, 0)\}$ **c.** third quadrant; $\{M({}^-4, {}^-2), N({}^-2, {}^-2), P({}^-3, {}^-4)\}$ **d.** fourth quadrant; $\{Q(1, {}^-3), R(5, {}^-3), S(5, {}^-1)\}$ **3. a.** $\{(5, {}^-3), (4, {}^-2), (3, {}^-1), (2, 0), (3, 1), (4, 2), (5, 3)\}$; no **b.** $\{({}^-5, {}^-4), ({}^-4, {}^-3), ({}^-3, {}^-2), ({}^-2, {}^-1), ({}^-1, 0), (0, 1), (1, 2), (2, 3), (3, 4), (4, 5)\}$; yes

Chapter 12. Addition and Subtraction of Polynomials

Page 347 Written Exercises A 1.–5. Answers will vary. **7.** $w^3 + 2w$ **9.** $3y^2 + 5y + 6$ **11.** $k^3 + k^2 - 2k$ **13.** $7a^5 + a^2 + 8a$ **15.** $-d^6 + 3d^4 + 4d^3$ **B 17.** $m^2 + mn$ **19.** $12r^5 - 7r^4s + 2r^2s^2$ **21.** $x^2 + xy + y^2$ **23.** $x^3y + x^2y + x + y^2$ **25.** $2x^3 + 0x^2 + 0x + 3$ **27.** $t^4 + 0t^3 + 0t^2 + 5t + 1$ **29.** $y^4 - y^3 + 0y^2 + 0y + 8$ **31.** $h^6 + 0h^5 - 2h^4 + 0h^3 + 0h^2 + h + 25$

Pages 350–352 Written Exercises A 1. $13z + 7$ **3.** $5x^2 + 4$ **5.** $2x^3$ **7.** $2y^2$ **9.** $4r^2 + 5rs + 4s^2$ **11.** $3r^2 - 5s^2 - t^2$ **13.** $10k - 8$ **15.** $3x - 2y^2$ **17.** $3a^2 - 6b - 1$ **19.** $2p^2 + q - 2r$ **21.** $10.1r + 2.5s + 9$ **23.** $2k + 2\frac{1}{4}n - 6\frac{3}{4}$ **25.** $11r + 7s$ **27.** $10x - 12y$ **29.** $16x - 5y^2$ **31.** $2a^2 + 2b^2$ **33.** $4t^4 + 20$ **35.** $y^3 + 3y^2 + 2y - 2$ **37.** $25x^2 + 10x + 3$ **39.** $z^2 - 20z + 65$ **41.** $2k^3 - 5k^2 + k - 3$ **B 43.** $x^2 + 3xy + 2y^2$ **45.** $a^3 + 5a^2b + 5ab^2 + b^3$ **47.** $7s - 10$; 25 **49.** $14x - 12$; 16 **51.** $9r + 8$; 44 **53.** $8r + 6$; 38 **55.** $x^2 + 4x - 9$; 3 **57.** $2.0x - 3.2w$ **59.** $4ab^2 - 3ac^2$ **61.** $1\frac{1}{3}r^3 - 1\frac{1}{2}s^3$ **63.** $-k$ **65.** $2.4rs + 4$ **67.** $-10ab + 8$ **69.** $9a + 2$ **C 71.** $13b^2 - 8.4c - 1.4c^2$ **73.** $k^3 - 6k^2 + 6k - 8$ **75.** $t^5 - 3t^4 + 7t^5 + 7$

Pages 354–356 Written Exercises A 1. $3y + 11 = 3y + 11$ **3.** $7 = 7$ **5.** $6r - 2 = 6r - 2$ **7.** $2x + y - 1 = 2x + y - 1$ **9.** $3a^2 + 2ab - 7b^2 = 3a^2 + 2ab - 7b^2$ **11.** $7x - 3; 7x - 3$ **13.** $-4; -4$ **15.** $6r - s + 19; 6r - s + 19$ **17.** 19 **19.** -10 **21.** 33 **23.** 11.2 **25.** 0 **27.** -4 **29.** 47; 47 **31.** 871; 871 **B 33.** 120; 120 **35.** 3; 3 **37.** $4\frac{3}{8}; 4\frac{3}{8}$

Pages 358–359 Written Exercises A 1. $-3x^2 + x - 10$ **3.** $-4x^5 - 2x^3 + x - 5$ **5.** $-15x^7 - 10x^6 - x - 3$ **7.** $25x^4 + 10x^3 - 7x^2 + x - 19$ **9.** $x^7 - 3x^5 + x^2 - 2x + 1$ **11.** $-2b^2 - 5b - 6$ **13.** $-2m^2 + 3m - 2$ **15.** $a^2 - a + 1$ **17.** $-3.7w^4 + 2.5w^2 - 0.3w - 10$ **19.** $x^3 - 6x^2 + x - 10$ **21.** $7a^2 - 16$ **23.** 0 **25.** 0 **B 27.** $8a^2 - 7b + c + 2$ **29.** $x^2 + \frac{1}{4}x + \frac{1}{6}$ **31.** $-4t^7 - t^6 - 3t^2 + 3$ **33.** $a^4 + a^3 + a - 5$ **35.** $b^3 - b^2 + 3b + 8$ **37.** $5x^6 + x^5 - x^2 + 3$

Pages 361–363 Written Exercises A 1. $2m + 10n$ **3.** $x + 22$ **5.** $9 - r$ **7.** $-6y - 2$ **9.** $11k + 1$ **11.** $9b^2 - 3c$ **13.** $x - 16 - 2x - 5$ **15.** $x^2 + 5x - 3x^2 + 10x$ **17.** $7a^2 + 8a + 4a^2 - a$ **19.** $a^2 - a + 5 - 2a^2 - 2a + 3$ **21.** $8x - 14$ **23.** $r^2 - 5r - 8$ **25.** $7s - 9t - 5$ **B 27.** $11x + 2$ **29.** $3a + 12b$ **31.** $1.8k^2 - 0.6k - 7.8$ **33.** $14x^2 + x + 15$ **35.** $-3m^2 - 4m - 15$ **37.** $4w^2 + 3w - 14$ **39.** $-3 - 9b + 7b^2$ **41.** $2z^2 - 1$ **43.** $m^2 - m - 2$ **C 45.** $5t + 8 - 2t - 3$ **47.** $-x - 2x + 8 + 3x$ **49.** $n + 4 + 3n - n - 5$ **51.** $10k - 2k + k + 5 - k^2 - 6$ **53.** $-3a^2 + 19a + 17$ **55.** $-5x + 5$ **57.** $x - 7$

Pages 363–365 Problems A 1. $7k - 2$ **3.** $3k^2 + 5k; 3k^2 + k; 3k^2 + 5k$ **5.** $33h^2 + 8h - 33; 18h^2 + 13h + 1; 33h^2 + 13h - 29; 3h^2 + 8h + 27; 3h^2 + 3h + 23$ **B 7.** $4at^2 + 4g + 5$ **9.** $2x + 6; 2\frac{1}{2}x^2 + 17x + 28\frac{1}{2}; \frac{x^2}{2} + 3x + 4\frac{1}{2}$

Page 366 Written Exercises A 1. $\{3\}$ **3.** $\{12\}$ **5.** $\{3\}$ **7.** $\{\frac{7}{2}\}$ **9.** $\{-3\}$ **11.** $\{-14\}$ **13.** $\{2\}$ **B 15.** $\{2\}$ **17.** $\{1\}$ **19.** $\{\frac{1}{3}\}$ **21.** $\{30\}$ **23.** $\{3\}$ **25.** $\{-\frac{3}{2}\}$ **C 27.** $\{-2\}$ **29.** $\{-\frac{10}{7}\}$

Pages 368–369 Written Exercises A 1. 6 **3.** 13 **5.** 0 **7.** 9 **9.** 15 **11.** 13 **13.** 3 **15.** 2 **17.** $(-4, 24)$ **19.** $-4; (0, -4)$ **21.** $-4; (3, -4)$ **B 23.** $\{(-3, 8), (-2, 4), (-1, 2), (0, 2), (1, 4), (2, 8), (3, 14)\}$ **25.** $\{(-4, -5), (-3, -7), (-2, -7), (-1, -5), (0, -1)\}$ **27.** $\{(-1, -9), (-\frac{1}{2}, -10), (0, -9), (\frac{1}{2}, -6), (1, -1)\}$ **29.** $\{(-\frac{1}{2}, \frac{3}{8}), (-\frac{1}{4}, \frac{3}{16}), (0, \frac{1}{8}), (\frac{1}{4}, \frac{3}{16}), (\frac{1}{2}, \frac{3}{8})\}$ **C 31.** $\{(-2, 17), (-1, 8), (0, -1), (1, -10), (-2, -19)\}$ **33.** $\{(-3, 7), (-2, 2), (-1, -1), (0, -2), (1, -1), (2, 2), (3, 7)\}$

Pages 370–371 Chapter Test 1. a. monomial **b.** binomial **c.** monomial **d.** monomial **e.** trinomial **3. a.** $5x^2 - x + 4$ **b.** $4x^2 - 8x + 2$ **c.** $-3ab - 2b^2$ **5. a.** $-x + 8$ **b.** $-a - b$ **c.** $5x - 5$ **d.** $-x^2 - 4x - 8$ **7. a.** $t = 4$ **b.** $t = 6$ **c.** $x = 6$ **d.** $k = -\frac{1}{2}$

Pages 371–374 Chapter Review 1. Answers vary. **3.** $-2x + 5$ **5.** $-5x^4 + 2x^2 + 3x$ **7.** $-a^3 + 3a^2b - 2ab^2 + b^3$ **9.** $3x^2 - x - 2$ **11.** 0 **13.** $1\frac{1}{8}x + 1$ **15.** $2t^2 + 3t + 3$ **17.** C **19.** B **21.** D **23.** 4 **25.** $-2 + 5x$ **27.** $x^2 - 5x - 12$ **29.** $5x^2 - 3x + 2$ **31.** 0 **33.** $6a$ **35.** $15 - 2x$ **37.** $2x - 11$ **39.** $3x - 5$ **41.** $-4b$ **43.** $7x + 11$ **45.** $t = -\frac{3}{2}$ **47.** $x = -6$ **49.** $x = 3$ **51.** 0 **53.** 2 **55.** $\{(-2, 9), (-1, 3), (0, -1), (1, -3), (2, -3)\}$ **57.** $\{(-3, -8), (0, 1), (1, 8), (2, 27)\}$

Pages 374–375 Review of Skills 1. H **3.** E **5.** B **7.** F **9.** r^3 **11.** t^7 **13.** 8 **15.** $2x$ **17.** -8 **19.** 9 **21.** 64 **23.** $-2y + 4k$ **25.** $5x - 5y + 5z$ **27.** $tm + tn - tp$ **29.** assoc. prop. **31.** comm. prop. **33.** mult. prop. of one **35.** 8 **37.** 8 **39.** 27 **41.** 3 **43.** 2^2 **45.** 11 **47.** 10

Page 376 Questions 1. 4 **3.** 8 **5.** 16 **7.** $-\frac{1}{2}$ **9.** 2 **11.** true **13.** false **15.** false **17.** $\{-25, 25\}$ **19.** $\{-4, 4\}$ **21.** $\{-9, 9\}$ **23.** $\{-7, 7\}$ **25.** 9 **27.** 1 **29.** -4

Chapter 13. Multiplication and Division of Polynomials

Pages 380–381 Written Exercises A 1. 8 **3.** 36 **5.** 16 **7.** 40 **9.** 36 **11.** 56 **13.** 27 **15.** 18 **17.** 36 **19.** 10^8 **21.** m^4 **23.** r^7 **25.** $-3t^3$ **27.** $21k^3$ **29.** $(a + b)^2$ **31.** $-15x^3$ **33.** $-14w^2$ **35.** $35t^2$ **37.** $20x^3$ **39.** $33a^5$ **41.** $-10b^3$ **43.** $-x^4$ **B 45.** 6 **47.** 2 **49.** 3 **51.** 3

Pages 382–383 Written Exercises A 1. $48x^3y^4$ **3.** $-4a^7b^5$ **5.** $48x^3y^2$ **7.** $-10a^2b^6$ **9.** $-80r^5$ **11.** $-36x^2y^2$ **13.** $-3x^2y^2$ **15.** $16ab^3c$ **17.** $\frac{1}{8}a^2b^2$ **B 19.** r^4t^3 **21.** $10a^4b^7$ **23.** $-90x^4y^9$ **25.** a^4b^3 **27.** $-15p^2q + 7pq^2$ **29.** $-17x^3y^4$ **31.** 0

ANSWERS TO ODD-NUMBERED EXERCISES

Page 385 Written Exercises A 1. 625 3. 1,000,000 5. 1 7. 10,000 9. 9 11. 0 13. 900 15. 25 17. $36a^4b^2$ 19. $0.09r^2s^4$ 21. $36b^6c^4$ 23. $64x^2y^2z^2$ **B** 25. $-37a^3b^3$ 27. 0 29. $-16x^5$ **C** 31. $0.1t^4z^5 + t^6z^3$ 33. $-4b^3$

Pages 387–388 Written Exercises A 1. $-6x^2 + 12x - 6$ 3. $a^3 - 2a^2c + ac^2$ 5. $-ab^2 - b^3 - b^2c$ 7. $-5x^5 + x^4 - 2x^3 - 3x^2$ 9. $-14v^5 + 21v^4 + 28v^3$ 11. $15x^2y - 6x^3y^5 + 9x^5y^3 - 3x^2y^6$ 13. $-2a^7b^3 + 2a^6b^4 - 4a^5b^6 + 2a^3b^7$ 15. $-21a^4c^2 + 28a^3c^3 - 35a^2c^4$ 17. $4y^3 - 8y^2 - 20y$ 19. $4.5m^3 + 13.5mn^2 - 36m$ 21. $-12a^3b + 8a^2b^2 + 11ab$ 23. $8r^2 + 8rs$ 25. $-m^2 + 6mn - 7n^2$ 27. r^2 29. $-6s^2t + 18st^2$ 31. $-4y^2 - 5yz$ 33. $-5r^3s + 2r^2s^3$ **B** 35. $a = 14$ 37. $b = -1$ 39. $x = -4$ 41. $r = 10$ **C** 43. $k = -16$ 45. $n = \frac{5}{4}$

Pages 388–389 Problems A 1. $30t^2 + 15t; 22t + 6$ 3. $20w + 4; 6w^2 + 2w; 24w^2 + 8w$ **B** 5. $30t^2 - 24t; 180t^3 - 144t^2; 45t^3 - 36t^2; 135t^3 - 108t^2$ 7. $5m^5 - 10m^4 + 50m^3$ 9. $3.8r^3 + 13r^2 - 0.08$

Pages 391–392 Written Exercises A 1. $x^2 + 9x + 14$ 3. $3n^2 + 20n + 12$ 5. $2a^2 - a - 15$ 7. $m^2 - 5m - 36$ 9. $y^2 - 5y - 24$ 11. $56n^2 + 95n + 36$ 13. $12y^2 - 34y + 24$ 15. $a^2 + 2ab + b^2$ 17. $2n^2 + 8n + 6$ 19. $2x^2 + 3x - 20$ 21. $x^2 + x - 6$ 23. $x^2 + 2xy + y^2$ 25. $r^2 - s^2$ 27. $r^2 + 2rt + t^2$ **B** 29. $9a^2 + 6ab + b^2$ 31. $2a^2 + 9.7ab - 1.5b^2$ 33. $3m^3 + 3m^2 - 2m + 8$ 35. $a^3 - b^3$ 37. $2n^3 - 15n^2 + 16n + 3$ **C** 39. $y^4 + 7y^2 - 30$ 41. $3 - 2ab - a^2b^2$ 43. $2a^4 + 7a^3 + 10a^2 + a - 4$ 45. $x^5 - 1$

Pages 392–393 Problems A 1. $24n^2 + 20n + 4; 22n + 10$ 3. $12k - 6; 2k^2 - k; 2k^2 + 11k - 6$ **B** 5. $4a^3 + 12a^2 + 5a - 6; 16a^2 + 28a + 2$ 7. $x^2 - 2x - 3 + 45 = x^2$

Pages 395–396 Written Exercises A 1. $x^2 + 2x + 1$ 3. $x^2 - 4$ 5. $x^2 + 12x + 36$ 7. $s^2 - 2st + t^2$ 9. $100x^2 - 240x + 144$ 11. $81 - 4k^2$ 13. $9n^2 - 25m^2$ 15. $25c^2 + 60cd + 36d^2$ **B** 17. $y^4 - 22y^2 + 121$ 19. $64 - 16xy + x^2y^2$ 21. $c^4 - 6c^2d + 9d^2$ 23. $25k^2 - 2k + \frac{1}{25}$ 25. $a^4b^2 + 10a^2b + 25$ 27. $x^2 + x + 0.25$ 29. $a^4 + 2a^2cd + c^2d^2$ 31. $2.56 + 3.2k + k^2$ 33. $0.0004s^2 - 0.12st + 9t^2$ **C** 35. $4y^2 + 3.2y + 0.64$ 37. $0.09y^4 + 0.6y^3 + y^2$ 39. $2.25n^4 + 1.2n^2 + 0.16$ 41. $324a^2b^2 + 360abc + 100c^2$

Page 396 Problems A 1. $16x^2 - 80x + 100$ 3. 12 in. × 12 in.; 15 in. × 15 in. 5. 51 and 52 7. 13 and 14 9. 16 and 17

Page 399 Written Exercises A 1. $\frac{m^2}{2n}$ 3. $\frac{3a+b}{2a}$ 5. $\frac{3}{3a^2+b}$ 7. $\frac{2}{m^2}$ 9. $15x^3y = 15x^3y$ 11. $16xy = 16xy$ **B** 13. $x^4y^4 = x^4y^4$ 15. $3x^3 + 11x^2y + 9xy^2 + y^3 = 3x^3 + 11x^2y + 9xy^2 + y^3$ **C** 17. $4x^5$ 19. $54t^3$ 21. $x^3y^2z^4$

Pages 402–403 Written Exercises A 1. a^6 3. $\frac{1}{m^7}$ 5. $-3n^5$ 7. $-q^2$ 9. $15x$ 11. $-7x^2y$ 13. $\frac{144}{11}bc$ 15. $-34z$ **B** 17. $\frac{a}{3b^3}$ 19. $-\frac{1}{2n}$ 21. $\frac{r^2}{3s^4}$ 23. $\frac{y^2}{3x}$ 25. $\frac{-3}{a^2b}$ 27. $\frac{-1}{2mn^2}$ 29. $\frac{r}{10s}$ 31. $\frac{p^2}{10r}$ 33. $\frac{b^4}{7a^{10}}$ 35. $\frac{-a^3}{b^2}$ **C** 37. $\frac{-3a^{x-1}}{2}$ 39. 4

Pages 404–405 Written Exercises A 1. $2m - 4$ 3. $4y + 1$ 5. $-3c^2 - 6$ 7. $-6a^2 - 2a + 7$ 9. $-4k + 3$ 11. $2x^3 + 7x^2$ 13. $8 + \frac{3}{n}$ 15. $r - 7$ 17. $4 + m^2$ **B** 19. $3x + 2$ 21. $2a^2 + 5a + 4$ 23. $10r^2 + 2r - 7$ 25. $-2n^3 + 3n^2 + 4n$ **C** 27. $-5a^3 - 4a^2 + 8a - 2$ 29. $\frac{2m^2}{-3} + \frac{mn}{3} + \frac{n^2}{6}$

Page 407 Written Exercises A 1. $x + 3$ 3. $x + 4$ 5. $2x - 5$ 7. $r - 7$ 9. $m - 3$ 11. $2t + 5$ 13. $2b - 3$ 15. $2t + 1$ **B** 17. $z - 8$ 19. $-4x - 7$ 21. $w + 6x$ 23. $5n - m$

Pages 408–409 Chapter Test 1. x^7 3. $4a^2b^4$ 5. $16t^6$ 7. $7r + 8r^2$ 9. $3t^3 - 15t^2$ 11. $9p^2 - t^2$ 13. $3y^2 + 10y - 8$ 15. $x = 5$ 17. $11t^2$ 19. $\frac{x+2}{3}$ 21. $2p + 1$ 23. $27t^3$

ANSWERS TO ODD-NUMBERED EXERCISES

Pages 409–412 Chapter Review 1. k^5 3. r^7 5. t^4 7. 6 9. 7 11. $15x^2y^3$ 13. $60a^3b^4$ 15. $-8y^4$ 17. a^6b^3 19. $25x^2y^2$ 21. $14x^3$ 23. $11k^2$ 25. $x^5 - 2x^4 - x^3 + 5x^2$ 27. $-3x^3y^3$ 29. $3r^2 + 4rs - 2s^2$ 31. $x = 3\frac{1}{5}$ 33. $6x^2 - 15x$ 35. $k^2 - 4$ 37. $-y^2 + 10y - 25$ 39. $10m^2 - 13mn - 3n^2$ 41. $6x^2 + 5x - 4$ 43. $-x^2y^2 - xy + 2$ 45. $x^2 - 10x + 25$ 47. $4t^2 - 2t + \frac{1}{4}$ 49. $x^2y^2 + 4xy + 4$ 51. $a^2 - 9$ 53. $10x^2 - 8x - 26$ 55. $\dfrac{k}{t-k}$ 57. $4x^4$ 59. $a^4b^4c^4$ 61. $\dfrac{1}{t^3}$ 63. z^{a-1} 65. $\dfrac{1}{x}$ 67. $\dfrac{1}{a^{x-1}}$ 69. $\dfrac{1}{4y^2}$ 71. $\dfrac{9x}{-y^2}$ 73. $\frac{3}{2} + s^2$ 75. $a - a^2$ 77. $3t + \dfrac{5}{s}$ 79. $3b - 2 + a$ 81. $x + 2$ 83. $n - 1$ 85. $x - 3$

Pages 412–413 Review of Skills 1. D 3. E 5. J 7. C 9. A and J 11. 360 13. 6 15. angles 17. longest 19. 144

Page 414 Questions 1. 1 3. 1 5. 1 7. 1 9. 1 11. 1 13. 1 15. c^6 17. $\frac{1}{5}$ 19. x^4 21. 7 23. 1 25. -8 27. 0

Chapter 14. Geometric Figures in the Plane

Pages 419–421 Written Exercises A 1. do not intersect 3. do not intersect 5. intersection is point C 7. $\overline{ST}$ or $\overline{TS}$ 9. $\overrightarrow{AB}$, $\overrightarrow{BC}$, or $\overrightarrow{AC}$ **B** 11. $\overline{KT} \cap \overline{TM} = T$ 13. $\overline{EF} \cap \overline{BC} = D$ 15. $\overline{CD} \cap \overline{JK} = \emptyset$ 17. $\overrightarrow{TR} \cap \overrightarrow{AQ} = \emptyset$ **C** 19. G 21. $\emptyset$ 23. $\emptyset$ 25. G 27. $\emptyset$

Pages 423–424 A 1. $\overline{AB}$, $\overline{BK}$, $\overline{KL}$, $\overline{AK}$, and $\overline{BL}$ 3. $\overline{AP}$, $\overline{PF}$, $\overline{FH}$, $\overline{AF}$, and $\overline{PH}$ 5. $\overline{SD}$, $\overline{DP}$, $\overline{GD}$, $\overline{DZ}$, and $\overline{GZ}$

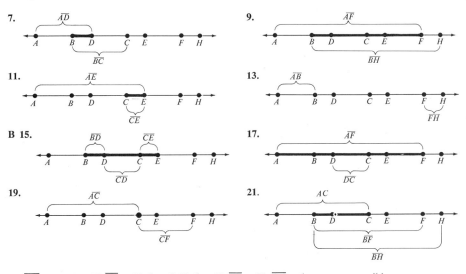

23. $\overline{AD}$ 25. E 27. $\overline{TF}$ 29. $\emptyset$ **C** 31. $\emptyset$ 33. $\overline{TF}$ 35. $\overline{DF}$; other answers possible

Pages 426–427 Written Exercises A 1. curve 3. simple closed curve 5. simple closed curve 7. curve 9. closed curve 11. simple curve 13. simple closed curve 15. simple curve 17. simple closed curve 19. simple closed curve 21. simple curve 23. on the curve 25. on the curve 27. interior 29. interior

Pages 430–431 Written Exercises A 1. $\overrightarrow{ST}$ and $\overrightarrow{SR}$ 3. $\angle RST$, $\angle TSR$, and $\angle 1$ 5. $\overrightarrow{PQ}$ 7. $\angle PQR$ 9. $\angle 2$ 11. G is in the exterior; T is in the interior, and R is on the angle 13. A is on the angle; R is on the angle; M is in the exterior **B** 15. $\angle CHB$, $\angle BHA$, and $\angle CHA$ 17. $\angle HDM$, $\angle MDB$, and $\angle HDB$ 19. $\overrightarrow{AB}$, $\overrightarrow{BC}$, and $\overrightarrow{AD}$ 21. $\overrightarrow{RA}$, $\overrightarrow{CA}$, and $\overrightarrow{HA}$

Pages 434–436 Written Exercises A 1. 42° 3. 45° 5. 45° 7. 125° 9. 35° 11. 115° 13. 80° 15. 150° 17. 45° **B** 19. $x = 43$ (degrees); $b = 58$ (degrees) 21. $t = 60$ (degrees); $k = 30$ (degrees)

ANSWERS TO ODD-NUMBERED EXERCISES

C 23. 25. 27. 29. 31.

Pages 439–440 Written Exercises A 1. $\overline{BG}$, $\overline{GH}$, $\overline{HB}$; $\angle 2 = \angle BHG$ 3. $\overline{EF}$, $\overline{FD}$, $\overline{DE}$; $\angle 3 = \angle EFD$
B 5. 90°; 65°   7. 45°; 45°; HG

Pages 443–444 Written Exercises A 1. true; is a Pythagorean triple   3. true; is a Pythagorean triple
5. true; is a Pythagorean triple   7. false; is not a Pythagorean triple   B 9. $3^2 + 5^2 = 6^2$; is not a right
triangle   11. $4^2 + 5^2 = 6^2$; is not a right triangle   C 13. 10, 24, 26   15. 20, 99, 101   17. 30, 224, 226
19. 34, 288, 290   21. 28, 195, 197

Pages 446–447 Chapter Test 1. M   3. D   5. J   7. G   9. B   11. 140°   13. 5   15. 135°   17. $\overline{BA}$
19. circle, or other answers

Pages 447–449 Chapter Review 1. $\overleftrightarrow{AB}$ 3. $\overline{JK}$ 5. $\overline{BC}$ 7. D   9. ∪   11. B   13. D   15. B
17. exterior   19. interior   21. exterior   23. ray   25. vertex   27. $\angle 1$ 29. obtuse   31. 23°   33. 71°
35. right   37. acute   39. isosceles   41. equilateral   43. 5

Pages 450–451 Questions 1. 5   3. 9   5. 5   7. 6   9. 14   11. $\overline{RT}$ and $\overline{GH}$; $\overline{RS}$ and $\overline{GK}$; $\overline{TS}$ and $\overline{HK}$
13. $\overline{RS}$ and $\overline{RT}$; $\overline{RV}$ and $\overline{RW}$; $\overline{SV}$ and $\overline{TW}$ 15. 145°; 35°; 75°; 105°; 75°

Pages 452–456 Cumulative Test Part I (Chapters 1–7) 1. 7521   3. 7512   5. 1257   7. 2715   9. $2 + 4 + 6 + 8$ 11. $18 + 2$ 13. even   15. odd   17. 350   19. 100   21. 250   23. $5\frac{3}{8} - 2\frac{1}{8} = 3\frac{1}{4}$;
$5\frac{3}{8} - 3\frac{1}{4} = 2\frac{1}{8}$ 25. $0.29 - 0.03 = 0.26$; $0.29 - 0.26 = 0.03$ 27. $\frac{m}{b} = c$; $\frac{m}{c} = b$

29.   31.   33. finite   35. finite   37. infinite   39. E
41. A   43. C   45. F   47. D   49. E   51. 1; (1, 1)   53. 2; (5, 2)   55. 4; (1, 4)   57. 16; (3, 16)   59. yes
61. yes   63. (3, 2)   65. (0, 2)   67. (0, 0)   69. 4   71. 80   73. $3rst$ 75. 56   77. 225   79. 109
81. $\{6, 12, 18, \ldots\}$ 83. $\{18, 9, 6, 3, 2, 1\}$ 85. $\{1, 3\}$ 87. $\{1\}$ 89. $A = 184$ sq. cm.; $P = 62$ cm.   91. B
93. A   95. D   97. $a = 25$ 99. $k = 5$ 101. $t = 15$ 103. $w = 12\frac{1}{2}$ 105. $c = 6\frac{3}{7}$ 107. $b = 5$
109. $x = 4\frac{2}{3}$ 111. I   113. A   115. E   117. K   119. C   121. J   123. 9 ft. × 45 ft.   125. 4

Pages 456–459 (1970 edition) **Cumulative Test Part II (Chapters 8–14)**   1. right; >   3. left; <   5. negative; positive   7. 70   9. -21 11. 101   13. -80 15. E   17. F   19. A   21. E   23. A   25. F
27. true   29. false   31. true   33. true   35. true   37. $\{-6\}$ 39. $\{-4.4\}$ 41. $\{50\}$ 43. $\{-3\}$
45. $\{-6\}$ 47. $4t^2 + 4t - 6$ 49. $13a^2 - 7ab - 2b^2$ 51. $11x - 8$ 53. $10n^2 + 3n - 5$ 55. $2b$
57. $-4rs - 2$ 59. $\{1\}$ 61. $x^3 + 6x^2 + 14x + 15$ 63. $35y^2 - y - 6$ 65. $2x^2 + x - 15$ 67. $4x^2 - 4xy + y^2$ 69. $-5ab$ 71. $-5s^2 - 2rs$ 73. $x - 4$ 75. 6; (−3, 6)   77. -2; (−1, −2)   79. -2;
(1, −2)   81. 10; (−4, 10)   83. -2; (0, −2)   85. 50; (4, 50)   87. $\{(-6, -20), (-3, -11), (0, -2),$
$(3, 7), (6, 16), (9, 25)\}$ 89. $\{(-\frac{1}{2}, 3\frac{3}{4}), (-\frac{1}{4}, 3\frac{13}{16}), (0, 4), (\frac{1}{2}, 4\frac{3}{4}), (\frac{1}{4}, 4\frac{5}{16})\}$ 91. 6 ft.   93. $2x^2 - xy + y^2$

(1973 edition)   37. $\{-5\}$ 39. $\{-3.7\}$ 41. $\{52\}$ 43. $\{-1\}$ 45. $\{-7\}$ 47. $2t^2 + 3t + 3$
49. $11a^2 + 3ab - 3b^2$ 51. $6x - 11$ 53. $8n^2 - 4n + 7$ 55. $2b$ 57. $4rs - 10$ 59. $\{\frac{5}{3}, -1\}$
61. $2x^3 - 4x^2 + 5x - 3$ 63. $6y^2 + 5y - 4$ 65. $3x^2 - 4x - 1$ 67. $9y^2 - 6xy + x^2$ 69. $-8a^2$
71. $-3rs + 1$ 73. $x - 2$ 75. -1; (−2, −1)   77. -5; (0, −5)   79. -1; (2, −1)   81. 50; (−6, 50)
83. -2; (−2, −2)   85. 10; (2, 10)   87. $\{(-5, 28), (-4, 19), (-3, 12), (0, 3), (3, 12), (4, 19), (5, 28)\}$
89. $\{(-1, 3), (-\frac{1}{2}, 3\frac{3}{4}), (0, 4), (\frac{1}{2}, 3\frac{3}{4}), (1, 3)\}$ 91. $288   93. $s^2 + 3st + 2t^2$